Green Energy and Technology

More information about this series at http://www.springer.com/series/8059

Nicola Labanca
Editor

Complex Systems and Social Practices in Energy Transitions

Framing Energy Sustainability in the Time of Renewables

Springer

Editor
Nicola Labanca
European Commission, Directorate-General
Joint Research Centre
Unit C.02 Energy Efficiency and
Renewables
Ispra, VA
Italy

Disclaimer: The information and views set out in this publication by the authors working for the Joint Research Centre of the European Commission are those of these authors and do not necessarily reflect the official opinion of the European Union. Neither the European Union institutions and bodies nor any person acting on their behalf may be held responsible for the use which may be made of the information contained therein.

ISSN 1865-3529 ISSN 1865-3537 (electronic)
Green Energy and Technology
ISBN 978-3-319-33752-4 ISBN 978-3-319-33753-1 (eBook)
DOI 10.1007/978-3-319-33753-1

Library of Congress Control Number: 2017935837

Printed on acid-free paper

This Springer imprint is published by Springer Nature
The registered company is Springer International Publishing AG
The registered company address is: Gewerbestrasse 11, 6330 Cham, Switzerland

Foreword

The social, ecological, and economic effects of a radical transition from fossil energy to renewable energies cannot be known, but there is a history of human civilization that can inform our thinking. As a hunter/gatherer society, we accessed high-gain food energy and low-gain fuel (wood). Moving about the land allowed a renewal of both once a local area was depleted. As human population in an area expanded to the point where moving was no longer an option, we switched to agriculture. When grains became our dominant, low-gain food energy source, the sun became society's dominant energy source for food and fuel. Land ownership greatly altered the structure of society, and excess energy moving through the system supported a more complex social hierarchy. That additional complexity demanded more energy to maintain it, so empires went next door and took over the lands of others. Yet another substantive change in society, and the beginnings of a global economy, based on taking sun-based fuel from others to support city-state structures. With the switch from renewable energy for food and fuel to fossil sun for both (coal and oil, and machine-based crop harvests) we hierarchically complexified society yet again, working from both a high-gain fuel energy and a high-gain food energy system (though still based primarily on four low-gain grains).

We are nearing the end of this present form of society, as the technology to extract and burn fossil fuels is peaking, as is the amount of food energy that can be extracted from a finite land base, despite fossil fertilizers and gene manipulation. The complication in all of this is that it is not a simple path back to hunting and gathering, as both the global population reliant on these energy sources is increasing, and the once-stable climate that provided for our complex global society is becoming less constant. The experts assembled for this book explore the ways, means and implications of just how a transition from high-gain fossil fuels back to a primarily low-gain renewable energy sources might unfold. What does it mean to return to our roots, to our initial condition of renewable energy as a sole-source fuel?

Despite being presented as simply a matter of technological innovation that will not substantively alter our societal structures, this energy source transition very likely entails dramatic and permanent changes in our social-ecological systems. The specifics of these changes are unknowable, but a plausible narrative of future scenarios and system trends can be woven. Narratives about how close to the rising oceans most humans live, how and where food is grown, how much of the land base will be needed just for renewable energy infrastructure (windmills, solar arrays, wave energy capture), and what level of complexity individual societies will be allowed to maintain. Who gets to keep a middle-class? Who has first rights to contribute to the greenhouse gas reservoir? Who pays for relocations of villages, cites, or whole nations caused by sea level rise? These are wicked questions, and the nexus of food-water-energy provision in light of an increasing population and a changing climate is at any level a wicked problem. The sections of this book tease apart and look at those possible levels from many viewpoints, and reach some actionable conclusions.

Increases in food production result not in better nutrition alone, but also in more humans on the planet. Increases in locating, extracting and burning fossil fuels lead to greater use of such energy sources, not less. This is Jevon's Paradox—that increased efficiency results in a greater use of a resource, not a reduction in use. When there is excess from increased efficiency, either someone will come along and take it, or the producer will seek out new markets to enhance overall profits. More food, more people. More fossil energy, a larger middle-class. A larger population living a middle-class lifestyle, a greater consumption of goods and services. Increased consumption demands greater efficiency in production, and we start all over again. The transition from high-gain, energy-dense fuels to low-gain, highly processed and organized energy carriers, will materially affect this cycle, the underlying economy, the societies from which the economy emerges, and the landscapes that support it all. It is not a mere change in where our energy comes from, but a material change in how we as a society use our landscapes, and that material change in demand and supply of ecological services will have substantive and permanent effects on our current, unsustainable global society.

By using the science of complexity and complex systems theory as the reference framework to study our transition to renewables, the contributing authors produce a series of relevant insights concerning what can be anticipated from this transition. They provide new insights, and some very interesting and non-conventional observations, that students, researchers and policy makers involved in the current energy transition will find useful in understanding the issues at hand, as well as the opportunities and pitfalls offered by a total transition to renewable energy sources.

Ron McCormick, Ph.D.*
Bureau of Land Management
U.S. Department of the Interior
Washington, D.C., USA

*The views expressed here are the author's and do not reflect the policy of the U.S. Department of the Interior.

Preface

When in 2015 we resolved to organize the experts' round-table discussions that led us to decide to write this book,[1] we were aware of two central problems affecting current mainstream policy and research activities for a sustainable energy transition.

The first problem relates to a very controversial dichotomous approach still mostly adopted by scientists and policy makers when carrying out these activities. On the one hand, they indeed still mostly aim at identifying and implementing solutions that may increase the sustainability of human activities by fostering the substitution of *single* technologies with assumed equivalent models functioning with less energy inputs and causing less harmful emissions in the atmosphere. On the other hand, they aim at finding and stimulating the adoption of policy approaches that can change the behaviour of technologies end-users. In doing so, they assume that end-users can somehow be individually persuaded to buy these alternative models or be induced to modify their conduct when employing single energy consuming technologies and do not take into account systemic factors may impede achieving expected policy impacts.

This dichotomous approach can certainly contribute to improve the energy efficiency of single technical applications in important ways, this result representing a very relevant result. When assessed against the possibility that it can lead to an overall reduction in the consumption of natural resources caused by human activities, it results nevertheless highly problematic in so far as it misses to take into account how individuals and technologies are nowadays interlinked within complex systems which evolve according to logics that it cannot capture. Indeed agency (i.e. the power to generate change) has to be considered nowadays as distributed over large series of human and non-human actors,[2] including a variety of technological products, institutional settings, rules and habits that co-determine people behaviours and all together induce a trend of energy consumption growth which neither

[1]See http://iet.jrc.ec.europa.eu/energyefficiency/round-table/experts-round-table-practice-theory-and-complex-adaptive-systems-theory for further information on this round-table.

[2]On this point see, for example, Latour (2005).

individuals, nor more energy-efficient technologies can reverse. Individuals taken alone cannot for example change the social constraints obliging them to commute every day or to buy houses where it is impossible to live without air conditioners. Moreover, the dynamics of growth triggered by the above-mentioned complex systems make often practically inevitable that the energy saved by one technology is then used as input for another technology in order to sustain this growth. In addition, the high level of power output that people composing societies can presently generate through these complex systems could be hardly achieved when renewable energy sources substitute on a large-scale non-renewable energy sources.

Although capable of determining relevant reductions in the consumption of energy inputs and in the production of greenhouse gas emissions that can be associated with the employment of single technical applications, the above-described dichotomous approach is hence affected by important limitations, which have to be ultimately considered as a consequence of two facts: it cannot significantly alter the overall energy consumption dynamics developing within current complex socio-technical systems, and it does not consider that a radical transition to renewables entails a radical reorganization of societies.

These are the main considerations that convinced us about the absolute relevance of alternative research and policy approaches that can take these complex dynamics into account.

At the same time, however, we were also aware of a major reflexivity problem affecting policy and research approaches informed by complexity. This problem relates to how social aggregates are mostly erroneously identified with kind of motors and information processors simultaneously maximizing their power output and energy efficiency while consuming abstract units of energy, time, information, money, etc. that are taken as actual ontological entities. Researchers and main stakeholders involved in the design and implementation of policies for energy sustainability tend indeed to identify socio-technical systems with input–output systems, while often forgetting that behind the abstract flows of resource units that they take as real entities and try to change there are very concrete and specific social habits, and there are people made of flash who can, on the one hand, potentially actively contribute to face contemporary sustainability challenges and, on the other hand, may not react as expected to implemented policies.

This problem, however, does not only affect policy approaches and solutions developed by specialists and experts. It actually concerns societies at large and the way in which people currently imagine the world around them. We think that this problem is the result of a large-scale social construction that has led motors and computers to become central metaphors whereby the functioning of societies and human beings is explained and being reorganized. The complex systems resulting from this social construction can generate enormous material benefits but are also responsible for an increased dependence on the technological supply and efficient utilization of given homogenized and standardized resource units while causing the disappearing of a variety of alternative practices established by people to provide for their necessities.

The dynamics of growth that can be triggered by these complex systems certainly contribute to increase material well-being in important ways. Moreover, the complexification of energy systems that might accompany the ongoing massive transition to renewable energy sources can generate huge environmental benefits. It is for example the possibility of generating an organized complexity that makes possible to conceive that highly distributed renewable energy sources can substitute non-renewable sources and be used to supply the energy needed by present large social aggregates. At the same time, however, the dynamics of these complex systems seem to obey abstract logics escaping any form of social control, whilst an increased complexification of existing energy systems can determine more frequent cases of crash and disengagement from rules and principles established by societies to regulate themselves due, among others, to an associated increased dependency on energy flows that can change unpredictably.

It becomes hence extremely relevant to understand how these dynamics are generated by existing social practices and how the development of new practices can possibly allow accomplishing the above-mentioned transitions in a more sustainable way while allowing preventing unwanted systems crashes or coping with these generally very unpleasant situations whenever they may occur.

These are the considerations that led us to conclude that the sustainability challenges posed by complex systems have to be necessarily also addressed by trying to take existing social practices and related theories as main research and policy target.

The policy approaches that can be designed and implemented in this way are generally radically different from approaches informed by complex systems theories. Whilst policy and governance strategies informed by these latter theories are inevitably based on considerations concerning existing and future energy, material and monetary flows, strategies informed by theories of social practices are supposed to take existing possibilities to reorganize the outputs of concrete actions undertaken by people as main starting point. Whilst the former strategies are informed by abstract considerations concerning inputs needed and outputs produced, the latter strategies can be designed based on considerations concerning what people say and do and how they organize and can concretely change own habits in a given context. Finally, whilst the former strategies are mostly based on technical considerations and do generally foresee a very limited active involvement of people in their design, the latter strategies are more genuinely political in so far as they relate to aspects that people can actively contribute to modify. In the case of future large-scale transitions to hypothetical renewable energy distribution networks, the former strategies are, for example, often focused on technical and economic interventions allowing an automated and mutual adaptation between energy demand and supply, whilst the latter strategies target people practices in their entirety and can be focused on whether and how these practices can be actually changed or reorganized by people in order to make them compatible with the increasingly intermittent energy availability that might be expected from these networks.

We are convinced that the different characteristics of these two strategies reflect a fundamental and unescapable complementarity that can be identified in the approaches that can be followed when developing or employing rules, material artefacts, institutional settings and know-how whereby societies are organized. On the one hand, these societies can develop or rely on general and abstract rules and principles that can be blindly applied to all of its members who are in this way mostly identified with kinds of passive users (this might happen for example in case of rules and technical solutions that can be implemented to allow that aggregated electricity demand and supply can be balanced in future smart grids). On the other hand, they must cultivate a particular practical sensibility allowing that these general rules and principles can be adapted and subordinated to the initiatives undertaken by individuals and to their specific conditions (in the previous example of the smart grids this might for example entail a subordination of these rules and principles to practices developed by people who could in this way be made collectively responsible for the management of the energy resources, the technical apparatus and the institutional settings whereby these grids can be administered). We think that the insights provided by social practices theorists can help policy makers and researchers to cultivate this particular sensibility, given the fact that a suitable way to combine the two above-mentioned approaches has always to be found and the fact that the prevailing of one out of the two approaches within policy making is generally destined to cause disasters of various nature.

Based on the above considerations, we decided to gather around a table a series of acknowledged scientists working on complex systems and social practices. Given the interdisciplinary character of the questions we wanted to address, very different competences were represented. The invited scientists are indeed acknowledged sociologists, physicists, engineers, economists, anthropologists, biologists, ecologists and policy analysts. During the two-day event we organized we managed to discuss some of the above questions with them, whilst other experts that could not be with us were sent the proceedings of our meeting and were involved in the e-mail discussions that took place during the following weeks. Altogether we then decided to produce a publication that could hopefully serve to make the scientific community and policy makers more aware of the relevance of the analysis approaches discussed and of the insights that can be gained through their application.

The present book is the result of this interdisciplinary effort.

Nicola Labanca
Paolo Bertoldi
Isabella Maschio
Daniele Paci
European Commission, Directorate-General Joint Research Centre
Unit C.02 Energy Efficiency and Renewables
Ispra,
Italy

Reference

Latour, B. 2005. Reassembling the social: An introduction to actor-network-theory. Oxford: Oxford University Press

Acknowledgements

The Editor would like to thank Heinz Ossenbrink and Paolo Bertoldi for the support they have given to the production of this publication and to the organization of the experts round-table where it was initially conceived. Many thanks also go to the acknowledged experts who have contributed to the various chapters and to the elaboration of the book concept.

Acknowledgements

The Editor would like to thank Heinz Ossenbrink and Paolo Bertoldi for the support they have given to the production of this publication and to the organization of the experts round table where it was initially conceived. Many thanks also go to the acknowledged experts who have contributed to the various chapters of this [illegible]

Contents

Contributors

Timothy Allen Department of Botany, University of Wisconsin, Madison, USA

Osman Arrobbio Department of Culture, Politics and Society, University of Turin, Turin, Italy

Françoise Bartiaux Institute of Analysis of Contemporary and Historical Societies, Université Catholique de Louvain and National Fund for Scientific Research, Louvain, Belgium

Thomas Bauwens École Polytechnique Fédérale de Lausanne. EPFL ENAC IIE HERUS, Lausanne, Switzerland

Paolo Bertoldi European Commission, Directorate-General Joint Research Centre, Unit C.02 Energy Efficiency and Renewables, Ispra, VA, Italy

David S. Byrne School of Applied Social Sciences, Durham University, Durham, UK

Kajsa Ellegård Department of Thematic Studies, Technology and Social Change, Linköping University, Linköping, Sweden

Brian D. Fath Towson University, Towson, USA

Mario Giampietro Universitat Autonoma de Barcelona (UAB), Institute of Environmental Science and Technology (ICTA), Barcelona, Spain; ICREA—Catalan Institution for Research and Advanced Studies, Barcelona, Spain

Zora Kovacic Universitat Autonoma de Barcelona (UAB), Institute of Environmental Science and Technology (ICTA), Barcelona, Spain

Nicola Labanca European Commission, Directorate-General Joint Research Centre, Unit C.02 Energy Efficiency and Renewables, Ispra, VA, Italy

Isabella Maschio European Commission, Directorate-General Joint Research Centre, Unit C.02 Energy Efficiency and Renewables, Ispra, VA, Italy

Mithra Moezzi Portland State University, Portland, OR, USA

Daniele Paci European Commission, Directorate-General Joint Research Centre, Unit C.02 Energy Efficiency and Renewables, Ispra, VA, Italy

Dario Padovan Department of Culture, Politics and Society, University of Turin, Turin, Italy

Jenny Palm Department of Thematic Studies, Technology and Social Change, Linköping University, Linköping, Sweden

Franco Ruzzenenti Department of Management and Quantitative Sciences, Parthenope University of Naples, Naples, Italy; Institute of Sociology, Jagiellonian University, Krakow, Poland

Duncan Shaw Nottingham University Business School, Nottingham, UK

Elizabeth Shove DEMAND Centre/Department of Sociology, Lancaster University, Lancaster, UK

Joseph Tainter Department of Environment & Society, Utah State University, Logan UT, USA

Abbreviations

ANT	Actor–Network Theory
ATP	Adenosine Triphosphate
CBE	Community-Based Energy
DNA	Deoxyribonucleic Acid
EROI	Energy Return on Investment
GDP	Gross Domestic Value
GVC	Global Value Chain
IEA	International Energy Agency
IT	Information Technology
KWAPT	Kilowatt at Peak Time
KWOPT	Kilowatt Outside of Peak Time
OECD	Organisation for Economic Co-operation and Development
OM	Order of Magnitude
OPEC	Organization of Petroleum Exporting Countries
PVC	Polyvinyl Chloride
RE	Renewable Energy
REE	Rebound Effect
RES	Renewable Energy Sources
SES	Social-Ecological System
SPT	Social Practice Theories
TMC	Toyota Motor Corporation

Abstract

This book is the result of an interdisciplinary effort undertaken by a series of sociologists, physicists, engineers, economists, anthropologists, biologists, ecologists and policy analysts who participated or were involved in the discussions that took place during a round-table organized by the Joint Research Centre of the European Commission in February 2015. That gathering allowed discussing fundamental issues at stake with policies for energy sustainability that are largely neglected by stakeholders involved in policy making. The participating scholars decided hence to produce a publication that could hopefully serve to make the scientific community and policy makers more aware of the relevance of the analysis approaches proposed and of the insights that can be gained through their application.

The book takes complementarity seriously and presents a double analysis perspective by taking complex systems and social practices as complementary references. It does so by acknowledging that Western societies have quite recently entered the age of *complex systems* and that our ideas and material artefacts are for this reason being shaped by a relatively new paradigm whereby complex systems are being extensively built. The *first part* of the book is indeed dedicated to discuss how complex systems are socially constructed and how they are framing the issue of energy sustainability within mainstream research fields. The *second part* analyses the ongoing transition to renewable energies and policies that can be generally implemented to conserve energy in the light of theories informed by complexity. On the other hand, the *second* and the *third part* discuss how complex systems take with them important drawbacks for energy sustainability that are linked to some phenomenological principles regulating their evolution. These drawbacks are mainly identified by adopting the alternative and complementary analysis perspective offered by social practice theories. Practice theories complementarity stems principally from their acknowledging of the primacy of *practical knowledge* over the abstract notions of energy, time and information that constitute some of the main elementary bricks whereby complex systems are being built. Axiomatically, this means that, rather than by very abstract concepts, the elementary units of the analyses performed under a practice theory perspective are given by the actions undertaken by people, by what people *do* and what people *say*. A series of contributions collected in the *third part* discusses therefore some main lessons for policy making that can be learnt by this complementarity and by integrating social practice and complex systems theories. Overall, the adopted analysis approach has then allowed drawing a series of relevant conclusions and indications for researches and policy makers involved in the ongoing energy transition that have been summarized in the *fourth and final part* of the book.

Introduction

Several countries in the world are currently engaged in an energy transition entailing a massive shift to renewable energies and a progressive increase in the efficiency of processes whereby energy inputs are used by economies. According to existing projections, in a few decades renewable energy sources will make most of the world's electricity production,[3] will provide almost 50% of the heat needed by buildings,[4] will provide a consistent share of the fuels used in the transport sector[5] and, above all, will allow markedly reducing anthropogenic emissions of greenhouse gases. Energy efficiency is then supposed to substantially contribute to this energy transition by reducing the burden of an ever-increasing energy demand on the existing natural resources system.[6] Researchers and policy makers rightly describe the realization of these scenarios as highly necessary and capable of contributing to the environmental, economic and social sustainability of human activities in important ways.[7]

There are, however, two very relevant areas of methodological improvement that are usually not sufficiently considered. The first one relates to the need of adopting an actual complex system perspective when performing the above-mentioned studies or designing and implementing energy transition policies, whilst the second one concerns the need for a better understanding of the role that people can have in the realization of this transition.

Concerning the first point, it cannot pass unnoticed how agents of resources consumption and associated emissions are nowadays mostly identified *either* with existing technological instruments *or* with individuals using these instruments and resources. As a consequence, adopted research and policy approaches mostly exclusively appeal either to the substitution of single technologies with more

[3]See, for example, IEA (2011a).

[4]See IEA (2012).

[5]See IEA (2011b).

[6]See, for example, C2E2 & IRENA (2015).

[7]On this point see, for example, UNEP (2015).

energy-efficient and less polluting ones, or to future so-called smart grids allowing continuous and automated exchanges of energy and information among all points of the energy network, or to behavioural changes expected from individual persons acting within competitive market settings. These approaches unfortunately miss recognizing that the actual agents of resources consumption are often represented by large and complex socio-technical systems wherein technologies and persons are nowadays integrated and made dependent on mutually reinforcing flows of energy, material and monetary resource units. They certainly can allow achieving a reduction in the energy inputs as well as a better integration into the environment of single technologies, but they are typically inadequate to face the dynamics of resources consumption growth that can be triggered by these complex systems and do not allow identifying suitable policy strategies and measures to counteract them.

Concerning the second point, the issue at stake relates to how the current energy transition is mostly envisioned as a problem of technological substitution where people will have to adapt to new technologies without having to significantly change their ways of life. The problem associated with this type of vision is that social practices reproduced by people are actually deeply embedded, co-evolve with and deeply affect the possible development of current energy systems. Research and policy approaches exclusively focused on technological substitutions or on individuals' behavioural changes around single technologies are hence problematic at least for two orders or reasons. Firstly, because they do not consider that existing social practices might not be as adaptable as expected and might hence represent an insuperable obstacle to the energy transition envisaged. Secondly, because social practices can provide an innumerable amount of alternative solutions and approaches that can better adapt to the ever changing local conditions that can be expected from this transition.

The objective of this book is therefore twofold. On the one hand, it wants to illustrate to researchers and policy makers the necessity to move from an instrumental to a complex system approach when studying socio-technical systems and policies that can be implemented to increase their sustainability within the current energy transition. On the other hand, it aims to show how relevant it is to perform this move by studying these complex systems by combining a positivist perspective with a constructivist one focused on the social practices wherefrom these systems emerge. Somehow, the book invites researchers and policy makers to perform a double change of gear when addressing problems linked to the finitude of existing energy and material resources and to greenhouse gas and polluting emissions generated by technologies and human activities.

While illustrating the research and policy insights that can be gained by studying the current transition by focusing on the complex systems dynamics that this transition can generate, it wants also to show that these complex dynamics do not have to be considered as an inevitable natural phenomenon. The constructivist perspective being proposed in several chapters of this publication aims indeed at showing that these dynamics are actually the outcome of social practices reproduced by people and that, due to this fact, research and policy approaches which are alternative and complementary to those taking complex systems dynamics as the ultimate reality can be devised to possibly counteract them.

Whilst positivist approaches informed by complexity tend to take phenomenological principles related to how natural systems optimize resources consumption and outputs production as a benchmark to study and design suitable policies, this latter perspective allows in principle *disaggregating* the energy, material, information and monetary flows circulating within complex systems into the myriad of human practices generating these flows and permits to take these practices as the starting point to possibly design and implement suitable energy transition policies. The study of the environmental sustainability of a transport system in a city can for example be informed by complexity and be focused on vehicles and persons flows and on associated emissions and energy consumption. This typically implies that the solution of optimization problems concerning how flows density can be increased while maximizing energy efficiency and minimizing polluting emissions of involved socio-technical systems becomes the main target. When studied under a social practices perspective, these flows may be disaggregated into trips made by people to go to school, trips to go to the supermarket, trips to commute, etc. Rather than being based on technical and abstract optimization problems, research and policy approaches that can be developed in this way can devise a reorganization of these different mobility practices which is subordinated to the specificity of the social context at stake and can therefore allow people doing better while reducing the environmental impacts of their activities. Why, for example, children go to school by cars in this city? Why people prefer cars to bikes or public transport to go to work in that municipality? What can be done to change these practices? By starting from this type of questions, these approaches acknowledge the primacy of what people concretely do and say over solutions exclusively informed by technical considerations and implicitly assume the irreducibility of the outcomes of these doings and sayings to what can be predicted by any type of modelling.

The constructivist perspective being presented offers therefore the possibility to effectively complement policy and research strategies treating the fluxes generated within complex systems as actual and ahistorical ontological entities and aiming at changing the associated dynamics by suitable technical solutions. In addition, this perspective allows interpreting the consumption of energy and material resources by complex systems as the outcome of an at least partly negotiable social construction that transforms standardized resources units supplied through specific technological artefacts into the necessary input needed for the reproduction of any kind of practice. How is it that nowadays we need kilowatt-hours supplied by utilities or by micro-generation systems installed in our houses to do anything? How is it that most of our daily tasks are being progressively associated with the transmission of bits of information throughout computer technologies? How is it that a temporary interruption in the supply of electricity or in the internet can nowadays potentially inhibit most of the activities performed in a city? Can our social practices be rearranged to reduce this dependency whilst improving the quality of our life?

All in all, the proposed constructivist perspective allows to not see the consumption of the standardized resources occurring within current complex socio-technical systems as the inevitable outcome of any human activity and in so doing allows conceiving valid policy alternatives resulting from the active involvement of people while permitting to understand the social dynamics whereby dependency on these resources can be unnecessarily reinforced.

The decision to combine a positivist perspective with a constructivist one is neither casual, nor opportunistically due to the additional research and policy insights that this combination seems to allow gaining. The complementary approach adopted in this book reflects in my opinion the presence of a fundamental complementarity and separation existing between know-how, institutional settings, social norms and rules validated through technologies and science, on the one hand, and the social tissue on which these know-how, technologies and social settings operate, on the other hand.

Complex socio-technical systems are not generally seen in this book as natural and ahistorical entities whose presence and properties can be completely understood by science. They are rather interpreted as the result of a social construction based on a series of implicit and usually undisputed assumptions concerning the nature of energy, information and monetary value which are shared among scientists, technologists, economists and within society at large. The fact that science has necessarily to proceed by means of assumptions and working hypotheses is nothing new. What however is generally neglected is that some of these abstractions and the technological instruments through which their properties are validated are at the same time the result and a reinforcement factor of a social imaginary concerning what we and the world around us are that informs current social changes. For example, energy, information and related technologies are nowadays the bearer of messages concerning the nature of our world which propagate through societies while causing their reorganization. The complex systems they constitute have therefore to be studied also in relation to how they act on societies and in relation to the needs for reciprocal adaptation and possible tensions that can arise between them and the social tissue on which they develop. If the properties of complex systems can be studied by referring to the dynamics of associated energy, information, matter and monetary flows, this process of mutual interaction however generally resists the reductions and the reification processes that science has to operate to possibly capture them. These two entities (i.e. the wide complex systems that are being constructed and the social tissue made of concrete and lively persons) have for these reasons to be considered as constituting a duality made of two separate and complementarity parts.[8] Technologies, institutional settings, know-how, social rules and norms constituting current complex systems have to be

[8]To a certain extent, this duality and the tension that may arise among its two parts is the same that can be found whenever people have to develop, learn or employ languages, artefacts, institutional settings and know-how whereby societies are organized. While rules and norms can be established to allow generating, using and understanding these material and conceptual artefacts, these artefacts actually result from and act upon a preexisting substrate made by what people practically do and say during their everyday life.

constantly enacted and lived by people through their bodies and their personal and collective experience.

Within this enactment process, these material and conceptual artefacts have to be adapted to all the specific cases represented by people to whom they are applied, whilst people have to conform their feelings, particular situations and inclinations to them. Their use requires a continuous process of confirmation and mutual adaptation which can never be taken for granted or be considered as achieved once and forever. The two dimensions being discussed and the tension and complementarity existing between them can be identified at all levels of societies, from the level of single persons to small social groups, to cities and countries.[9] Either people are engaged in artistic activities, or in the employment of technological artefacts, or in the enforcement of laws and policy measures, they are always called to personally live and possibly resolve the tension existing between the general principles and ideas that may inspire their action and the particularity of the case they have to face. Either complex systems at stake are represented by the general principles, know-how, institutional settings, technologies, social norms and rules whereby energy is (or will be) produced and consumed, or are constituted by the conceptual and material artefacts whereby the current global economic market is organized or by the information systems being created to timely respond to emerging threats (wars, nuclear accidents, environmental accidents, etc.), there are basically two options to deal with them. Their evolution can either be subordinated to people decisions and be adapted to various social circumstances, or they can become abstract entities whose evolution is passively determined and accepted by people despite the very high social and environmental pressure this may determine. The vital and sometime violent force exhibited by social phenomena has to be ultimately found within the long and short term interactions taking place between aggregates of people and the material and conceptual artefacts they put in place to carry out their daily lives also in case these artefacts end up constituting the very complex systems addressed in this publication.

The binocular perspective proposed in the book to study the current energy transition reflects at a speculative level the presence of this duality and tension. This publication has indeed been structured into four parts.

The *first chapter* included in the first part describes the social construction of present complex systems and discusses the transformations they are inducing in how human artefacts are conceived. The *second chapter* illustrates instead the role of energy in the dynamics exhibited by these complex systems and how this social

[9] When, for example, people have to learn a language, to play an instrument or dance a music, they can refer to general rules and methods established within grammars, music or dance scores. These general rules and methods, however, do not determine the practices of speaking, playing or dancing. They have to be confirmed by and adapted to the practical knowledge that people use for their creation and generally develop around them. The fact that languages or play and dance arts can be learnt also without these rules is, among others, an index of the primacy of this practical knowledge over standards and methods that can be established to facilitate their creation and reproduction.

construction frames the issue of energy sustainability within mainstream research fields. Moreover, it discusses some main implications of this social construction for policies that can be implemented to foster the current energy transition.

The chapters included in the second part of the book are instead informed by a positivist perspective and are specifically dedicated to analyze the ongoing transition to renewable energies in the light of complex systems theories. The *third chapter* of the book included in this part discusses whether the ongoing transition to renewable energies that is taking place worldwide is leading to a higher complexification of associated energy systems. The *fourth chapter* employs instead the concept of energy metabolism to discuss whether the scale at which the present economy has developed is strictly dependent on the energy intensity of fossil fuels and whether this scale can be sustained by renewable energy sources. Its authors use this concept also to discuss the proper scale of governance to be developed for ecosystems and whether circular economy is attainable at the scale of the present global economy. The *fifth chapter* then analyses the role of hierarchies within complex systems. It discusses to what extent the advent of renewable energy sources will lead to a new hierarchical organization of matter and energy and how to cope with it under the viewpoint of governance and policy. The *sixth chapter* included in the second part of the book is instead dedicated to discuss how complexity theory can allow understanding the role of community-based energy initiatives in increasing the resilience of energy systems within current low-carbon transitions. Finally, the *seventh chapter* of this part highlights the urgency of revisiting the role of energy efficiency within the current energy transition and of accompanying energy efficiency policies with policies aiming to achieve an absolute reduction in energy consumption also within the current transition to renewable energies.

Chapters belonging to the third part of the book reflect instead a constructivist perspective. They discuss the problems caused by the processes of energy, time and information reification (i.e. the processes whereby these abstract entities come to be considered as a concrete thing) occurring within current complex systems while showing how a practice theory perspective can serve to very effectively complement research and policy approaches informed by the positivist perspective adopted in the second part of the book. The *eighth chapter* of the book is dedicated to discuss how energy and information play the role of central metaphors that are constantly taken literally in the present age of complex systems. The author of this chapter discusses how people and societies are being constantly identified with motors and information processors and which are the consequences of this aberration for policies that can be implemented to increase the sustainability of the current energy transition. The *ninth chapter* focuses on how current methods of knowing and managing energy that depend on techniques of abstraction, standardization and equivalence (like those leading to reduce different energy sources and end-uses to time independent representations of quantities measured according to a same metrics) prevent researchers and policy makers from engaging effectively with the multiple dynamics of energy demand or with the fundamentally different characteristics of renewable and fossil fuels. Along a similar line of thinking, the *tenth chapter* discusses the problems generated by energy and time reification, the fundamental differences existing between renewable energy sources and fossil fuels and the risks and problems generated by the fact that they are often treated in equivalent terms by policy analysts and stakeholders dealing with the issue of a low-carbon energy transition. The *eleventh chapter* shows instead how the current energy transition requires a deeper knowledge about the relation between people's daily activities and their electricity use and how to increase existing knowledge through time-use surveys and the visualization of aggregate activity patterns.

The *twelfth chapter* explains how to achieve a deeper understanding of smart grids and analyses them a) as technological zones where metering standards, communication infrastructures and socio-technical evaluation assemble and b) as apparatuses made of asymmetric lines of power, knowledge, information, decision making, energy intensities and artefacts. The *thirteenth chapter* aims to help expand current demand response thinking to include a fuller appreciation of what actions can provide demand response and how changes in technological regimes, policies, social structures and expectations could increase demand response capacity. Finally, the *fourteenth chapter* generally discusses how practice and complex systems theories can be profitably integrated and can inform policies implemented to foster the ongoing energy transition.

The fourth and last part of the book is then dedicated to summarize indications for research and policy making and conclusions drawn by the authors of all book chapters. It has been conceived to facilitate researchers and policy makers in accessing information concerning key research and policy aspects that the authors of this quite voluminous publication have mainly produced for policy makers dealing with the ongoing energy transition at the international, national, regional or even city level.

I hope the book can render with sufficient clarity the importance of the complementary perspectives proposed by its authors.

Nicola Labanca
European Commission, Directorate-General Joint Research Centre
Unit C.02 Energy Efficiency and Renewables
Via E. Fermi 2749 IT-21027
Ispra, VA
Italy

References

IEA. 2011a. Solar Energy Perspectives. ISBN: 978-92-64-12457-8. Report available at https://www.iea.org/publications/freepublications/publication/.

IEA. 2011b. Technology Roadmap. Biofuels for Transport. Report available at http://www.iea.org/publications/freepublications/publication/Biofuels_Roadmap_WEB.pdf.

IEA. 2012. Energy Technology Perspectives 2012. Pathways to a Clean Energy System. ISBN: 978-92-64-17488-7. Report available at https://www.iea.org/publications/freepublications/publication/ETP2012_free.pdf.

C2E2 & IRENA 2015. Synergies between Renewable Energy and Energy Efficiency. Working paper available at http://www.irena.org/DocumentDownloads/Publications/IRENA_C2E2_Synergies_RE_EE_paper_2015.pdf.

UNEP. 2015. UNEP Environmental, Social, and Economic Sustainability Framework. Report available at http://www.unep.org/about/eses/Portals/50272/Documents/UNEP_Environmental_Social_and_Economic_Sustainability_Framework.pdf.

Part I
The Social Construction of Complex Systems

Chapter 1
Complex Systems: The Latest Human Artefact

Nicola Labanca

Abstract Complex systems are presented in this chapter as an emergent social and historical phenomenon related to the making and the using of artefacts. Rather than as the result of scientific discoveries, these systems are mainly seen as the product of a social construction which has affected any department of knowledge and human activity. The proposed account revolves around the idea that the intensive scientific and technical reflections that have taken place in specific historical periods in relation to specific human artefacts have transformed the concepts associated with the creation of these artefacts into central ideas and metaphors around which societies have started being organized while leading to their massive technological reproduction. By building on an historical enquiry on instrumentality developed by a series of acknowledged scholars, this chapter discusses how the nature of human artefacts has changed starting from the twelfth century. In particular, it shows how these artefacts have been mainly seen during subsequent historical phases as organa, instruments, motors and, more recently, as complex systems. In addition, it illustrates how these transformations have been accompanied by as many radical changes in the social imaginary concerning the meaning of human action and in the way in which delegation to machines and agency (i.e. the power to generate a change) has been conceived. The chapter also illustrates how the ongoing transition to renewable energies can reinforce the social construction of complex systems and represents an introduction to the second chapter where the implications of this construction for the energy sustainability of this transition are discussed by the author.

N. Labanca (✉)
European Commission, Directorate-General Joint Research Centre,
Unit C.02 Energy Efficiency and Renewables,
Via E. Fermi 2749, 21027 Ispra, VA, Italy
e-mail: Nicola.labanca@ec.europa.eu

N. Labanca (ed.), *Complex Systems and Social Practices in Energy Transitions*,
Green Energy and Technology, DOI 10.1007/978-3-319-33753-1_1

Introduction

I have had the privilege of being part of the scientific community who contributed to the detection of the top quark at the *Tevatron* collider at the *Fermilab* of Chicago. It is thanks to this opportunity that I could follow the scientific discourse that has developed around the detection of this elementary particle since 1995. That was the first time I could experience what nowadays seems to me a very curious and somehow misleading approach to present and interpret scientific advancements. I was quite young at that time and my readings had been until then mostly focused on the application of the scientific method to physics. My interests widened a bit afterwards and I could in this way realize that the strangeness I will describe probably affected the interpretations proposed by most scientists for the new conceptual artefacts they develop and divulge to the public. What I could identify was the presence and the relevant implications of a misleading interpretation typically arising among physicists, biologists, chemists and other natural scientists, when they explain the detection of new entities resulting from the interaction of given material objects with suitably prepared experimental apparatuses in terms of a *discovery*. The misleading character of this interpretation is generated by a methodological issue linked to a problem of *reflexivity* that can cause confusion among scientific communities and in the general public concerning *what* has been actually discovered. Researchers (notably researchers involved in so-called hard science) seem indeed particularly prone to neglect the fact that what they discover is actually the result of what they have partly contributed to create. Rather than signalling that their discoveries concern the interactions of measurement instruments with material objects prepared under the assumptions of theories developed in a given historical period, they often tend to present the outcomes of their experiments as the unveiling of absolute and eternal truths which could not be revealed beforehand because of still underdeveloped or limited cognitive and experimental capacities.

Because of this it happens, for example, that the Higgs Boson recently observed at the "Large Hadron Collider" (LHC) at CERN in Geneva is not just interpreted and presented as the result of the interaction of portions of matter (prepared according to the assumption of the so-called *Standard Model*) with the very sophisticated experimental apparatus that could be set up by the scientists of the twentieth century. This boson is rather assumed to be as old as the universe itself and capable of explaining how the mass of elementary particles has emerged in an extremely remote past. When generally referred to contemporary science, this apparently slight semantic shift concerning what should be meant by *discovery* is the cause of a series of misconceptions that ultimately result in the cancellation of the historical character of given conceptual artefacts and in the disregard or misinterpretation of the fundamental role played by the historical context for their discovery. Rather than as the result of a social construction, these conceptual artefacts are usually presented as eternal entities whose discovery just results either from a mere accident, or from a linear process of knowledge accumulation, or from the geniality of given scientists. This downplaying of history and of the social context in the processes leading to the creation of scientific concepts and artefacts is not so difficult to be verified. It can suffice to observe how these concepts are usually explained to students at schools and universities. Rather than as the outcome of a social tissue that creates, keeps them alive and can possibly decree their death, these concepts are either presented as the logical implication of assumptions taken within given undisputable theories or as entities whose nature can be easily inferred through intuition or induction. I could make this quite estranging experience several times during my university courses when I have been, for example, introduced to the concepts of time, space, mass, acceleration, speed, etc., through operational definitions whereby it was implicitly assumed that the measurement methods being presented for these physical quantities just served to quantify the extension or the intensity of manifestation of entities actually populating the real world.

What a surprise it has been for me to discover after my university courses that the existence of and the self-confidence of the professors introducing these physical quantities was not so unquestionable, that social communities could and actually had developed a variety of alternative conceptions and ways of life around these entities and that the wide scale application of their operational definitions could sometimes result in a very questionable reorganization and homogenization of societies.[1] The misleading and curiously seductive approach experienced by a university student is a common practice very often adopted by scientists and consists in presenting the *scientific construction* of physical entities as the detection and measurement of *natural phenomena*. The retroactive and distorting impact on the role played by history caused by this practice could hardly be overestimated. It transforms history in a kind of laboratory where all the activities being undertaken are seen as guided or constrained by the presence of recently discovered entities assumed to have always and incontrovertibly constituted reality.

Energy provides a nice example of how everybody is still nowadays trained to this distorting vision of the past. As Ivan Illich noticed already in the 1980s, images, explanations and advertisements of scientists contributing to this distortion abound in the media.[2] Still nowadays, energy is presented as something arcane that everybody has always needed, from the Australopithecus to today's Mr. Smith. Compared to their ancestors, today's people would be the luckiest ones because they can get energy very easily by pushing a button and without unpleasant side effects, at least as long as it is supplied in the most efficient and greenest way possible. It is here not very relevant to question whether this very common understanding of scientific discoveries makes contemporary people actually feel as the luckiest ones (because they would have much easier access to the possibilities disclosed by natural resources compared to forebears) or as the unluckiest ones (because they would not still have the access possibilities that will be disclosed by

[1] On this point see, for example, Bauman (1998).
[2] See Illich (1983).

discoveries of their descendants). What appears much more relevant is that such a view projects human activities within an advancement process whereby old and current knowledge seem to be destined to be continuously superseded by new knowledge whereby an increasing number of artificial phenomena can be explained. *Truth*, or the best approximation of truth currently available, would be represented by the most recent theories of science just because older theories cannot explain the latest phenomenic manifestations observed within the latest laboratories settings. The fact that these manifestations might be just human artefacts and that, within a kind of auto-referential loop, the theories and the assumptions whereby these manifestations are explained are the same theories and assumptions whereby these manifestations are created is apparently deemed not very relevant. I am convinced that this approach to science and to related technical applications is actually a dazzle that implies, or at least facilitates, a progressive cancellation of collective memory while legitimating a continuous activity of destruction and reconstruction. Moreover, I think that this type of blindness impedes highlighting relevant limitations concerning the application of scientific findings to everyday life. After all, if constructions of science are seen as natural entities actually populating everyday life like the tree planted in our garden or the cat living in our house, how could the circumstances of everyday life where their presence should not be invoked be identified?

Yet, our views over the world and our interpretation of past events would radically change when the assumption that conceptual artefacts provided by science represent *eternal* truths is simply released. If these artefacts would constantly be seen as the creation of a given historical period, as something that has had an *origin* and could, therefore, achieve an *end*, then history would get suddenly highly re-evaluated. Previous theories and worldviews considered as something obsolete and not thrust worthy could probably in this way be seen and understood as something capable of disclosing the implicit and, why not, socially negotiable assumptions of apparently undisputable present worldviews. History allows looking at the origins of the present grasp over the world and permits in this way to take to the foreground its implicit assumptions and limitations while possibly offering some glimpse concerning what can be expected in the near future. Having access to past ways of life can allow discovering different ways of perceiving the world and re-discussing present scientific assumptions. This experience can be extremely liberating and can disclose new research avenues. Its possibility is a consequence of the fact that concepts, principles and laws formulated by science are typically constructed and rigorously applied within laboratories under very restricted and controlled conditions, whilst all the details of the dynamics of everyday life escape by definition the reductions and abstractions performed and created by science.

This being said, it would be a big mistake to assume that the above-mentioned possibility can detract from the solidity of the outcomes of the scientific method and from the reliability of technics developed by its application. Energy has, for example, proved an extremely powerful concept to study natural phenomena and its impact on science can be hardly overestimated. This concept and the associated conservation and degradation laws have however originated within laboratories only in the nineteenth century and assuming that in the future they could be complemented by new and alternative concepts and principles to study and reorganize our environment is not an act of irreverence to science. This possibility, however, does not necessarily imply, for example, that it will be possible to extract useful work from a fluid of given heat engines by violating the energy conservation law or the Carnot theorem on heat engines efficiency. As long as natural phenomena are analysed in the thermodynamic framework in relation to the amount of useful work that can be extracted therefrom, no evidence has been so far able to prove the violation of these laws and theorems. Possible new and alternative explanatory principles will probably be adopted, not because these laws and theorems will be violated within laboratories, but because for some reason it will be deemed socially relevant and useful to overcome the inevitable reductions and distortions that can be associated with the application of thermodynamics to study the dynamics of human affairs. Despite, for example, societies are nowadays mostly modelled and described as motors and input–output systems by scientists and policy makers concerned with their energy sustainability, it would be profoundly wrong to assume that the dynamics of resources consumption of human aggregates can be completely captured by thermodynamics laws and that alternative research approaches based on different assumptions cannot improve our understanding of these dynamics in the future.

The general considerations so far reported have very practical implications that I have decided to discuss in this chapter for the case of one of the latest creations of science: complex systems. One of the main reasons for this endeavour is the fact that the notion of complex systems and associated phenomenal principles, although still probably lacking of a common understanding within the scientific community, are becoming omnipresent. Every field of knowledge is being revisited through complex systems theories, this indicating that the fundamental assumptions behind the creation of these entities are becoming invisible. To use Hans Blumenberg vocabulary, they are becoming kind of "absolute metaphors" whereby everything is explained.[3] The fact that some of the notions associated with complex systems can be so powerful to be associated with a reorganization of every aspect of social life is in my opinion astonishing. As much (if not even more) astonishing is the fact that these notions are so abstract that nobody has a clear picture of their meaning. Another connected reason that has stimulated my interest in the topic concerns specifically the implications of the social construction of complex systems for policies that can be implemented for energy sustainability. It might be stated that my endeavour has been animated by the following research questions: How can it be showed that complex systems have been socially constructed? How can they nowadays shape every department of knowledge? How is it possible to become more aware of the biases generated by reflexivity when complex systems science is applied to social phenomena? If complex systems are being socially constructed, then what may be the unexpected impacts of policies for energy sustainability that

[3]Blumenberg (1988).

are designed and implemented by assuming that, rather than being built on a massive scale in every department of human activity, these systems are actual entities obeying to universal and eternal laws? How becoming reflexive aware can generally help avoid unwanted impacts of these policies?

The best approach to address these questions is in my opinion represented by an historical enquiry on instrumentality as first attempted by scholars like Ivan Illich, Carl Mitcham, Jean Robert, etc.[4] Contrary to what is typically assumed, the origins of human artefacts generally named instruments are indeed not prehistorical. They probably have an origin that dates around the twelfth century and have subsequently undergone a series of fundamental transformations leading to the creation of so-called complex systems around the mid-twentieth century. These material transformations have been accompanied by as many transformations in the central metaphors whereby human action has been explained and natural entities have been imagined. By briefly describing these transformations, I would like to take to the foreground the implicit assumptions of present complex systems views and discuss the implications of their massive construction for energy sustainability and for policies that are informed by these views.

How to Intend the Social Construction of Complex Systems Outlined in This Chapter

In order to avoid possible misunderstandings, it is probably better to start by spending some words to clarify how the process of social construction of complex systems is being intended. Complex systems are primarily seen as an emergent social phenomenon related to the making and the using of artefacts. They are seen as the result of a non-deterministic co-evolution occurring within a bundle made of material objects, human habits, technical skills, ideas and narrations about reality and human action. Their construction is therefore not intended as the result of a linear sequence of transformations whereby new ideas and material objects are produced in given historical periods and completely replace preceding ones. On the contrary, it is assumed that, as happens with technologies becoming obsolete, previous ideas and material arrangements generally recede to a kind of background whilst sometimes serving as entry or leverage point for the creation of new material and conceptual artefacts which become dominant for reasons which may be often purely contingent. It is usually very hard, if not impossible, to understand and collect all the evidences needed to describe the dynamics whereby these transformations take place and such description is certainly not an objective of the author of this chapter. The huge difficulties often associated with a causal description do not nevertheless impede to identify the presence of relevant points of discontinuity in

[4]For a detailed account concerning how this historical enquiry has been conceived and developed see Cayley (2005).

the evolution of the mentioned bundle and to study the necessary changes that had to occur in relation to how tools of physical nature were conceived in order to allow the social construction of complex systems. Despite the ultimate reasons that have led to the emergence of these discontinuities might remain obscure, this type of study remains possible. The assumption made by the author of this chapter is that these points of discontinuity have been generated during historical periods when human tools have become objects of a particularly intensive philosophical and scientific reflection and that some of these historical periods coincided with specific periods of development, namely: (1) the time of the invention of mechanical science at the beginning of the twelfth century; (2) the time of the invention of the steam engines and the energy concept around the mid-nineteenth century; and, (3) the time of formulation of cybernetics as a discipline around the mid-twentieth century with its subsequent reformulation of the so-called second-order cybernetics lasting until the 1980s. The impressive technological developments that occurred during these periods have been accompanied by as many radical changes concerning how the making and the using of artefacts have been conceived. These radical changes are assumed to have substantially contributed to the social construction of complex systems and will therefore be described in this chapter in order to discuss underlying assumptions, potentialities and possible drawbacks associated with the massive diffusion of these quite recent artefacts. The proposed account revolves around the idea that the intensive scientific and technical reflections that have taken place in relation to human artefacts during the above-mentioned periods have transformed the concepts and ideas associated with the creation of these artefacts into central ideas and metaphors around which society has started being organized while leading to their massive technological reproduction. In this way, it could happen, for example, that the ideas developed around the technical instruments that were produced starting from the twelfth century made it possible to conceive the world and societies as a gigantic clock mechanism during the following centuries; it could happen that the massive production of steam engines and the thermodynamic principles established since the mid-nineteenth century made it possible to conceive the universe and human beings as consumptive and dissipative energy motors, or that information theories and technologies transformed ourselves and things out in the world into computer processors since the mid-twentieth century. Clearly, specific types of human artefacts and ideas developed around them might certainly have been in circulation before they become an object of social attention and scientific reflection and can continue being used also when largely superseded by new types of conceptual and material artefacts. Instrumental tools have, for example, been in use and described in all cultures since antiquity and continue existing also in the age of complex systems. It is however the fact that scientific and technological thought has transformed their presence into an issue of fundamental theoretical importance that has made their massive reproduction possible and has changed the concepts accompanying this reproduction into central metaphors whereby societies have been and still are being reorganized. When it comes to study how they impact on our environment, how they inform our ideas concerning sustainability of human activities, and how alternative ideas can be formulated, the study of this interplay between scientific reflection and massive production of given types of artefacts becomes much more relevant than any discussion concerning exactly when and how these artefacts and the social imaginary accompanying their reproduction have originated or have disappeared.

From Organa to Instruments

It is not difficult to realize how the distorting effect caused by projecting concepts and views that have been elaborated in specific historical period to previous and remote epochs of the past occurs also in case of the notion of "instrument". The idea that instruments are probably older than the human being is profoundly rooted in the contemporary social imaginary. Examples provided by literature of first humans using instruments typically refer to beings very similar to apes grabbing tree branches, stones or various kinds of bones to pick up fruits, broke nutshells or defend themselves from the assault of wild beasts. The idea that instruments date back to human prehistory is also supported by modern cinematography. Stanley Kubrick's 2001 *Space Odyssey* depicting an ape casting a bone into air that suddenly transforms into a spacecraft illustrates exemplarily how the social imaginary conceives instruments and how they are assumed to have been always present and to just evolve in their shapes and functions within societies. With a few frames, this director has managed to render a supposedly historical evolution of human instruments by reducing the forebears of the first types of utensils and weapons and the complex devices employed to travel into the space to a same matrix. This type of imaginary might perhaps appear very realistic to a hypothetical distant observer, a kind of extraterrestrial being having the privilege to observe from a large distance how the interactions of human beings with their environment have evolved during millennia. When observed from a large distance, the evolution of these interactions may indeed seem to keep some basic characteristics unchanged. Men and women of the ancient Mesopotamia ploughed their fields with oxen to produce the food they needed. Contemporary men and women may have substituted the plough and the oxen with tractors to get their food from the Earth. Overall, the same *end* seems to be achieved by using instruments that apparently evolved to alleviate as much as possible the burden of labour while increasing productivity. Some basic objectives seem to remain unchanged. Occurred changes seem to be limited to the *means* whereby these objectives are achieved.

Unfortunately, these kinds of descriptions and explanations completely neglect how perceptions and interpretations concerning the nature of human relationships with the material world may have changed with time and may have affected the way in which material and conceptual artefacts have been conceived and produced. According to several scholars, important historical discontinuities can indeed be identified concerning the way in which people perceive their relationships with the material objects they use. These discontinuities cannot be noted without considering how the ideas that men and women have about themselves and the surrounding environment have changed. This can be done only by adopting an analysis perspective that, rather than from a large distance, studies human societies from the inside. To identify and understand these moments of discontinuity, it is necessary to analyse cultures and how narrations and assumptions whereby people explain their actions and perceptions change.

This is the endeavour that the scholars I have previously mentioned have attempted in relation to instruments. Ivan Illich in particular has maintained that it is not possible to find any evidence confirming that western societies could conceive human artefacts as instruments before the twelfth century, i.e. there is no evidence dating before the twelfth century and indicating that human tools were seen as means designed and created to allow *any* person achieving predefined ends in the same way as, e.g. a typewriter can be seen as a device designed for any person to print letters of the alphabet on a sheet of paper. Writings by Plato, Pliny and Aristotle show that before that period it was not possible to distinguish, even verbally, between, for example, a hammer, a pencil or a sword and the hand that held them. The hand, the hammer and the hammering unit made of the hammer and the hand were all named *organon*.[5] The perception of human artefacts existing before that century induced to assume that only a particular type of hand could grab a particular type of artefact to perform a particular type of action. What could be defined as an inter-specificity existing between the person using an artefact and the artefact itself was so high to make a distinction between these two elements impossible or irrelevant. These elements were completely integrated and described by a same word. The possibility that a blacksmith, a knight or a baby could, for example, hold a sword to accomplish a same action was simply unconceivable. In order to understand how this could happen, it is necessary to realize that activities accomplished by persons were seen as activities whereby their soul showed its nature, i.e. they showed what this soul was and what it could be. Human activities were seen as activities of their souls. They did not aim at transforming the world. They were rather seen as aiming at transforming human souls according to their destinations. Rather than as autonomous entities, human artefacts were conceived as at the service of a body that was in its turn at the service of its soul. When compared to modern ways of thinking, this kind of imaginary certainly looks quite exotic. As pointed out by Marianne Gronemeyer,[6] a contemporary person asking for a job allowing his/her soul to find its destination would probably nowadays not be left in circulation. Nevertheless, it must have been exactly the fact that artefacts were at the service of persons' body and of their soul that determined this high integration between persons and artefacts (every person is indeed supposed to have his/her own soul and this soul differs from the soul of any other person). Because of this high integration with the person, human artefacts could not cause or be separated from the developments associated with the manifestation of a particular soul. They were somehow perceived as the reflection of this soul and it was mainly for this reason

[5]See Cayley (2005).

[6]See Gronemeyer (2012).

that mass production of artefacts was not conceivable before the twelfth century. Rather than by some technical limitation, this type of production was probably mainly prevented by the social imaginary developed around persons and their material environment. This kind of social imaginary must certainly have also had deep implications concerning how agency and *responsibility* for the effects of actions accomplished by employing organic tools was imagined. As these tools were completely integrated into and at the service of persons' body and of their soul, responsibility had to be necessarily circumscribed to the person who mastered them and could certainly not be ascribed to the tools themselves. In the following paragraphs of this section, it will be discussed how subsequent radical changes occurred in the social imaginary developed around tools will completely reconfigure the problem of agency and responsibility attribution. It has finally to be mentioned that, when analysed under the point of the view of the duality constituted by *persons* and the *material things* they employ to provide for their necessities, the relationship existing at the time of organic tools between the two poles of this duality has to be interpreted as a relationship where the pole constituted by material things and their possible conceptual representations were always submitted and adapted to the pole made of a particular person and the particular soul that manifested itself through the use of these things. In the remainder of this section, it will be shown that subsequent transformations occurred to the human tools can be very usefully characterized in terms of as many transformations occurring in the relationship existing between the two poles of this duality.

As a consequence of a radical change in the social imaginary that occurred most probably during the twelfth century, human artefacts indeed became separated from the body and were not any more principally seen as at the service of persons' souls. Starting from that century, human tools underwent a metamorphosis that changed them into *instruments* that could be used by *any* person to achieve abstract and predefined *ends*. Various hypotheses deserving further investigations have been formulated to explain the nature of this metamorphosis. One of these hypotheses is that this transformation occurred when mediaeval theologians started assuming that God had delegated to the Angels the task of acting upon the world by means of instruments named *corpora coelestia* that were moved around the Earth. Illich maintains that the new type of causation associated with this new version of a myth could have made possible for the first time to conceive specific types of artefacts as a *means* that can be used by *anybody* to achieve given *ends*. The utilization of the *corpora coelestia* as neutral instruments transmitting Angels' intentionality would have led to conceive that also human intentionality could be transferred to neutral artefacts[7] and the idea that men could share with the Angels this capability of administering the world by creating and using instruments would have come for the first time to the mind of the Saxon canon regular Hugh of Saint Victor. This leading

[7]See Cayley (2005). Within the interviews documented in this book, Illich points out that the notion of an instrument whose functioning is mostly independent from the capacity, the will and the intentions of its users may be also closely linked and is coeval to the birth of the idea that sacraments are God's instruments for man's salvation.

theologian and teacher of the twelfth century, whose books became mandatory reading for people seeking for a liberal education until the seventeenth century, would have been one of the first investigators on the nature and origins of tools for manual labour.[8] People have always used tools and reported about their use since the antiquity, but their presence was somehow taken for granted whilst their shape and nature changed from culture to culture, as it happens, for example, with language, until the twelfth century. It would have been only around the year 1120 that tools of manual labour were recognized as a social and philosophical theoretical problem by scholars like Hugh of Saint Victor, Honorius of Augsburg and Theophilus the Priest. As Illich explains to us, the twelfth century was indeed a period of intense technical innovation in north-western Europe with an impressive increase in the consumption of iron, in the number of mills and in the variety of machines that these mills could activate. It is in this period that Hugh of Saint Victor's ideas concerning the possibility of improving tools for subsistence appeared and tools started being studied by science in terms of *means* that can be used by any person to perform specific and predefined actions. The transformations occurred in the social imagery during the twelfth century, would have led to conceive human tools as objects which can embody human intentions and remain clearly detached from the body of the persons using them. This newly perceived separation or *distality* (Cayley 2005) between the instrument and their users would be at the roots of the separation between an objective reality and the subjects who know and act on it by using tools. Whilst persons and their organic tools were seen as highly integrated and inter-specific in the previous centuries, a detachment between these two entities was instead created with so-called "*instrumenta separata*". Organic tools were seen as utensils whose presence was taken for granted. Their fabrication did not result from a conceptual representation of their functions by their users, and handling and usage were probably the main patterns whereby their nature of tools was discovered. There are indeed very good reasons to believe that the description of a material thing as, for example, something "for hammering" is much "more primordial than any conceptual description of a hammer as being of some particular size, shape, weight and colour".[9] Contrary to organic tools, instruments can instead be the result of and have paved the way for engineering design while creating an object–subject dichotomy.

The new perception that developed around human tools would have ultimately resulted from a change in how causation was intended. Causation was indeed mainly explained in terms of the Aristotelian *causa materialis*, a *causa efficiens*, a *causa formalis* and a *causa finalis*[10] until the twelfth century. Whilst persons and their tools could not be distinguished within the *causa efficiens*, the birth of

[8]See Illich (1981), pp. 75–95.

[9]See Mitcham (1994), p. 256.

[10]In his Metaphysics, Aristotle distinguishes among four types of causa: *causa formalis, causa materialis, causa efficiens, causa finalis*. The difference among these can be grasped by the classical example of the sculptor. To make a statue the sculptor (causa efficiens) is supposed to produce changes in a block of marble (causa materials) with the aim of producing a beautiful object (causa finalis) having in mind his idea of the statue to be carved (causa formalis).

instruments and the way in which human intentionality can be transferred to them would have to be associated with a fifth type of causation (named by Illich *causa instrumentalis*) generated within the causa efficiens.[11]

The consequences of the separation established between persons and their instruments are huge and manifold. Large-scale standardization of artefacts became, for example, possible only once this separation was created. At the same time, it became possible to assume that human intentionality could be transferred to objects and two contrasting views concerning agency and the responsibility for the consequences of instruments mediated actions could be generated. It became indeed possible to assume that instrumental tools could be employed by any person provided with sufficient skills and information background without affecting or redefining his or her intentions. For this reason, a kind of neutrality and objectivity was generally ascribed to them, whereas the full responsibility of the consequences of the actions they allowed to perform had to be attributed to the will of their users. Paradoxically, however, it was exactly because of this separation that it became possible to conceive that agency and responsibility could also be entirely attributed to instruments that appeared as able to deeply redefine human intentions with unexpected and often disastrous consequences for humans and their environment.[12] These contrasting assumptions and perceptions, still largely present in contemporary society, have deeply influenced any field of knowledge and human activity since they entered diffusely the public discourse. With instruments, the two poles of the duality made of the persons and of the material things they use during their everyday life became more independent and autonomous.

As pointed out by Marianne Gronemeyer,[13] the artificial separation created by instruments has also radically changed the sense of the existence of human artefacts and of human beings. Contrary to *organa*, instruments are not artefacts at the service of the human soul. With instruments, human artefacts can become independent entities generating effects that can in principle be completely unknown and deserve investigation. At the same time, however, this separation is what makes it possible to conceive for the first time an idea of *delegation* of human tasks to machines. It is this separation or *disembodiment* that makes it possible to think of human artefacts as a kind of automata that can be activated, for example, by pushing a button. With instruments, human artefacts can be changed into autonomous entities to which human action can be delegated and their autonomy is exactly what makes it possible that the effects of their employment escape human control and foresight. Despite their birth makes it possible to think of the world and of

[11]Aristotle's *causa efficiens* did not indeed make possible to distinguish between the artefact and the hand handling this artefact.

[12]The current debate on increasing access limitations to weapons for US citizens is an example of this dichotomous perception. Part of the public opinion attributes the responsibility for the increased number of murders being registered in US to the wide presences of weapons among US citizens. Another part (weapons manufacturers especially) maintains that the responsibility for murders has to be ascribed to the will of murderers and not to the weapons themselves.

[13]See Gronemeyer (2012).

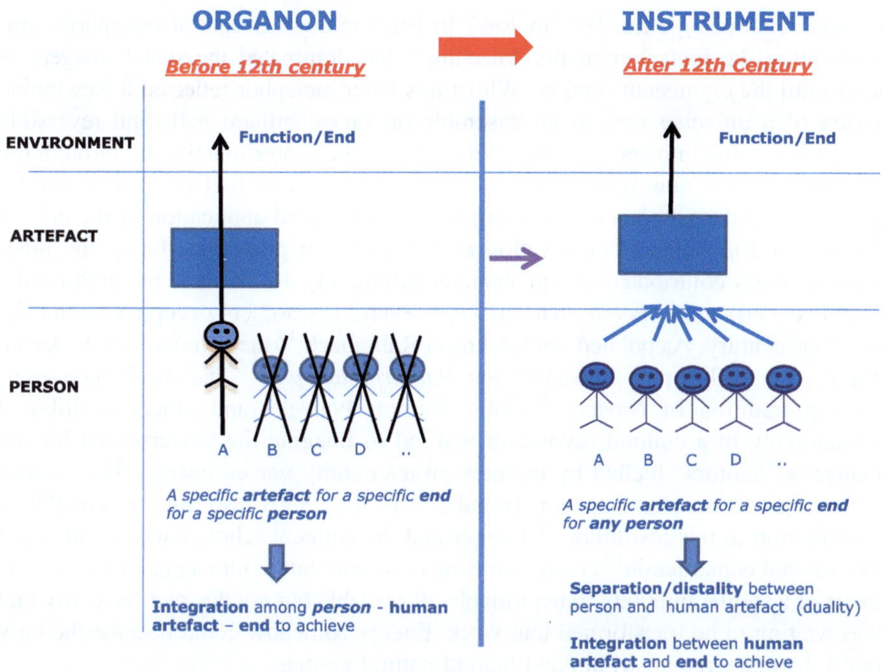

Fig. 1.1 Transformations associated with the birth of instruments

human beings as machines obeying deterministic laws, instruments project on human artefacts a shadow of unpredictability that was unknown before their creation.

The transformations associated with the creation of instruments are schematically illustrated in Fig. 1.1 by distinguishing among persons, artefacts and functions accomplished thereby. However, it has to be stressed that this distinction has just a descriptive function and did not hold, for example, in case organic tools. It has been introduced to illustrate some major metamorphoses occurred in the way in which human artefacts were conceived afterwards.

From Instruments and Machines to Motors

The central metaphors whereby the world and human beings have been imagined have been affected by another radical change that occurred in the mid-nineteenth century. As happened with the transformations that led to the birth of instruments, this later change has taken with it a transformation in the way in which natural phenomena were conceived and delegation to human artefacts was imagined and realized. As briefly discussed in this section, the invention of the energy concept has had a fundamental role in a cultural change that still deeply affects contemporary society. This change has led "motors" to become another central metaphor complementing the metaphor of the "machine" that dominated the social imagery at least until the eighteenth century. Whilst this latter metaphor reflected a Newtonian vision of a universe seen as an ensemble of forces, billiard balls and reversible mechanisms, the universe became a kind of gigantic motor functioning through the degradation of a new type of natural resource serving as fuel input. It is not by accident that the scholars contributing to the widespread application of the energy concept and associated conservation and degradation principles during the nineteenth century contributed also to abandon definitively the chimeric research for the so-called *perpetuum mobile* that has kept several researchers occupied during the previous century. As pointed out by Anson Rabinbach,[14] the invention of the steam engine, the philosophic impact of the *Natuarphilosophie*[15] and the French engineering tradition of Navier, Coriolis, Carnot, Poncelet and others contributed substantially to a cultural revolution that led to imagine the universe and human beings as "motors" fuelled by the new protean entity named energy. The famous lectures given by Hermann von Helmholtz in the 1840s also gave a remarkable contribution to this revolution.[16] Energy and the eminent scholars who contributed to its social construction[17] changed the universe and nature into a gigantic *reservoir* made of a single, infinitely transformable, degradable but not destructible entity that was waiting to be transformed into work. Energy somehow could become the only real substrate existing within and behind natural entities.

It however passes often unnoticed how, despite that common parlance implicitly acknowledges an indisputable ontological concreteness to energy still today, this concept has actually undergone a series of profound metamorphoses within laboratories of physicists and engineers that actually started already in the seventeenth century. These metamorphoses have progressively led to associate energy with a magnitude remaining intact during collisions of rolling balls and springs

[14]See Rabinbach (1992).

[15]This philosophy drew on Shelling and Hegel and postulated the presence of an Urkraft or vis viva containing the secret of energy and life in the universe. It was particularly important in the work of Mayer and Helmoltz who contributed in important ways to the social construction of energy. For further information concerning the link between Naturalphilosophie and the energy conservation principle see for example Caneva (1993), p. 310.

[16]For further information on Helmoltz lectures see e.g. Rabinbach (1992).

[17]Mirowski (1989) points out that the energy concept has to be probably seen as the result of the joint and mutually reinforcing social constructions of invariants and conservation principles taking place in the fields of physics, biology and economics. According to this scholar, the structures of explanation produced in these three different fields have probably always been homeomorphic and would legitimize each other even in the face of possible disconfirming evidences produced in each field. The above-mentioned mutual reinforcement would have been already operating when the institution of money was disconnected from any reference to a particular commodity and became the abstract representation of pure value, when the dual concepts of the organism and of natural selection were established within the evolution theory of Darwin and when the energy conservation and degradation principles were established by physicists and engineers around the mid-nineteenth century.

oscillations, with a primordial entity obeying conservation and degradation principles, with states of electromagnetic fields, with fields symmetries, with time homogeneity.[18] On the one hand, energy has therefore been suggesting since the last two centuries that there is no free lunch, that the whole universe and all human activities are naturally regulated by conservation and degradation principles indicating that there is a cost to be paid for anything we do and that nothing can be created for free. On the other hand, the energy concept has evolved within laboratories in such a way that cosmologists admit nowadays that everything could have begun from a vacuum fluctuation and that the whole universe could actually be a free lunch.[19] Despite these quite recent evolutions and discrepancies in the interpretation of the energy concept, the organization of economies and societies remains entirely informed by the idea that everything happens in the universe thanks to the consumption of the protean entity that has been named energy. The implications of this misconception for how energy sustainability has been and is still being conceptualized will be discussed in a subsequent section. What deserves to be briefly specified here is rather how the rise of the energy concept has changed the way in which human beings and human delegation to machines is conceived. A description of the changes induced in the previously mentioned duality made of persons and the material things they use during their everyday life can be very insightful in this respect. Once again, the modifications occurred in how material tools were conceived mirrored as many modifications in the way in which persons and action delegation to machines were imagined.

As long as human tools were principally seen as instruments, persons and the outside world were identified with clockwork reversible mechanisms. With motors, human action became dependent on the provision and on the optimized consumption of suitable and quantified resources inputs. Delegation to machines assumed in this way a connotation of human *empowerment* to be achieved and/or maintained through the consumption of various forms of energy. Either actions were accomplished by using motors or by human bodies, agency and the reproduction of these actions were in this way associated with and subordinated to the consumption of quantifiable energy resources units. If the disembodiment and separation between tools and human bodies that were generated by instruments led to conceive actions in terms of mechanisms, motors subordinated these actions to the consumption of an abstract entity named energy. Material and physical infrastructures whereby these actions could be realized became a kind of energy *stock* transformers that may work and produce expected outputs more or less properly or more or less efficiently and this change was perfectly reflected in how the shapes of these infrastructures and their integration into the environment changed. Whilst previous machines, like windmills, tended to fit into landscapes and to put into relief specific features of this landscape, the shape of the plants that have started being built since the nineteenth century to extract and store the energy

[18]For further information on these transformations, see Mirowski (1989).

[19]For further information, see Tryon (1973) and Akatz and Pagels (1982).

serving as motors fuel were and still are completely abstracted from the landscape and just stand-up as ready to use objects without any specific aesthetical connection with the surrounding environment.

Moreover, the energy concept has taken with it also a completely new relationship to be entertained with *time*. Energy and the devices relying on energy use that started to be massively built contributed indeed to interpret time as a quantifiable resource that is consumed at a constant pace and whose measurement and consumption can be used to re-organize and control human activities.[20] Energy, time and the associated conservation principles allowed in this way to put human activities under a scarcity paradigm, according to which the consumption of energy and time units needed to perform given activities inevitably causes that less energy and time is available both at the individual and the social level to perform other activities. Needless to say that this scarcity paradigm has been alone the reason for a hugely intensified and more energy efficient delegation to machines whereby people were supposed to liberate their time to perform additional activities. This type of imaginary, however, was fundamentally based on an idea of energy derived from fossil fuels. Despite the thermodynamics laws that have been established within laboratories may induce to think differently, the nature of energy can indeed change and this change can generate different types of social imaginaries and different types of perceptions concerning how human activities can be organized. In the following sections, it will be discussed, for example, why renewable energy is a fundamentally different energy type compared to fossil fuels energy and how a transition to renewable energies implies, among others, a different relationship with time and therefore a different organization of human activities. Energy and time are somehow the two sides of a same coin and a modification in the nature of one side inevitably induces a modification in the nature of the other side. The transformations entailed by the social construction of energy are schematically represented under Fig. 1.2.

From Motors to Complex Systems

The separation or *distality* that instruments have created between persons and their artefacts has made it possible to conceive that any end can be achieved by fabricating means that can be used by the arbitrary hand of an arbitrary actor.[21] Instruments, however, are still entities deeply integrated into the ends they allow achieving. They are indeed conceived and fabricated to perform specific functions and their structure and shape are deeply dependent on these functions. A further fundamental transformation takes place when it becomes possible to assume that a *same* material object can be produced and used by *any actor* to perform *any kind* of function. Human artefacts get so separated in a very particular way from the ends

[20]For further information on this transformation see Perulli (1996).

[21]See Gronemeyer (2012).

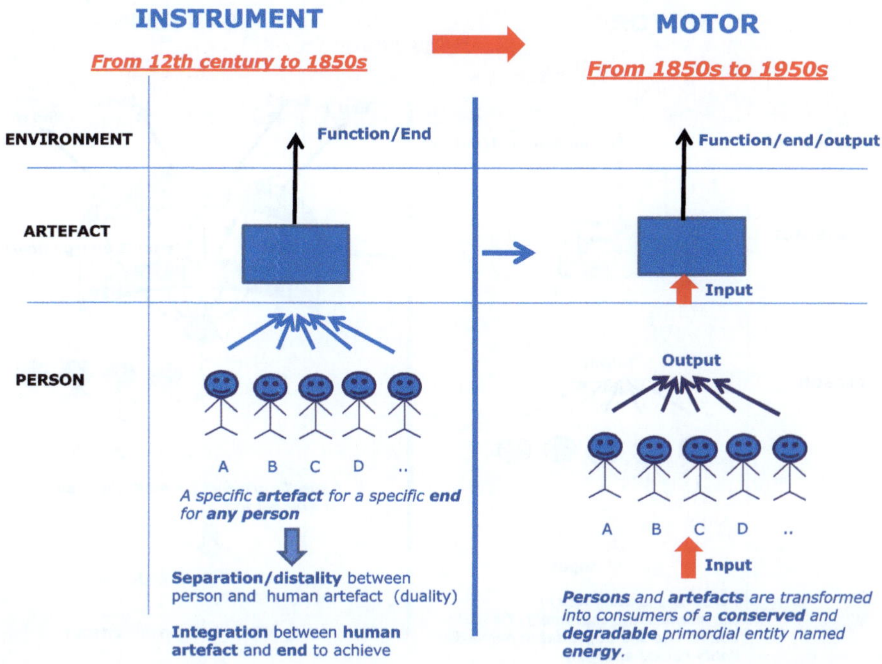

Fig. 1.2 Main transformations associated with the transition from instruments and machines to motors

they allow achieving. This is a fundamentally new and common characteristic of a large series of quite recent conceptual and material artefacts that can be identified with so-called complex systems. The main phases of the social construction of these systems can be found by identifying the main knowledge advancements that have made their large-scale production and employment possible. As the next paragraphs will try to illustrate, the latest phases of this construction occurred probably after the mid-twentieth century and the previously described passages to instruments, machines and motors somehow represent the necessary preliminary conditions for this latest transition. Some of the basic characteristics of complex systems are schematically represented in Fig. 1.3.

The implications of this third type of metamorphosis occurred to material artefacts can be understood by referring to a large series of nowadays very familiar devices embedding human beings within complex systems. Personal computers, smart phones, computer servers, audio-visual systems and all devices generally subsumed under the category of computer and information technology are the most common examples of these types of device. It is indeed quite easy to realize how they allow or are supposed to allow people performing an increasing number of functions. By interacting with a computer, a person can nowadays, for example, send mails, write a text, purchase products, call other persons, etc. With the increase in the number of functions they allow performing, these material objects cannot

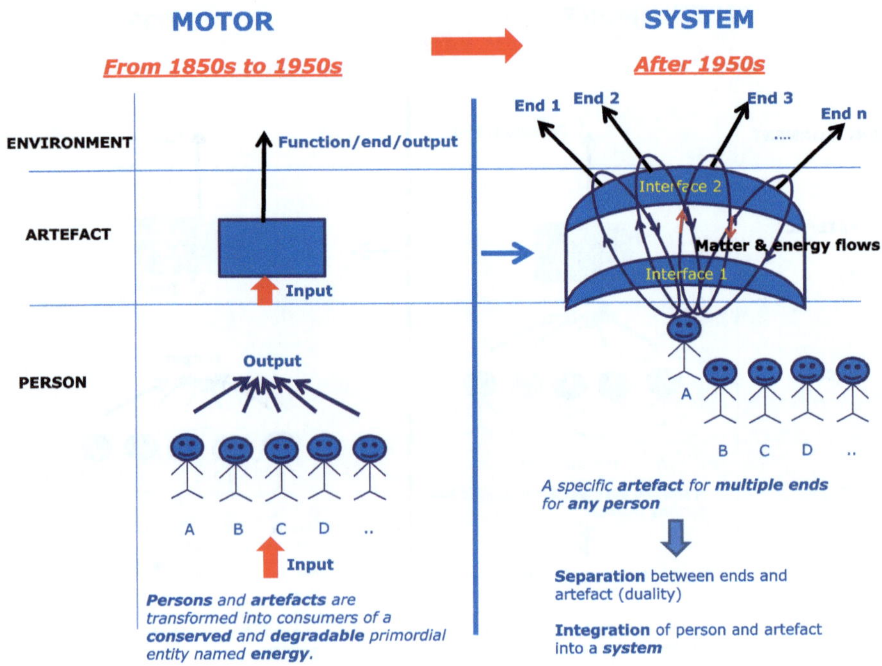

Fig. 1.3 Main transformations associated with the creation of complex systems

anymore be considered as instruments, i.e. as means conceived to allow any person achieving a *specific* end. When they are used by people for most of their interactions with the external environment, they rather become ends in themselves and get in this way separated from each of the specific ends they allow achieving. At the same time, they become more and more integrated in the human body because the increased number of functions they allow performing implies that people have to stay "attached" to them for longer time.

It is for this reason that the transition to complex systems makes human artefacts constituting these systems very similar to kinds of human *prostheses*. Human prostheses are indeed generally assumed to allow disabled people performing the highest possible number of functions compared to normally endowed ones. A prosthesis replacing a missing arm should, for example, allow grabbing, writing, feeding, driving and performing all the other functions that normally endowed people can perform. The higher the number of functions that can be accomplished through it, the better the prosthesis and the higher its integration into the human body. The same principle applies to the types of artefacts previously described. The higher the number of their functions, the closer and the more integrated into the human body they become. The *distality* between user and the used artefact that characterized the age of the instrument gets lost with systems. A man can still decide whether to use or to leave a hammer and the hammer remains the tool of a man as long as this hammer is conceived as an *instrumentum*. When the unit made by a man and a hammer is conceived in terms of a system, this unit becomes a strange entity made of quasi-objects and quasi-subjects, to use Bruno Latour terms,[22] wherein information flows. One part of the system defines and could not exist without the other. Systems are units integrating special types of devices into persons' bodies. If the analogy with a human prosthesis would not be sufficient to illustrate this fact, the quite common experience of the relationship we have started entertaining with computers and cell phones can certainly provide everybody with further insights concerning the nature of this progressively increasing integration. Complex systems represent the tangible realization of the myth of the *cyborg*. It is not certainly necessary that chips are implanted in human bodies for this to happen. When the number of functions accomplished through these new artefacts increases, they start constituting a sort of membrane that inserts itself between our senses and the outside world. Whatever the physical distance existing between us and them, they function as a kind of very thin plastic bag that can perfectly adhere to our body and mediate any relationship undertaken with the external world. Given the high number of functions that they allow accomplishing, they end up shielding and impermeabilizing the body from the outside world. At the same time, however, they can perfectly adhere to the body. Contrary to instruments, these types of artefacts can indeed be extremely flexible and adaptable, this adaptability being due to the fact that their functioning relies on the standardized transmission of an extremely immaterial and protean entity. Whilst instruments standardization relates to their shapes and functions, systems standardization relates indeed to the information codes they employ.

The type of integration between person and artefacts realized within systems should however not be confused with that realized by the organic tools previously described. Systems rely on a double interface whereby a double translation is constantly and actively performed. As it can be probably understood by Fig. 1.3, the first interface translates and reduces acts accomplished by the user into codes and messages that can be processed by the artefact and translates codes generated by the artefact into inputs and messages that can be understood by the user. The second interface translates instead the inputs from and the outputs to the external world. Systems can ultimately be seen as units made by persons integrated into artefacts whose functioning is based on the elaboration of information within very complex feedback loops. Change and stability become in this way the result of positive and negative information feedbacks which generate along system feedback loops following external perturbations. The distinction between action and reaction becomes often meaningless because circular causation loops are the only ontological entities of systems. It follows that the loss of distality associated with systems makes the conceptual category of the person and the distinction between subjects and objects also meaningless. The only elements needed to describe systems dynamics are indeed the above-mentioned information feedback loops circulating between persons and material objects. This loss of distality somehow

---

[22]See Latour (1993).

also implies a loss of persons' control over the material objects they are interacting with. When the interaction between a person and a material object is described in terms of a system, the interacting parts can indeed constitute a whole pursuing own ends.

Systems can inscribe persons' intentionality into their workings. Heinz von Förster described, for example, a man walking a dog[23] as a system with the man, the leash and the dog forming a unit processing informational signals that manages to make its way down the sidewalk. In the same way, the system made of a man interacting with a modern Internet-connected computer can be described in terms of a two-component unit processing signals to achieve its own ends in the surrounding environment. The possibilities that persons can exercise some type of control over the evolution of "their" systems are therefore markedly reduced or even nullified. As already happened with the creation of instruments, the notion of agency and responsibility for human actions are hence once again profoundly redesigned. A description of the transformations induced in the duality made of the persons and of the things they use to provide for their necessities can be once again very insightful in this and in many other respects. It is indeed not very difficult to realize how the new imaginary associated with the new type of material artefacts constituting complex systems has mirrored a change in the imagery associated with persons and their psychological and organismal dynamics. It can, for example, hardly pass unnoticed how complex systems have contributed to reformulate psychological problems in terms of communication problems linked to how information is processed among and within persons.[24] At the same time, bodies of persons have been progressively identified with immune systems capable of keeping the value of its vital parameters (e.g. blood pressure, glycemic rate, etc.) within predefined variation ranges in a changing environment while body health has been identified with a risk profile, i.e. a list of numbers representing the conditional probabilities that the measured values of its vital parameters may correspond to a system evolution towards a status threatening its own existence. Genetics is then another research field where human and not human organisms have been progressively identified with the information processors representing the central metaphor around which complex systems are being socially constructed, i.e. computers. However, it has to be stressed that, although the transformations that have accompanied the creation of these systems might seem to integrate and completely abolish any distinction between the two polarities of the previously mentioned duality, a more attentive analysis reveals instead that, rather than disappearing, this duality moves from persons and their artefacts to the unbridgeable gap and separation artificially established between complex systems functions and their material infrastructures. The puzzle posed to computers programmers having to find suitable algorithms whereby specific human functions can be reproduced by technological devices is exemplary of the nature of this separation and of how the

[23]See Cayley (2005).

[24]See for example Watzlawick et al. (2014).

approaches that can be elaborated to overcome this separation actually redefine the nature of the problem without resolving the duality at stake. As also mentioned at the beginning of this section, this problem is always and can only be formulated in terms of a *translation* problem consisting in finding suitable information algorithms obeying specific internal logics that can serve to faithfully reproduce these functions. This exercise is undoubtedly stimulating and often provides with extremely useful solutions. Nevertheless, it also progressively and increasingly contributes to create and maintain a separation between an underworld obeying the rules of information theory and an upper world where people continue conducting their everyday life. Either referring to the entities integrating humans and their tools, or to the entities studied by biology, sociology, physics, linguistics, informatics, etc., complex systems take always with them an inescapable and irreconcilable separation between observable functions and meanings and their underlying material or conceptual infrastructures supposed to generate these functions and meanings by following own internal rules. As also discussed in the next section, despite complex systems and science informed thereby seem to propose an holistic view of the world, these artefacts and the associated body of knowledge remain profoundly dualistic.

The type of integration achieved within complex systems entails however also a change in the way in which agency, human delegation and disembodiment are realized. It has been previously mentioned that the social construction of instrumental tools presumably made it possible to conceive human artefacts as autonomous entities to whom human intentionality can be transferred because of the distality created between them and the persons using them. Moreover, it has been pointed out that the subsequent social construction of motors has led instead to conceive delegation in terms of *empowerment* achieved thanks to the optimized consumption of a natural resource named energy. Strange as it may seem, with complex systems, delegation and disembodiment are instead the result of an *integration*. Persons integrated within complex systems have to be imagined as nodes of very wide and highly interconnected information networks. The spatial extension and the strong coupling of these connections enhance incredibly the geographical area that can be covered in very short time and the power capacity that can be activated by single human actions. These same characteristics however render complex systems similar to entities which follow own logics and escape the control of individuals and makes often practically impossible to track the ultimate consequences of single human actions and ascribe some kind of personal agency and responsibility for these consequences. With complex systems, the views of the extremely wide regions of the world that become achievable through artificial prostheses and the actions that can be accomplished thereupon by individuals located in a given place are inevitably filtered by the artefacts these individuals are integrated into and are shaped by a necessarily limited number of circuits and information feedbacks constituting the whole system. Given the wide spatial extension and the strong couplings operating within complex systems, changes occurring within them can however also be dramatic and completely unpredictable. They can occur after a long period of stasis or can repeat after a short time according to unknown frequencies. Too rigid infrastructures can represent a serious obstacle to properly face these types of events in this type of environment. Complex systems require extreme flexibility and adaptability to individuals integrated into them. Because of their characteristic dynamics, they entail therefore a different relationship with *space* and *time*. Motors have, for example, contributed to interpret time as a quantifiable resource that flows, uniformly and uni-directionally, whilst the perception of space that accompanied the diffusion of motor-like devices was informed by material infrastructures symbolized by the huge silos where energy and material resources could be stoked. With complex systems, time becomes instead a discontinuous and punctuated entity whilst assuming qualitative and relational characteristics. Its flowing is marked by single and sudden events whose occurrence in a place depends on a series of ever-changing spatial relationships with other places. Material infrastructures might have to be rapidly disassembled and reassembled and have to become more and more flexible and liquid in order to allow coping with unexpected challenges. All these changes depend substantially on a change in the nature and in the role played by energy sources. Complex systems undergo indeed intensive material and non-material exchanges with the external environment and are *open* by definition; this fact makes their dynamics naturally dependent on exogenous rates of energy supply. Put in other words, energy sources involved in the dynamics of complex systems can more hardly be described only in terms of available and predeterminable amounts of resources *stocks* that can be used at any time. They have often to be seen as *funds*[25] of resources whose utilization occurs according to not pre-establishable rates. The combined changes occurring in how time, space and energy are perceived should not come as a surprise. As already mentioned, these are closely linked and inter-dependent physical quantities and changes induced in the nature of one quantity mirror the changes occurring in the others and vice versa.[26]

The Role Played by Information in the Social Construction of Complex Systems

The short explanations offered in the previous paragraphs may be sufficient to hint that *information* is one of the main building blocks of complex systems. The type of information at stake is however very particular and its peculiarities have been put into evidence by the seminal works of scholars like Claude Shannon,

---

[25]For a definition of stocks and funds see for example Georgescu-Roegen (1971). This aspect will be further discussed in a following chapter section.

[26]The fact that changes in the nature of space, time and energy are being inferred through changes incurred in the nature or in the interpretation of observed phenomena should not come as a surprise either. If, for example, the events reproduced by thermodynamics have led to conclude that time flows uniformly in one direction, events reproduced by complex systems can led to conclude that time is discontinuous and punctuated.

Gregory Bateson, etc., starting from the mid-twentieth century. Because of one of the strange curiosities I have referred to in the introduction to this chapter, it is generally assumed that this relatively new type of information has always been present in nature, for example, within the DNA of any biological organism. Contrary to what is generally assumed, there are instead at least three important transformations that had to occur in what has been traditionally meant by information, before this notion could be conceived as a kind of natural entity and contribute to the social construction of complex systems. The main transformation that had to occur has consisted in assuming that *information can exist and play a role in nature without the presence of a human reading it*. Although it could at first sight appear quite irrelevant, the assumption that any (part of any) biological organism and even machines can somehow be regulated by the transmission of something named information is a huge step towards abstraction. This reinterpretation of information has caused a deep change in the nature of this entity and would most probably have been unthinkable without the advancements that occurred in computation science during the first decades of the twentieth century.[27] This has led, for example, to conceive that a particular type of information is available in any biological organisms and regulates their epigenesis and phylogenesis.

A second closely related important change concerning information that has also taken place in the past century is instead responsible for having made of information something that can be calculated. Today, we are probably not very surprised in hearing that information can be reduced to *numbers*. For this to happen, it has been however necessary to think that the information that can be found within any sign and natural manifestation is actually *one* manifestation among a *finite and prefigured number* of possible manifestations. To a certain extent, this is equivalent to assume that anything that can be read or written in the book of nature actually corresponds to one or more combinations of a finite number of letters constituting the alphabet employed by nature. By translating natural manifestations into numerable combinations of a limited number of signs, this change has led to create classifications that may somehow resemble to Classical Age taxonomies[28] whereby all possible identities and differences detected in nature were arranged into ordered tables. Unlike taxonomies that were created at the end of Renaissance, the ordered tables created by the modern notion of information are huge tables written in the binary language of computer technologies. For each of the messages that can be written in this language, it is nowadays possible to assess its so-called information content by calculating the ratio between the number of binary combinations corresponding to this message and the total number of totally possible combinations. Whereas the first mentioned transformation has made possible to think of an abstract entity at work within the natural world, the second one has hence led to reduce it to ordered tables that can be studied and manipulated by using computer

[27]For a detailed description of the evolution of this concept see for example Poerksen (1995).

[28]See for example, Foucalt (1966).

technologies. A fundamental contribution to these changes has come also from Alan Turing and a series of other eminent scholars during the first decades of the nineteenth century.[29] By demonstrating that any function that can supposedly be calculated by humans can be calculated also by a machine, these scholars have contributed to transmogrify also the human part of the natural world into autonomous computational systems whose functioning is based on the elaboration of information.

A third and very relevant metamorphosis in what has to be meant by information relates finally to the nature of the elementary units that constitute it. This nature can be grasped through the definition provided by one of the fathers of the cybernetics: Gregory Bateson. By defining information as "*a difference which makes a difference*"[30] within any natural system exhibiting a mind-like behaviour, Bateson has contributed to give information a purely *relational nature*. Through information, any entity of the natural world comes ultimately to be made of infinite chains of relationships (i.e. differences) with other entities. Rather than from some kind of intrinsic characteristic, objects are defined by their relationships with the surrounding environment. To explain this, Bateson provides a variety of examples ranging from phylogenesis, to phenomenology of perception, to linguistics. When, for example, he disserts on what an *elephant's trunk* is phylogenetically, he concludes that what defines the meaning of the trunk is nothing but the context where the trunk grows from within the elephant's embryo. It is the fact that what we call trunk "stands between two eyes and north of a mouth".[31] The trunk would hence not result from an intrinsic characteristic of a specific embryo part. It would rather be the result of an internal process of communication during embryo growth. Bateson offers a series of experimental evidences that can prove this conclusion. He mentions, for example, the experimental evidence provided by a study of unfertilized frogs' eggs demonstrating that for these eggs "the entry point of the spermatozoon defines the plane of bilateral symmetry of the future embryo". The parts of the frog's egg that can become the frog's nose would hence be defined by their relationships with other egg's parts based on the spatial relationship of all the egg's parts with the axis fixed by the spermatozoon entry point. There would not be any specific internal characteristic that can predestine any specific part of the

[29]See for example Teuscher (2004), p. 216.

[30]See for example Bateson (1972). Terms and expressions like *information, information about a difference, difference that makes a difference* are used interchangeably by Bateson. In order to produce information, two (real or imaginary) entities are needed such that the difference can be immanent to their reciprocal relationship; moreover this difference must be such that information about this difference can be represented as a difference within some information processor (e.g. a brain or a calculator). Each of the two entities producing information is a non-entity if taken alone. A relationship between two parts or between a part at time 1 and the same part at time 2 is needed in order to activate some third component that could be defined as the *receiver*. This receiver (e.g. a terminal sensor in an organism) reacts only to a difference, to a change. As the reaction of the receiver is in its turn nothing but a difference, this reasoning implies that *information is just a difference producing another difference.* See the original explanation in Bateson (1979).

[31]See Bateson (1979).

unfertilized egg to become the frog's nose. Another example is then provided for sensory perception when Bateson illustrates how human eyes require motion to see anything. In this respect, he explains how static objects would disappear from our sight without the tremors that move our pupils along objects borders. It is the variation, the difference generated during our perception by the perceived differences between the static object being observed and adjacent objects and surfaces that makes the perception of the former object possible. A further example to explain the relational nature of information is then taken from linguistics. Bateson wonders in this case what gives meaning to letters, words and sentences and provides the following reasoning to answer this question and to support his thesis. He maintains that the letter "p" would have no meaning if, for example, it were not part of the word "perhaps". The word "perhaps" would have in its turn no meaning if, for example, it were not part of the sentence "perhaps this is soap". This sentence would in its turn have no meaning without the context where it is stated and this meaning would be different if the sentence were mentioned, for example, in a bathroom, on a stage or within the reasoning presented in this chapter. Meaning and information content would therefore be purely relational and depend on a series of piled contexts. In agreement with linguists like Ferdinand de Saussure, Bateson concludes that, rather than from an objective relationship between the sign and the thing this sign refers to, meaning emerges from a series of relational contexts that can be established with other signs. *Signifier* and *signified* are in this way completely separated. Moreover, as the last example would prove, the contexts at stake would always be hierarchically organized and it would never happen that the smaller context determines the characteristics, the evolution and the meaning of the larger context. According to Bateson, hierarchies necessarily cross and entirely organize complex systems, either these systems are constituted by biological organisms or by the aggregates studied by linguistics. Within complex systems, hierarchical organization actually appears already at the level of the two irreducible entities that constitute each elementary sign, each information unit, whatever this information unit may represent. The difference between these two irreducible entities is the ultimate elementary brick where hierarchies are built upon. In other words, systems hierarchies are built upon a duality which is already present in the two irreducible entities whereby the modern notion of information is constructed and which is closely connected to the type of duality mentioned in the previous paragraph. Interestingly, the presence of hierarchies itself would make the evolution of most complex systems highly unpredictable and counter-intuitive. Considering that these systems are *open*, the possibility that higher hierarchies and wider information feedback loops are not taken into account when analysing them is very concrete. This may cause that systems' evolution results the opposite of what can be forecast, especially when assessed in the long term. Paradoxes and unexpected evolutions within complex systems are indeed everyday practice.

The three transformations just mentioned have substantially contributed to the social construction of complex systems. Through them, it has become possible to conceive of single artefacts potentially capable of performing any type of function by integrating any person within larger units. The material world and the human beings had to be integrated and reduced to the common denominator represented by the kind of just described information before these systems could be massively constructed. Any phenomenological manifestation had to be reduced to complex information flows that can be (re)constructed and analysed by calculation machines before complex systems could be presented as one of the latest discoveries by science. The next chapter will further explore these transformations in order to hopefully allow better understanding their nature and the connections existing between the massive construction of these types of artefact and the ongoing transition to renewables.

References

Akatz, D., and H. Pagels. 1982. Origin of the universe as a quantum tunnelling event. *Physical Review D* 25: 2065.

Bateson, G. 1972. *Steps to an Ecology of Mind.* Chicago: The University Chicago Press.

Bateson, G. 1979. *Mind and Nature: A Necessary Unity (Advances in Systems Theory, Complexity and the Human Sciences).* New York: Hampton Press. ISBN 978-0-553-34581-0.

Bauman, Z. 1998. *Globalization: The Human Consequences.* Cambridge: Polity Press in association with Blackwell Publishers Ltd.

Blumenberg, H. 1988. *Work on Myth.* Cambridge: The MIT Press.

Caneva, K.L. 1993. *Robert Mayer and the Conservation of Energy*, 310. Princeton: Princeton University Press.

Cayley, D. 2005. *The Rivers North to the Future: The Testament of Ivan Illich as Told to David Cayley*. Toronto: House of Anansi Press Inc.

Foucalt, M. 1966. *The Order of Things.* Abingdon: Routledge.

Georgescu-Roegen, N. 1971. *The Entropy Law and the Economic Process.* Cambridge, MA: Harvard University Press.

Gronemeyer, M. 2012. *Wer arbeitet, sündigt: Ein Plädoyer für gute Arbeit.* Darmstadt: Primus Verlag.

Illich, I. 1981. *Shadow Work.* Marion Boyars Inc., 75–95.

Illich, I. 1983. *The Social Construction of Energy.* Published in New Geographies 02: Landscape of Energy (2009). Massachusetts: Harvard University Press.

Latour, B. 1993. *We Have Never Been Modern.* Cambridge: Harvard University Press.

Mirowski, P. 1989. More *Heat than Light. Economics as Social Physics, Physics as Nature's Economics.* Cambridge: Cambridge University Press.

Mitcham, C. 1994. *Thinking Through Technology: The Path Between Engineering and Philosophy*, 256. Chicago: University Chicago Press.

Perulli, A. 1996. *Il Tempo da Oggetto a Risorsa.* Franco Angeli Editore.

Poerksen, U. 1995. *Plastic Words—The Tyranny of a Modular Language.* University Park: The Pennsylvania State University Press.

Rabinbach, A. 1992. *The Human Motor: Energy, Fatigue, and the Origins of Modernity.* New York: Basic Books.

Teuscher, C. 2004. *Alan Turing: Life and Legacy of a Great Thinker*, 216. Berlin: Springer.

Tryon, E. 1973. Is the universe a vacuum fluctuation? *Nature* 246: 396–397.

Watzlawick, P., J.B. Bevelas, and D.D. Jackson. 2014. *Pragmatics of Human Communication. A Study of Interactional Patterns, Pathologies, Paradoxes.* New York: W.W. Norton & Company Inc.

Chapter 2
Energy and Complex Systems Dynamics

Nicola Labanca

Abstract This chapter discusses the role played by energy within complex systems dynamics and compares this role to that played by information. In this respect, it briefly shows how information theory can confirm and incorporate thermodynamics and illustrates how given energy flow principles become unifying principles allowing studying the evolution of any complex system under a same phenomenology. This evolution can be characterized in terms of a proper balance to be achieved between improvements in the efficiency whereby systems inputs are converted into outputs (in a situation of resources scarcity) and a diversification/intensificaton in systems outputs production (in a situation of resources abundance). The ongoing transition to renewables is then presented as a very relevant reinforcing factor of the large-scale construction of complex systems and of the manifestation of the above mentioned dynamics. These considerations are employed by the author to discuss how the role of energy efficiency policies, although still fundamental, becomes ultimately functional to an intensification and diversification of outputs production in the age of renewables and how new types of policies have therefore to be devised and implemented to ensure the sustainability of the ongoing energy transition. To do so, it is necessary to acknowledge that the construction of complex systems is based on a particular and very abstract commodification of natural resources and human activities. This construction relies on the assumption that functions accomplished by people within societies can be reproduced and sustained through an underlying network wherein energy, matter, information and monetary values circulate and it reflexively validates this assumption by contributing to the materialization of this network and by creating a situation of increased dependency thereon. The final part of the chapter is therefore dedicated to discuss how new policies questioning this assumption and allowing escaping the increasing dependence on complex systems dynamics of growth can be devised.

N. Labanca (✉)
European Commission, Directorate-General Joint Research Centre,
Unit C.02 Energy Efficiency and Renewables,
Via E. Fermi 2749, 21027 Ispra, VA, Italy
e-mail: Nicola.labanca@ec.europa.eu

N. Labanca (ed.), *Complex Systems and Social Practices in Energy Transitions*,
Green Energy and Technology, DOI 10.1007/978-3-319-33753-1_2

The Apparently Vicarious Role Played by Energy Within Complex Systems

The relation between energy and information within complex systems is quite intricate. It might seem that complex systems relegate energy to play a vicarious role with respect to information and difference. As mentioned in the sections of the previous chapter, energy has been seen as an extremely powerful explanatory instrument since the mid nineteenth century when it came to represent a kind of primordial cause whereby the dynamic properties of matter could be understood. In the world of complex systems that has come to life after the 1950s, effects are however not propagated through energy exchanges any more: they are brought about by "differences." Nothing—that which is *not*—can be a cause within complex systems in the same way as, for example, "the letter that you do not write can get an angry reply"[1] in the social system where you live. In the world of information a "zero" can generate huge impacts just because it is different from a "one." Although still fundamental (difference can indeed propagate only along the pathways where energy is available) the energy concept cannot apparently serve to explain the dynamics of complex systems. It is the structure and the difference that can be found within systems that ultimately determines their evolution. Bateson explains this by providing the example of a chain stretched by two equal and opposite forces applied at its two extremities. If the chain would not have a weakest link, he states, the chain would never break whatever the intensities of the opposite forces. It is the structure, the difference existing between some parts of the chain that allows understanding and generates the dynamics of the system under study. The tension applied by these forces cannot serve to explain how a particular link came to be the weakest link.[2] The presence of this weakest link has to be considered as a given whereby the evolution of the system can be explained. According to Bateson, energy would belong to the world of quantity, whilst information belongs to the world of structure and it is the latter that drives systems evolution. By looking at how these two concepts are conceived and used within complex systems theories, it is however possible to verify that they are closely interconnected and reciprocally dependent. Information drives energy flows in so far as energy flows can be generated only where some kind of difference is maintained. At the same time, however, information cannot be maintained, created and/or transmitted without energy flows and can actually be seen as a driver of energy flows only within a static description of complex systems. In so far as the evolution of complex systems is at stake, energy remains the ultimate fuel whereby information is produced and destroyed. The way in which information content is statically associated with energy flows within complex systems represents however an important point of conjunction between these two entities and can be grasped by looking, for example, at how this notion is used within ecology.

---

[1]See Bateson (1972).

[2]See Bateson (1979).

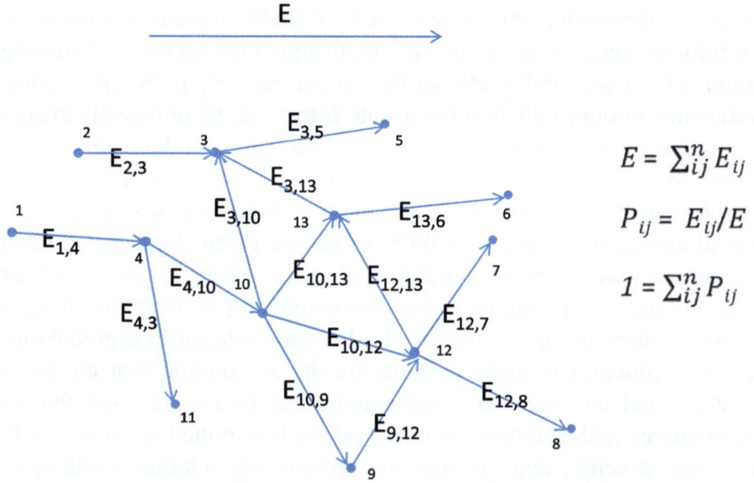

Fig. 2.1 Schematic representation of energy flows and associated probabilities whereby the mutual information exchange within nodes of ecosystems is typically defined

To illustrate this point, the following approach to associate energy and material flows with information within ecosystems can be considered as an example. The example presented here makes exclusively reference to energy but can be very easily generalized to matter flows within any complex system. Very broadly speaking, ecosystems can be represented as networks made of nodes whereby energy and matter flow. Each node (i) of an ecosystem can be assumed to have its own inputs from and outputs to other network nodes. The information content of this ecosystem can be defined based on the probabilities *Pij* representing the fraction of the total energy passing through the ecosystem that flows from the node (i) to the node (j). The probabilities that can be associated with all of its *n* nodes fix all the energy flows travelling through the ecosystem and the function $H = -\sum_{ij}^{n} Pij * lnPij$ can be assumed to represent the information content of the ecosystem (see Fig. 2.1). It can indeed be shown that the value assumed by the function *H* is lower when the distribution of the total energy flux over the different nodes is more even (e.g. in case of two nodes *H* is lower when $P_1 = 50$ and $P_2 = 50$, than when $P_1 = 90$ and $P_2 = 10$) this indicating that the more even the flux distribution, the lower its information content.[3]

Although explained by paying more attention to the substance than to the rigor of mathematical formulas, this example illustrates a very general approach whereby information is generally associated with an energy distribution and, as in a snapshot, with all the energy fluxes existing within a system. The function *H* has been indeed used both by Boltzmann and Gibbs in the 1870s to statistically define the

[3]A lower information content corresponds to a situation of lower predictability of the flows taking place within the ecosystem.

entropy of any thermodynamic system and by Claude Shannon in 1948 to define an average information content of any communication process made of messages that may occur with a probability *Pij*. In the former case *Pij* is the probability that a thermodynamic system falls in a microstate *ij* that can be univocally characterized by the position and the energies of all of its constituents. In the latter case *Pij* is the probability to get a given combination of characters within a fixed number of possible combinations. Whether we call them probabilities of getting a given distribution of energy flows or probabilities of getting given characters combinations, *Pij* represent a ratio between a number of given (energy values or characters) combinations and a total number of possible combinations. Both the energy and the information content of any system are in this way reduced to a probability distribution. This reduction is made possible by the assumption that all the possible energy states and all character combinations can be counted and that the total amount of energy and of information that can be transmitted are conserved. It is in this way that structure and quantity are reciprocally interlaced within complex systems. The amount of energy and information content is just determined by the number of combinations whereby a given state can be achieved as assessed against the number of all possible combinations corresponding to all the possible states. However, when it comes to assess and explain the evolution of thermodynamic systems which are open and far from equilibrium, further phenomenological energy principles need to be invoked to explain how new information can be created and order may emerge from disorder. As also described in the subsequent chapter sections, the Belgian Chemist Ilya Prigogine has showed in the 1970s how energy drives and maintains organization changes within complex systems which are far from equilibrium. It is energy that causes the gradients/pressures driving organizational change. New structures can be created and maintained within complex systems when small initial fluctuations determined by these gradients can amplify within the system and establish new and comprehensive stationary paths whereby additional energy can circulate and be dissipated. Despite the relevance acknowledged to information within systems, energy remains therefore the ultimate driver whereby order can be created and energy principles remain the principles whereby this creation can be explained. Within complex system theories, energy remains the *reservoir* made of a single, infinitely transformable, degradable but not destructible entity that is awaiting to be transformed into work that is supposed to have been discovered already in the 1850s. The explanations so far provided show nevertheless that information theory can confirm and incorporate thermodynamics. Energy flows and information content can be described by a same mathematical formalism relying on a conservation principle[4] whilst creation and/or destruction of information can be identified with creation and destruction of order and be quantified through the formula whereby entropy is statistically defined.

[4]In case of energy the quantity conserved is the sum of the amounts of energy entering and exiting the system at stake, whilst in case of information the quantity conserved is the sum of the probabilities whereby the information content of this system is calculated.

At this point, it may also be interesting to observe that energy and information have to a certain extent a same degree of immateriality. Information and energy flows are both present wherever change and difference occur or are detected. Like information, energy materializes only during transformations and change. Paradoxically, however, they are also seen as entities with a separate and self-contained existence.[5] In case of energy this has for example to be considered as a kind of paradox, because classical physics theories already indicate that the amount of energy that can be attributed to whatever isolated system is a physical quantity that can be defined up to an arbitrary additive constant and its absolute value is per se meaningless. Strictly speaking, this implies that the energy content of whatever matter or substance can never be properly defined in absolute terms.[6] What can be defined is instead the amount of energy that can be transferred from this matter to another matter under a given transformation and in a given amount of time. Energy materializes therefore only in terms of a variation and a transformation. It can only manifest itself during change as something that is transferred and flows from one part of a system to the remaining part of this system.[7] Despite this characteristic does not allow localizing it within any physical object, energy is nevertheless still typically imagined as the ultimate resource fuelling our economies.[8]

[5]The problems caused by the fact that both energy and information are considered at the same time as fluxes and stocks represent one of the interesting and problematic aspects of these types of conceptual artefacts that will be discussed by the author in more detail in another chapter of the book.

[6]Although under a different perspective, this aspect is analyzed also by Giampietro et al. (2013) and Diaz-Maurin and Giampietro (2013).

[7]This is due to fact that, rather than energy (E), the physical quantity that can actually be measured and assumed to have some degree of concreteness is always a variation of energy (ΔE) over a given amount of time (Δt). Whenever we deal with an isolated system, the notion of energy is of some utility in so far as this notion is employed by referring to a system transformation and is used under a conservation principle. As it can be easily realized by considering the example of an isolated system made of two colliding spheres, all that this principle allows establishing is just that, whenever the energy of one part of our isolated system (e.g. the energy of one of the spheres of the mentioned example) varies by ΔE over a given amount of time Δt, the energy of the remaining part of our system (e.g. the energy of the other sphere of the example) shall vary by $-\Delta E$ in the same amount of time. In other words, what can be defined and be measured unambiguously is not energy. What can be measured is a flow of energy ($\Delta E/\Delta t$) passing from a part of an isolated system to the remaining part of this system.

[8]It may here worth to briefly note that, rather than energy, the resources actually consumed to fuel our economies are, for example, coal, oil, biomass, etc. from which work is extracted through transformations that change the status of these resources and that make more and more difficult that further work can be extracted from them. I hypothesize that the misinterpretation mentioned in the text above is due to the fact that the energy concept has initially served as leverage for an industrial revolution which mostly relied on the utilization of resources which were available in the form of stocks of different materials and that had to be processed to produce the desired outputs. The depletion of material resource stocks caused by any industrial process has probably led to associate energy with the resources themselves. Further information on what has to be meant by *stock* can be found e.g. in Georgescu-Roegen (1971).

What Are Complex Systems Ultimately Made of?

The characterization of complex systems provided through the historical enquiry outlined in the previous chapter points to the fact that these systems appeared on a large scale within societies and in the scientific discourse when it became possible to massively produce particular types of single artefacts whereby an increasing variety and number of human functions could be reproduced. Moreover, this characterization has allowed showing how this has happened at the expenses of a progressive integration and loss of differentiation between persons and their artefacts realized through a reduction to information feedback loops equally circulating within and regulating the functioning of these two entities. In addition, it has been discussed how this integration relies on a bidirectional process of translation whereby persons' actions are translated into information that can be processed by machines and information elaborated by machines are translated in their turn into information and signals that can be understood by persons. These bidirectional translation processes have always also to be intended as a reconstruction process, in the sense that they represent also the means whereby functions and structures observed in biological entities are artificially reconstructed by an observer within a technological environment. It would indeed be a mistake to assume that complex systems are just a description of phenomena that pre-exist observation. It is the possibility offered by current theories and technologies to isolate and reconstruct underlying information flows that makes their observation and wide diffusion possible while permitting a variety of extremely useful applications. As suggested by Jacobs (2000), the nature of these information flows can nevertheless be of very various nature. Any not isolated system constituted through the circulation of matter, energy or of any other type of resource whose total amount can be assumed to be conserved during circulation can indeed potentially represent a complex system exhibiting the same characteristic dynamics of energy flow networks (more will be said about these dynamics in the following section). As also suggested by Goerner et al. (2015), either these flows are constituted by monetary flows occurring within economies, or by circulation of energy and matter within biological organisms, or by flows of oxygen, carbon, nitrogen, etc. occurring between organisms constituting an ecosystem, or by flows of cars and trucks circulating within road networks, these flows can be studied through same phenomenal principles observed in case of energy-flow networks. Complex systems become in this way a sort of underlying natural entity whose dynamics can be identified anywhere in physical systems, economies, living systems and ecosystems in general, whilst energy flow principles become unifying principles allowing to study the evolution of any complex system under a same phenomenology. The origin of the above mentioned flows are generally explained through causation mechanisms relying on the presence of two distinct and complementary elements. On the one hand, there are the mechanisms whereby these flows are generated and maintained. On the other hand, there are the circulating abstract units that remain unchanged during circulation. The mechanisms whereby these abstract units may be assumed to be put or maintained in circulation may change across systems. Moreover, they can be identified at different scales within complex systems and be described with the different languages of physics, biology, economics, or even psychology. Money may be for example exchanged because of specific needs and wants animating individual initiatives or because of some type of economic pressure assumed to act at a larger scale between national economies. Energy can be exchanged because of the presence of given spatial gradients observed in matter distribution. Oxygen can be supplied to the cells of an organism by passive diffusion or convective transport mechanisms and information related to a genetic modification occurring within a part of an organism may start spreading because of some type of pressure assumed to be exerted by the surrounding environment. Whatever the causation mechanism, there must then be some object (money, energy, oxygen, information, etc.) whose identity doesn't get lost whilst it circulates within complex systems and whose circulation is for this reason generally assumed to obey a conservation principle. At the same time, however, the underlying presence of complex systems and of their characteristic dynamics can be substantiated and possibly put in relation to everyday life human activities only if the circulation of the above mentioned entities is associated with some observed structure or with the reproduction of some function. It is at the point where the complex systems underworld made by matter and energy flows has to be jointed with observed structures and functions accomplished during everyday life that the characterization of complex systems proposed in the previous sections play an essential role. It is indeed at the interface between this underworld and the upper world of observed structures and functions that the social construction of complex systems comes into play. In the same way as the observation of given structures within complex systems depends on a selection performed by an observer establishing what is relevant and what is not relevant in the description he is producing for the phenomena under study, it cannot pass unnoticed how the construction of these systems is generally accomplished by massively and artificially joining structures and functions observed in natural entities to bits of information and, thanks to information, to energy and matter flows. These type of artificial joints are being established everywhere. They are being established, e.g. when it is attempted to merge molecular biology (studying life with a thermodynamic/informational posture) and organismal biology (studying life in terms of evolution and adaptation of functions). They are being established, e.g. when it is pretended that each action we accomplish can be associated with the consumption of given units of energy and matter. They are being established, e.g. when we, like cyborgs, act in the world through computerized prostheses thanks to the elaboration of information. Present possibilities to manage huge amounts of bits of information while observing nature from its most microscopic parts up to its most macroscopic aggregates make it even appear that these joints are something created by nature itself. Unfortunately, the establishment of these artificial joints always generate (or is generated through) a *discretization* and a reduction to *s-tandardized* functions and structures of an otherwise *continuous* spectrum of *unique* functions that nature and human beings can generate.[9] This discretization and standardization, certainly very often extremely useful, remains however the sign of the artificial character of an underworld made of energy and matter flows supposed to generate functions reproduced within complex systems. This being said, it is now necessary to briefly discuss how specific energy flow principles can be assumed to represent unifying principles explaining the dynamics observed for any kind of complex system.

Trade-Offs Between Power and Efficiency Within Complex Systems

The two key phenomenological energy principles whereby the evolution of complex systems is generally explained will be briefly described in this section. Based on what was previously discussed, it should not be extremely difficult to understand how these principles can be expressed in the language of information theory. The concepts of entropy and information content of complex systems can indeed be considered as synonyms, whilst the energy flow through a given pattern can be associated with the elementary probability of observing this flow as calculated against all the possible flows that can be observed through all the possible patterns of a complex system (see Fig. 2.1 in a previous chapter section). It should then not be extremely difficult to realize how these energy principles apply also to the circulation of money or of any other type of resource flowing in different types of complex systems (what generally changes in how these principles are described for the different systems is just the metrics whereby the circulating units are measured). Leontief (1951), Boulding (1981), Fischer-Kowalski et al. (1998), Odum (2007), Lindeman (1942), Hannon (1973) can then provide more detailed explanations concerning how these principles apply both to economic networks and to ecosystems.

[9]The type of *discretization* and *standardization* mentioned here can be seen as the result of an (at least partly) arbitrary resolution of an otherwise unsolvable allocation problem. The problem of having to establish how much energy (or e.g. time) one person consumes when he/she walks (i.e. when he/she accomplishes the function of "walking") can perhaps help clarify this point. Such apparently simple allocation problem actually involves a high level of arbitrariness and standardization. A person walking might indeed actually being also talking, looking at a landscape, making some kind of sport, etc. and all these activities can be assumed to require some type of "additional" energy input. We therefore might discover that in order to establish the amount of energy (or time) consumed while walking it is necessary to refer to a kind of reduced and standard version of walking (e.g. without talking, without exerting sight, etc.). On the other hand, we might discover that a given amount of allocated resources (whether these resources are energy, or matter, or time, or information) can serve to generate only very particular and specific aspects of the functions we are trying to reproduce. Complex systems somehow always invite to take decisions in relation to these types of unsolvable allocation problems and make people blind to the distortions they generate in this way.

This being said, the two above mentioned phenomenological principles will be hence described by using the language of energy. According to a series of scholars, the evolution of complex systems is indeed regulated by two different principles depending on energy and time availability.[10] Minimum entropy production or minimization of the input needed to obtain a given output are the expressions coined and most frequently used to refer to the first principle which dominates in a situation of energy scarcity and stable system boundary conditions. This phenomenological principle has been formalized by Prigogine (1961), Glansdorff and Prigogine (1971), Nicolis and Prigogine (1977) for energy-dissipating systems in a non-equilibrium steady state and applies to systems which are close to the thermodynamic equilibrium. Broadly speaking this principle establishes that, in a condition of energy supply limitation and quite stable boundary conditions, system structures and components requiring a lower energy input to produce a given output have a competitive advantage and will prevail over less efficient ones (i.e. over system structures requiring more energy to produce a same output) determining a system transformation that can be characterized in terms of an increased organization. This reorganization causes therefore a lowering in the diversity of options available to perform a same function in the short term and may put system survival at risk in case of a change in the boundary conditions. On the other hand, it contributes to liberate energy whereby the activity within more efficient structures can be focused and intensified so making the whole systems more robust and capable of generating new diversity in case a new condition of energy abundance will be achieved.

The second principle has been instead formalized in terms of maximization of energy flows and has been proposed for the first time by Lotka (1922). Several names have been proposed for it by different scholars. It has been defined, e.g. as "maximum power principle" by Odum and Pinkerton (1955), as "maximum exergy degradation" by Morowitz (1979), Jørgensen (1992), Schneider and Kay (1994). It establishes that in a situation of energy abundance and time scarcity complex systems tend to increase the speed of energy intake in order to speed up the activity of existing structures and to generate new structures so increasing diversity in how activities are performed at the expenses of system efficiency. The overall effect of this augmented energy intake can be described in terms of an increased intricacy, interconnection, diversification and intensification of outputs produced per unit of time accompanied by a decrease in system efficiency. The augmented system power output may determine a higher stress on the environment and on the boundary conditions. On the other hand its increased diversity and interconnections increases the possibility of a system reorganisation in case of a significant change in systems boundary conditions. System maximum power output corresponds therefore to a status of increased diversity which is a prerequisite for higher system adaptability and increases the chances of system survival through a system complexity leap towards increased efficiency whenever the conditions of energy resources scarcity

[10]These principles have been described by the author also in Labanca and Bertoldi (2013).

and minimum entropy production are possibly achieved. Overall, health and survival of complex systems would depend on a balanced interplay between these two principles. The proper balancing of these two principles can however be assessed only on a long temporal term whose actual length is impossible to establish given the intrinsically unpredictable evolution of the environmental conditions within which complex systems typically operate.

Polimeni et al. (2009) provide an example of household management to illustrate how the principle of efficiency and power output maximization co-operate in the evolution of complex systems. According to them, economies made by families during routine activities can be assimilated to the above mentioned minimum entropy production principle allowing to save money amounts that can be subsequently reinvested in additional activities. What is saved at the lower level of routine metabolism can indeed be transformed into investments enhancing social interactions and creating new activities at a higher level of household activities organization in accordance to the maximum power output principle. The final outcome of this co-operation process would be a better integration of families' metabolic systems with the environment during their evolution. Nevertheless, the reciprocal influence between efficiency and power output represents for Polimeni et al. (2009) an overall drive toward instability. Systems evolution seems to be a question of eliminating the least energy efficient practices in order to be able to employ the available energy to generate more diversity whereby increasing adaptability in a context of continuously changing system boundary conditions. These authors underline that the goal of increasing diversity per se collides with the goal of increasing efficiency as defined at a particular point of space and time, although these two goals co-operate in the long term. Moreover, they point out that the phase of increased diversity is a phase during which additional system outputs are generated and system efficiency cannot be properly defined. They illustrate, e.g. how energy efficiency improvements in cars have been associated to or have determined the introduction of new categories and variables in the formal identity of cars due to addition of many different gadgets and services and how this has represented an increase in the *diversity* of possible options available for consumers looking for a car. It is only during the phase of resource scarcity and system reorganization that an efficiency function can be defined and the different structural types can be mapped on this function in order to eliminate the least efficient and amplify the most efficient ones. Interestingly, these scholars consider identity redefinition as an intrinsic and fundamental property of systems that implies a continuous redefinition of what should be intended by systems output, systems power output, system efficiency and a continuous redefinition of the related metrics.

This important insight deserves further consideration. If the evolution of the technology of digital cameras is taken as an example, then it can be observed that when the first models of this new technology were put on the market the increasing of cameras' resolution was the main objective of R&D activities and their efficiency was therefore mainly assessed in terms of number of pixels/cm^2. After a period of about ten years during which digital cameras resolution grew exponentially and allowed in this way to generate new models with new functions and attributes, consumers' interest in this parameter started decreasing and drifted towards the speed of sensors so determining what could be called a complexity leap. This triggered a new growth in the performance of digital cameras with respect to this parameter that became the new driver of the evolution of this technology generating in its turn new diversity and determining a dumping in the growth of their resolution. The definition of systems efficiency seems hence destined to change during system evolution and the same destiny seems therefore to be reserved to the definition of system power output (i.e. to the metrics employed to measure system outputs, efficiency and number of outputs per unit of time). Despite their continuous redefinition, efficiency and power of systems seem to remain correlated as depicted by applying the thermodynamics principles briefly described above. However, it has to be pointed out that what allows power output increase during systems evolution is the peculiar nature of systems power output and the peculiar role played by *information* during system evolution. While evolving, systems would manage to increase their power output by continuously re-defining their outputs and this can happen only because the essence of systems outputs has the same material consistency of information. It is as systems could be endowed by an incredible level of vitality. Whenever the resource they consume to generate their outputs is abundant, they react by intensifying the activity of existing input–output structures and by generating new structures that can increase the possibility of system reorganization and survival in conditions of resources scarcity.[11] However, this increased power output will be generally achieved by reducing the amount of material resources whereby this power output is generated, rather than by increasing this amount, given the general scarcity of material resources typically available in the environment. This is confirmed, e.g. by the fact that the metabolic rate of small organisms (i.e. watt/kg produced) is higher than that of larger ones[12] and by the fact that in general the exponential power output increase achieved within materials relate to a scaling down of the dimensions of these materials.[13]

All in all, complex systems evolution would hence consist in a circular pattern whereby they grow and increase their power output and diversity (i.e. they add new activities and intensify existing ones at the same hierarchical level) while decreasing their overall energy efficiency as long as a condition of energy resources abundance persists. As soon as a situation of energy resource scarcity and system

[11]Clearly the possibility of a successful system reorganization cannot be established beforehand and efficiency improvements might also determine system collapsing due to the reduction in its adaptability caused by improved efficiency. Similarly, the system might collapse due to a stagnation caused by lack of organization and efficiency in its structures.

[12]Polimeni et al. (2009) point out for example that mice has a metabolic rate around 3.0 W/kg, whereas an elephant has a metabolic rate around 0.5 W/kg.

[13]Computer technologies are probably the most relevant example of an increased power output (as measured e.g. in terms of bit/sec/cm^2, or watt/cm^2) involving a scaling towards small dimensions of components.

stress is achieved, a complexity leap corresponding to a system reorganization and to an increased efficiency is realized in such a way that additional energy is liberated and the system can start growing again while increasing its diversity and power output. A recursive pattern that could then be depicted as growth-saturation-complexity leap-growth would hence be followed by systems. Within this recursive pattern, energy efficiency improvements in situations of time scarcity would be the necessary prerequisite for a continuous system and power growth. Ruzzenenti and Basosi (2008b) support this conclusion by examples illustrating, e.g. how in the aftermath of the second oil crisis of the 1980 (i.e. in a situation of energy scarcity) efficiency of trucks in the EU was maximized while trucks power increased slightly. As energy prices started decreasing (i.e. as a situation of energy resources abundance was somehow re-established), trucks power started increasing significantly on average while their efficiency started decreasing because of the higher average speed trucks were requested to achieve and of the additional functions they were requested to execute. At a larger scale, the increase in truck efficiency would have been accompanied by a structural change from the Fordian production system to the post-Fordian production system characterized by a much higher frequency and distance of shipments as well as by a much higher system power output.

Overall, a power output increase seems to be the main driver of complex systems development (whatever this system power output may represent) and an increased efficiency in the transformations of systems inputs into systems outputs seems to represent the natural consequence of a pressure exerted by the environment when the resource in term of which the system input rate is measured is scarce (either this resource is represented by time, or space, or bits, or Euros, etc.) and the necessary prerequisite for system survival through subsequent power output enhancements. When artefacts integrate human beings within complex systems, human beings and complex systems survival comes in this way to depend on a continuous process of growth to be achieved through a proper balance between efficiency and increased coupling and diversity among systems structures. Rather than being the result of intentional actions undertaken by persons, this balance would represent the manifestation of principles that con be observed everywhere in nature. Despite we might think that we are contributing to change the course of the events, our struggles to increase complex systems power capacity and efficiency would actually reflect the manifestation of a universal trend according to which existing biological and not biological aggregates have evolved (starting from galaxies, to stars, to the earth, up to plants, motors, animal bodies, human brains, cars, airplanes up to computer chips) by increasing their power outputs and energy densities through and augmentation of the level of their internal organization and efficiency.[14]

[14]See Chaisson (2001).

Complex Systems and Renewable Energies

If complex systems are the result of a social construction, a massive transition to renewable energy sources can certainly highly reinforce this construction and, if still possible, make complex systems and their characteristics dynamics much more present in our daily lives. There are multiple reasons justifying this conclusion and these reasons are mostly connected to how renewable energy sources are distributed over very large geographical areas and can constitute interconnected funds of low energy intensity supplying energy according to fluctuating rates.[15] Non-renewable energy sources like fossil fuels usually constitute well localized stocks typically generated in millions of years which can mostly be used at will without specific time constraints other than those dictated by associated identification, extraction and transformation processes. Renewable sources (like wind, sun but also biofuels, wood, etc.) are instead energy funds characterized by intrinsic regeneration processes that take place on infinitely shorter temporal scales and involve a series of variable interactions and extensive energy exchanges between the physical system where they can be localized and the external environment. Contrary to what happens for example with stocks of fossil fuels, renewable energy sources cannot generally be delimited spatially within fixed boundaries. Renewable energy sources like wind, solar radiation, biomasses, etc., cannot indeed be disjointed and conceived separately from their intensive and ever-changing interactions with an external environment, this external environment being represented by, e.g. the thermal sources whereby wind is generated, by the sun, by the ecosystem wherein biomasses are grown, etc.[16] The usually strong and highly variable energy coupling with the external environment that characterizes energy systems relying on renewable energy sources makes the dynamics observed within complex systems an everyday experience.[17] There is however another much more tangible reason why a massive transition[18] to renewable energy sources can reinforce the social

---

[15]These points have been discussed by the author also in Labanca et al. (2015).

[16]It might be argued that all energy sources are ultimately generated through the energy coming from the sun and that also renewable energy sources are hence located within the practically closed and isolated system including the earth and the sun. The temporal and spatial scales which are relevant for existing energy supply systems require nevertheless that the boundaries of these systems cannot be enlarged to include all the actual sources of renewable energies.

[17]It may be worth pointing out that complex systems dynamics can certainly be observed also with non-renewable energy sources. The point made here is however that these types of dynamics do not necessarily play a relevant role when energy has to be generated from non-renewable sources, whilst they become a fundamental characteristic of supply systems relying on renewable energy sources like wind, sun, biomasses, etc. As previously mentioned, these dynamics do not follow the physical laws so far formulated and verified for physical systems that are in thermodynamic equilibrium with their external environment.

[18]It may be worth mentioning that the author does not intend to maintain here that renewable can completely substitute non-renewable energy sources in developed countries. Putting aside extremely important social constraints, important physical and economic constraints to this possibility are for example represented by the rates of power output to be guaranteed in these countries, by the

construction of complex systems. Present ways of life in so called developed countries require such huge amounts of final energy and renewable energy sources have such an high spatial distribution and low energy density that a very high interconnection among these sources will be necessary in order to allow that they can sustain a consistent part of present energy end-uses. More specifically, the type of transition at stake entails a large-scale transformation from uni-located to multi-located and interconnected energy production centres where these centres can also possibly play the role of energy consumption centres.[19] This transition can be characterized in terms of a *complexification*, as defined for example by Ruzzenenti and Basosi (2008a) and resulting from the necessity of creating hierarchical control systems at multiple levels in the energy supply network because of the presence of distributed geographical energy gradients[20] leading to an increase in the average distance from the points where energy is produced to the points where energy may be consumed. Moreover, it involves the creation of more interconnections and more frequent interactions (i.e. an increased *connectivity*) among the different energy production and consumption points of the energy network. The resulting networks exhibit a higher connectivity primarily because a large number of their nodes are both points where energy can be conveyed from other nodes in order to be consumed and points where energy is produced and redirected towards other network nodes (it is indeed obvious that the possibility of redirecting energy inputs determines more potential connections with other nodes).[21] This aspect contributes to confer on the end-users located at the nodes of these energy networks a higher degree of *flexibility* and possibility for *self-organization*. This possibility however depends ultimately on the creation of additional hierarchical control systems whereby decisions can be taken concerning, e.g. whether to redirect the energy produced to the network or to consume it locally, whether to exploit one type of energy source or another, etc. Overall, a complex character is indeed ultimately conferred on these energy networks by the establishment of these additional hierarchical control levels.[22]

(Footnote 18 continued)

significantly different amounts of energy from renewable and non-renewable energy sources that are needed to produce same amounts of energy carriers (see the concept of energy return on energy investment—EROI), by the impacts on land use, etc.

[19]When this transformation takes place, energy end-users can decide whether to use renewable energy sources for self-consumption or to sell the energy produced in the energy networks, so becoming prosumers.

[20]Geographical energy gradients are spatial regions where energy flows pass from a condition of higher concentration and intensity to a most likely arrangement made of more diffuse and less intensive flows.

[21]Compared to other energy networks, complex electricity networks fed by renewable energy may however show a higher connectivity also because energy generated from more diversified energy source types can be conveyed to their nodes.

[22]Additional links to the nodes of a network do not per se make this network more complex. On this point, see e.g. the distinction between *complication* and *complexification* formulated by Allen et al. (2003).

These networks can be more *adaptable* to changing conditions within and outside the energy network in so far as energy end-users located at their nodes can decide to switch from an energy source to another or can decide whether to consume or to input into the network the energy they can possibly produce. At the same time, however, they are also exposed to more uncontrollable and unpredictable factors (linked, e.g. to the decisions that can be taken at the different network nodes, or to changing conditions in the wider geographical area where the energy sources used to provide energy inputs are located) compared to centrally managed energy networks. Interestingly, the energy supply that can be provided through these complex energy networks can fluctuate unpredictably not only because of the intermittent availability of renewable energy sources possibly used, but just because of the complex character of the energy supply network. As complexity of the energy network depends on how the energy gradients whereby energy is provided are *spatially* distributed,[23] it can hence be concluded that the *spatial distribution* of energy can determine unpredictable conditions solely generated by complexity. A complexification of existing energy networks can hence to a certain extent be considered as the vehicle whereby the *space* dimension affects the *time* dimension of energy, this type of mutual interaction being enabled by information technologies. It should indeed not pass unnoticed how this complexification can be enabled by and represent a formidable push to exploit the available technical capabilities for the reconstruction and monitoring of huge amounts of information concerning the energy flows taking place within energy networks. At the same time, however, it should also not pass unnoticed how the construction of these networks can lead to an intensive manifestation of the previously described mutual reinforcement mechanisms between energy efficiency improvements and power capacity increases. This might happen in several ways. Whenever more energy end-use efficient technologies would be installed at one node of the network, the energy saved thanks to these technologies might for example be made available for other nodes so allowing performing additional activities. Or, it might happen that energy prosumers operating at the nodes of these networks are highly incentivised by existing market rules to produce more energy by installing additional and more efficient energy production plants in such a way that they can either sell more energy to the network or consume this extra energy by installing additional energy end-use technologies. Complex energy networks and energy markets potentially associable with them unfold plenty of possibilities to establish these mutually reinforcing mechanisms between energy efficiency improvements and augmented power capacity also because these mechanisms can represent a way to increase complex systems adaptability and possibilities of survival. As further discussed in the following chapter section, these considerations point to the fundamental role to be played by energy conservation policies for a sustainable evolution of these networks.

[23]On this point see e.g. Ruzzenenti and Basosi (2008a).

Conclusions and Implications for the Design and Implementation of Policies for Sustainable Energy Transitions

Complex systems have been presented in the first two chapters of this book as the latest human artefacts resulting from a series of transformations concerning the way in which instrumentality has been socially intended. While doing so, the fundamental role played by science in their construction has been outlined and it has been discussed how this construction has been supported by a particular and fundamental misconception, i.e. the idea that the artefacts constructed through science are natural entities that have always existed and that can be experienced by people during everyday life. This endeavour has not certainly been undertaken to criticize the results achieved by scientists within their laboratories, or to contest the validity of the laws and the phenomenological principles they establish, or to deny the often extremely useful conceptual artefacts developed by science and their related technical applications. It has rather been undertaken to signal the important problems and negative implications of a particular attitude unfortunately often assumed and widely popularized by scientists as well as a series of relevant opportunities that can derive from its recognition. This attitude consists in pretending that the conceptual entities and experimental laws observed and verified under very restricted and controlled assumptions and conditions are eternal truths and guiding principles that can be used to interpret and act upon any society to hopefully improve its conditions. If this attitude is indeed somehow the necessary pre-condition for a massive multiplication of technical applications having often undeniable benefits, it also leads to forget the fundamental problems and alternatives represented by all those particular cases which escape categorizations and dynamics associated with the abstract and general principles which science must necessarily rely on and which are sometimes blindly applied anywhere, to anybody and at any time to explain how societies function and have functioned. The acknowledgment of this situation has made the adoption of a *two-faced* strategy necessary when presenting the social construction of complex systems. On the one hand it has indeed been necessary to try to understand the implications, the internal logic and the dynamics that can be expected from the enactment of complex systems dynamics. These dynamics are indeed what can be expected to be massively reproduced in the future also, but not only, because of the ongoing transition to renewable energies that is taking place in several parts of the world. On the other hand, however, it has been also necessary to highlight problematic aspects of and to hint to possible alternatives to what has been presented as an *artificial* generation of these dynamics. The best approach that could be conceived to prove this artificiality has consisted in the identification and description of the transformations that had to occur in some key concepts contributing to constitute the notion of instrumentality before the social construction of complex systems could become possible. These transformations have typically to be considered as the result of a non-linear and non-deterministic coevolution of material artefacts, ideas, discourses and technical skills emerging from a series of alternative evolution patterns that come to be discarded for reasons which are often contingent. The presence of these alternative evolution patterns is alone sufficient to legitimize the just mentioned two-faced analysis strategy and this strategy will be followed also in the remainder of this section to discuss the implications of the social construction of complex systems for policies that can be implemented to increase the sustainability of current energy transitions to renewables.

If complex systems will continue framing our social imaginary and will therefore continue to be massively constructed, it can reasonably be assumed that the phenomenological principles regulating their evolution will become more and more manifest and that a sustainable and healthy transition to renewables will be progressively identified by policy makers with the achievement of a proper balance between efficiency and power capacity improvements. A proper balance between these two complementary trends is what seems to have to be necessarily achieved by complex systems exhibiting a long term capability to withstand environmental challenges.[24] The possibility of a complex system break down can always be around the corner and nature has demonstrated that a long term survival capability can be identified with the capability of maintain this balance. A complex system that would increase the efficiency whereby some of its main inputs are transformed into outputs without maintaining a sufficient level of diversification and coupling among these outputs would be probably destined to collapse due to its scarce adaptability to possible changing conditions in the environment. Similarly, however, a system increasing its resilience through an augmented diversification and coupling among its outputs without a corresponding increase in the hierarchical organization and co-ordination among these outputs will probably be destined to collapse because of stagnation and lack of focus. The presence and the relevance of these phenomenological principles has been already acknowledged by scientists, policy makers and experts working in very different fields (e.g. ecology, energy sustainability, companies management, definition of national budget laws, etc.) and existing studies and literature hint already to a variety of policy approaches whereby such balance between efficiency and resilience can in principle be achieved. Goerner et al. (2015) have for example showed that factors like flexibility, diversity, small size and dense connectivity contribute to increase complex systems resilience, while factors like streamlining, large size and high capacity contribute to increase complex systems efficiency. When national economies are identified with complex networks converting resources and information into energy and products needed by societies, economies' resilience would entail a need for a diversity of options that can provide choice, competition and alternatives in case of failure by industrial activities. At the same time, however, economies' efficiency would be highly needed to generate robust flows, although the presence of extremely large, efficient and powerful organizations would tend to drain resources from smaller organizations so reducing an economies resilience and increasing brittleness. Interestingly, Goerner et al. (2015) deduce from the previously mentioned

[24]See for example Chaisson (2001).

energy principles a series of measurable characteristics which are in clear contrast with current and widely applied neoliberal competitive principles and that have to be guaranteed to ensure that economic systems can develop in a healthy way within a transition to renewable energies. They infer for example that rather than by exports, resilience is enhanced in the long term by the presence of as many as possible self-feeding return loops whereby energy and material flows generated by an economy are constantly redirected back into this economy to maintain internal productive capacities and processes. The same resiliency principle would also indicate that healthy economies have to constitute intricate networks wherein human expertise, material infrastructures and cultural systems grow together and play a mutually supportive role. On the other hand, hierarchical organizations would be absolutely necessary to regulate societies and economies beyond a certain size, but these organizations would have to operate according to a "subsidiarity principle" because the degree of flexibility required by complexes societies cannot be achieved exclusively through top-down administration approaches.[25] Along a similar line of thought, Elinor Ostrom has formulated and empirically demonstrated the validity of a series of design principles whereby she has acknowledged, among others, the need for subsidiarity and for a proper balance between self-organized initiatives by local actors and hierarchical organization within the complex systems made of the cultural arrangements, the institutional arrangements and the physical environment whereby people produce and exchange their goods and services.[26]

Although described quite synthetically, the above mentioned aspects impose a radical change of gear to policy strategies currently adopted to ensure a more sustainable transition to renewable energy sources. The complexification linked to the on-ongoing transition to renewables obliges for example policy makers and researches operating in the field of environmental policies to increasingly pay attention to the temporal dimension and rhythms of energy consumption. They have now for example to deal with research questions like the following ones: if an increased speed in the circulation of energy and matter within densely interconnected energy systems is what can actually enhance systems health and possibilities of survival, how can then this condition be assessed and be reasonably achieved in a transition to renewable energies? Can renewable energies provide the same amount of power output produced through fossil fuels in all energy end-uses? To what extent can current energy end-uses expected to be flexible and adaptable to energy sources which are intermittently available? Which policies can be implemented to change the temporal profile of current energy demand? Which market rules can be established for a transition to renewable energies where power, rather the energy, could become the commodity mostly traded?

---

[25]Interestingly, complex systems developments according to *fractal* patterns would represent a way in which a healthy balance between intricacy and hierarchical organization is achieved in nature. Fractal patterns could therefore be used to measure and assess complex systems health.

[26]The fundamental role that can be played by these principles and by community-based energy initiatives to ensure a sustainable low-carbon transition are just briefly discussed in the remainder of this section and will be the subject of a specific chapter of this book.

Another set of relevant policy research questions relates then, for example, to how to define and measure the balance to be maintained between energy efficiency and power output increases. How could indeed this balance be assessed in future renewable energy systems? On which temporal scale should it be established? Which are the parameters to be considered?

A further set of important questions comes then from the role played by self-organization in relation to hierarchies that have to be established across complex systems to ensure that they can endure un-expected environmental changes. How can indeed be guaranteed that a sufficient level of self-organization can be expressed and the necessary control hierarchies can be established? To what extent future renewable energy prosumers can be given the freedom to establish own governance systems whereby they establish own rules and sanction systems to administer energy and related infrastructures? How can these governance systems be integrated within a system of nested control hierarchies? Given the extremely high number of technical applications that in principle can be developed by anyone to exploit the extremely distributed and highly different types of possible renewable energy sources, how could a market of these applications be possibly regulated?

Changes induced in current policy strategies by all the above aspects are extremely relevant, especially when it is taken into account that these strategies are presently mostly implemented within competitive market settings and rely either on the stimulation of technical innovations capable of reducing energy inputs and/or CO_2 emissions of technologies or on initiatives aiming at changing individual behaviours. At the same time, it should not pass unnoticed how, contrary to what so far generally happened, energy efficiency improvement actions undertaken within complex systems cannot just be conceived as a means to reduce the energy consumption associated with human activities. Although continuing playing a fundamental role, energy efficiency improvements become indeed one part of a two-legged strategy where augmented power, diversification and coupling represent the other leg and a sustainable complex systems growth is the final objective to be achieved in order to guarantee systems survival. Rather than as a means to reduce energy consumption, energy efficiency becomes therefore a means whereby complex systems can continue growing and increase their survival possibilities by reallocating the energy saved for the production of given outputs to the production of the additional outputs that can increase their resilience. Complex systems survive indeed by using energy efficiency as a means to maintain and increase the density of their energy fluxes. They implicitly frame the problem of sustainability as a problem of sustainable *growth*.

All the above considerations apply in particular to existing electricity networks which are destined to expand and become densely interconnected within the ongoing transition to renewables energies. We are in a phase where the liberalization of the electricity market has separated the structures of production and distribution in many parts of the world and has made them more transparent. At the same time, however, this relatively new situation has certainly not managed to curb the increasing impact on existing resources by energy systems. If most of the electricity supply will have to rely on common resources like the sun, wind and water, then this further re-configuration implies that solutions to the new challenges determined by these energy sources cannot probably be provided by technical innovations operating within one of the two traditional and alternative institutional settings represented by a liberalized electricity market or by state regulated energy systems. As already mentioned, a series of studies has indeed already demonstrated that complex energy resource systems can be administered in a much more sustainable way when collaborative approaches, rather than competitive or authoritarian ones, are adopted.[27] Compared to institutional settings where resource systems and related technical equipment are owned individually (according to competitive market settings) or by a central authority (e.g. the state), commons-based institutional settings designed by establishing that these resource systems and technical equipment are owned in common by people can often achieve much better performances in terms of reduced environmental impacts and energy conservation. The reasons for this are quite intuitive. The self-interest of competing market agents can only achieve sub-optimal and short term solutions to solve the issues linked to the depletion of the energy sources possibly at stake, whereas centralised authorities can only rely on command-and-control and adopt unified and standardized solutions that do not fit optimally to all the local situations where they have to be applied. Local self-governing and self-organized institutions whereby equipments and resource systems are owned and managed in common by people could instead in principle exhibit the flexibility and adaptability required by the complexity of the problems at stake while being much more suitable to adopt strategies for long term sustainability.[28] The complexity of renewable electricity networks offer hence the opportunity to go beyond the conventional two binary usage structures based either on buyers and sellers (in case of competitive market settings) or on a central and unique owner and electricity customers (in case, e.g. of governmental settings). These structures can indeed in principle be replaced by a user community whose members are both electricity customers and electricity producers and can develop more suitable and flexible strategies to administer this resource. Lambing (2012) rightly mentions that the creation of these communities requires that consumers participate actively in the creation of rules and sanctions concerning electricity consumption and supply by taking into account local social, natural and technological conditions. Clearly, there are important barriers still hindering the establishment of these administration types. These barriers mostly include still too high costs associated with the installation of technologies and related infrastructures (e.g. windmills, PV panels, etc.) and negative impacts on a large circle of persons affected by the installation of these solutions (whose interests can however be integrated in the associated decision making processes). Lambing (2012), however, also mentions that the natural trend of electricity grids to aggregate and communalise electricity consumption (due to the fact that the larger the grid, the lower the additional power capacity needed to meet peaks in electricity

[27]On this point see Chap. 6 of this book and e.g. Ostrom (1990).

[28]See Chap. 6 of this book and Ostrom (1990).

demand) may lead to very large grids that may be quite difficult to administer according to a commons-based approach.[29] These grids, like any other complex systems, have indeed to be hierarchically organized and the conciliation between commons-based approaches and hierarchies may be hard to achieve. Despite these barriers, the implementation of these governance systems looks nevertheless very promising. Due to the present situation of existing energy infrastructures, only hybrid solutions where common-based types of electricity supply coexist with a liberalized electricity market can however be realistically hypothesized and the first examples of these types of governance systems are represented by energy cooperatives.[30] Although most of these cooperatives deviate from a "pure" form of communalised electricity consumption (i.e. a model where the cooperative is owner and operator of the production plants and the power grid and where the cooperative includes all the electricity consumers and the decision makers on its electricity infrastructures), the ongoing multiplication of these types of undertakings can already highlight the huge economic interests at stake when citizens self-organization in the field of energy consumption and production becomes a reality.[31] In principle, it cannot be excluded that a further deployment of multi-located renewable energy sources within electricity networks and the associated diffusion of electricity communalisation can even trigger movements in the civil society for a re-appropriation of power industry. These, however, are just speculations. The governance options sketched above have been just briefly described to explain how complex systems dynamics may contribute to create new and interesting governance scenarios.

Either partly administered through local and self-organized institutional settings or not, complex systems remain hierarchical systems resulting from a social construction based on a particular and very abstract commodification of natural resources and human activities. This social construction relies on the assumption that functions accomplished by people within societies can be reproduced and sustained through an underlying network wherein energy, matter, information and monetary values circulate and it reflexively validates this assumption by

[29]Very large grids could however still be managed based on a commons-based approach. Multistage control systems can indeed in principle be used to allow that overcapacity in one community compensate for demand peaks in other communities. This would certainly require the wide scale usage of smart grids and smart meters, but the resulting management system would be fundamentally different from the usually prospected solutions to the challenges posed by renewable electricity. These solutions propose indeed top-down management approaches mostly relying on price signals processed by automated systems regulating electricity usage in each consumption point.

[30]For a brief overview of existing energy cooperatives around the world, see e.g. ILO (2013).

[31]One important area where existing interests have started generating power conflicts concerns e.g. access rights to technologies for smart metering and smart grid management and access rights to personal data concerning consumption within households. Self-organized energy prosumerism relies e.g. on citizens sovereignty on data concerning their energy consumption and on the possibility of having free access to smart technologies. The actual realization of electricity commons will depend on the outcomes of existing and future conflicts in this area.

contributing to the materialization of this network. Societies can certainly be organized and policies can be designed and implemented based on the abstract principles regulating the evolution of this underlying network. Policies can, for example, be established to substitute given technologies with other technologies with reduced energy input and CO_2 emissions based on pre-established measurements and evaluations. Similarly, policies can for example be implemented to diversify and intensify activities for the production and exchange of goods and services in a given area and the optimal distribution and balance of these initiatives can be perhaps be assessed by using fractal branching patterns taken from complex systems theories. The usefulness of these methods cannot certain be denied and it would be probably insane to reject their adoption a priori based on some kind of ideological preconception. The point is, however, that these methods, when blindly applied to any social context without paying due attention to each specific case and circumstance, can produce serious damages and problems simply because they cannot take the effects they produce on every person and context into account. This problem is certainly not a minor issue and cannot certainly be solved or bypassed by Machiavellian considerations or by invoking the higher interests of the environment or of a not better specified collective with respect to individual persons. The current economic crisis is for example showing how a passive submission to the abstract principles and laws of the complex systems constituted by the international markets and monetary systems can escape any form of social control and be detrimental for millions of people. At the same time, however, what may seem to constitute an unfortunate impediment and obstacle to the application of abstract principles and laws to regulate societies is also what constitutes the lively force of these societies and the source of often more valid and alternative solutions to the problems at stake. The problem that is being delineated here is genuinely *political* and relates to how the blind and large-scale implementation of technical solutions can in some circumstances become a way to bypass people and aggravate the conditions that it should contribute to ameliorate. Policies cannot indeed be implemented exclusively based on abstract considerations related to reduction of energy inputs (or of associated polluting emissions) or multiplication of outputs. The experience that has been matured since the 70s of the last century with energy efficiency policies aiming at reducing the energy inputs of single technological instruments has for example showed how policies approaches exclusively fostering the diffusion of technologies with lower energy inputs can represent a way to bypass any political consideration concerning the actual utility of these technologies, how they often reduce the question of diminishing energy resource consumption into a mere question of individual choice (that can typically be influenced by economic considerations and incentives) and can actually end up reinforcing the need for these technologies without significantly reducing the associated energy consumption. To make an example, when the problem of reducing energy inputs and polluting emissions of private transportation in a city is primarily faced by implementing policies fostering the diffusion of more energy efficient and/or less polluting new vehicles, any political discussion concerning how urban mobility could be reorganized to reduce the amount of travelled kilometres by people is typically bypassed or is assumed to be of secondary importance compared to the diffusion of new technologies despite this reorganization could potentially result much more effective, more diversified and more tailored to the specific context at stake. Policies aiming at reducing energy *inputs* and polluting emissions by fostering the diffusion of more efficient and less polluting vehicles do not indeed call into question and may even stimulate a more intensive usage of vehicles, whilst the political decisions that people can effectively take around mobility relate to the reorganization of energy *outputs* and could for example concern the redistribution of shops and sales point in the different city areas, the implementation of solutions to reduce commuting, decisions concerning the need for new parking places nearby given sites, etc. In addition, the policy approach used as an example relies on private and economic decisions to be taken by the individuals supposed to purchase new and more environmental friendly vehicles, whilst the alternative approach being described relies on decisions to be taken by a community of people. Finally, the implementation of policies fostering the diffusion of specific new technologies results mainly from the active involvement of manufacturers supposed to produce these technologies, whilst the active involvement of so called end-users remain quite limited and the actual necessity of using these technologies (e.g. vehicles) is not put under discussion. This necessity is rather usually implicitly reinforced by the policy at stake and by all the activities that will lead to its implementation. For example, vehicle manufacturers are certainly interested in producing more environmental friendly solutions and will actively promote them, but will generally not be willing to put the necessity of using what they produce under discussion; at the same time all the forms of persuasion - including economic incentives - put in place to convince people to participate in the policy can reinforce the diffusion of vehicles and create lock-in effects that are very difficult to be eradicated.[32]

An important watershed lies between the two approaches just described. The former approach is designed by having measurable and reduced energy inputs and polluting emissions as primary and often exclusive objective. It is based on an hypothesis of calculability and measurability of its energy effects typically relying on the assumption that the policy being implemented will not produce any reorganization in the existing outputs.[33] Moreover, it is usually expected to produce

[32]The radical monopoly that cars can for example exercise on mobility (i.e. the way in which their extensive diffusion can lead to the elimination of existing alternatives to mobility) by inducing a redesign of urban landscapes and the lock-in effects that can often generally be generated by the extensive diffusion of given technologies due to how they co-evolve with other material and conceptual artefacts in the environment where they start being widely used should never be underestimated while designing and implementing policies.

[33]The energy effects of the policies being discussed are typically calculated under a often questionable *ceteris paribus* condition consisting in assuming that these policies will not produce any other change than the expected reductions in the energy inputs. Put in other words, the outputs produced by the energy end-users participating in the policy initiative are generally supposed to not be changed by the policy itself (e.g. it is assumed that all end-users buying more energy efficient cars thanks to the economic incentives received through a given policy instrument will not change their travelling behaviours).

predetermined energy impacts through a process of homogenization associated with the diffusion of standardized solutions and its success depends on individual choices supposed to be typically taken within and be driven by competitive market settings. The latter approach takes instead the reorganization of energy outputs as the starting point to be considered for a possible reduction of associated energy inputs and polluting emissions. It consists of collective actions and decisions whose effects in terms of reduced energy consumption and emissions can be very relevant but cannot generally be easily quantified beforehand. Moreover, it typically can generate solutions which are highly diversified and can in this way better fit to the different necessities usually at stake.[34] The relevance and the nature of the characteristics associated with this latter approach (e.g. active involvement of people through collective actions, more diversity and adaption to specific contexts and conditions, potentially higher effectiveness, etc.) are alone sufficient to understand the consequences of disregarding the primacy of the role that it should play within policies and how, contrary to what usually happens, it can be socially relevant to find suitable ways to subordinate the former approach to the latter.

It might be objected that the above considerations do not hold in case of policies possibly informed by complex systems theories as these considerations mostly refer to energy efficiency policies mostly targeting single technological instruments and energy end-uses (as energy efficiency policies focused, e.g. on cars, refrigerators, air conditioners, etc.). This objection, however, is valid only to a limited extent. Complex systems certainly reframe the nature of the policy issues to be faced to increase the sustainability of human activities. As mentioned, they even allow better understanding the actual implications of energy efficiency improvements on single systems functions and allow adopting preventive strategies to cope with the problems that may be caused by the decreased diversification and systems' resilience that may be associated with these improvements. Nevertheless, they cannot certainly represent a way to bypass communities of people and subordinate their views and the solutions that they can elaborate based on the exertion of their practical knowledge to the large-scale application of technical solutions. Complex systems and policy informed thereby can certainly nowadays stimulate the diffusion of plenty of different individual options to perform given functions more sustainably thanks to the possibilities disclosed by new technologies to manage and organize huge amounts of information. The intensity and diversification of fluxes that can nowadays be generated and managed in relation to mobility are for example astonishing. An individual moving in some cities with a smart phone can nowadays detect the presence and decide to use a bike made available by a system of bike sharing for a segment of his journey, then he can leave the bike wherever he

[34]In the above mentioned example of private transportation in a city, rather than causing that as many citizens as possible buy same types of new and more energy efficient vehicle, the political process whereby outputs are reorganized may for example cause that a consistent part of citizens will not have to use a vehicle anymore, while another part could be induced to use public transportation, another (hopefully minor) part could continue using inefficient vehicle because it cannot afford the usually more expensive energy efficient vehicles, etc.

wants and decide to take a bus, a tram, a train or a metro based on the information he finds within an online time table, then he can decide to take a car or to go by feet because his smart phone shows that this is the more convenient option to reach his final destination and can take him to this destination through an interactive map. It can be assumed that, in a hypothetical future, multitudes of persons and goods could be moved in this way and that these mobility practices could be extensively adopted within larger and larger geographical areas because of a series of associated advantages mostly linked to increased speed and convenience but also related to reduced emissions pollution by individuals. Thanks to the amount of information they can manage, individuals integrated within complex systems can indeed potentially identify plenty of alternative and optimized mobility options whereby the environmental impacts associated with their movements can be reduced. Nevertheless, it should not be forgotten that this multiplication of available options is alone responsible for an increased affluence and intensification of activities that overall may counterbalance the reduced impacts that can be associated with the mobility options adopted by single persons. Mobility systems organized as just mentioned are for example capable of generating and managing a density of fluxes that would have been impossible to imagine few decades ago.[35] Complex systems are naturally devoted to growth and, rather than the possibility to increase their efficiency, it is their growth and the increasing integration of people within them that constitutes a sustainability problem, either the outputs are generated by renewable or non-renewable energy inputs. Moreover, the multiplication and diversity of functions that they allow accomplishing is to a certain extent only apparent, as this diversification is generated through a progressive standardization and homogenization occurring at the level of the energy types, materials and information flows circulating in the underlying network supporting the generation of these functions. It should hence not be neglected how this increased homogenization, together with the increased extension and couplings existing among the nodes of the networks that are being constructed, makes these networks extremely vulnerable and exposed to breakdowns that may be caused by minimal and unpredictable perturbations occurring in some of their parts. Underneath the multitudes of people moving through their smart phones in the previous example there is a network that mostly works through electricity and electromagnetic fields which have invisibly colonized our public and private spaces and there are huge amounts of 0 and 1 s circulating through it. It is amazing to think of how a sneeze, a small accident and perturbation occurring in a remote part of a complex network can potentially and unpredictably put all people integrated into it in a kind of pneumatic vacuum without any point of reference. Recent cases of blackouts[36] already reveal how an increasing integration into complex electricity networks can put communities of people in very estranged and dangerous conditions due to the impossibility

[35]Descriptions of mobility solutions informed by complex systems theory can be found for example in Newman et al. (2009).

[36]See for example RAENG (2016).

of accomplishing the most elementary activities in case of temporary interruptions in the supply of electricity. Taking just people mobility into consideration, it is probably not so unrealistic to assume that the increasing dependence on electricity and on the internet that can be expected in a near future can cause that, in case of temporary blackouts, most of the vehicles available in a city result inaccessible and plenty of people do not know which direction to take or how to reach their destinations. This might happen either because part of these vehicles might not function without electricity supplied by a network, or because the information system regulating their flow and/or providing information to end-users (e.g. within bus and train stations) could not be activated without electricity, or because most of the pumps whereby vehicles are refuelled might be electric pumps, or because electric lighting services might not be provided by night, or because communication devices like radios, cell phones, etc. might not work without electricity, or even because practical knowledge[37] of people related to how to move within and between cities might have been partly lost due to prolonged delegation of mobility related tasks to complex technical systems. The description of this extreme situation is not dictated by any kind of technophobia. It has been provided just to explain (a) how the diversity of options that can be identified within complex systems can be only apparent; (b) how these systems can potentially weaken the social tissue whereby people have always provided for their necessities and (c) how it can become politically relevant to ensure minimal conditions allowing that people can, at least temporarily, live disconnected from the increasing series of complex systems wherein they are being progressively embedded and can collectively identify own alternative spaces and ways of life. Despite it becomes more and more difficult to explain why these spaces are so important per se, it should not be difficult to understand how important it is that they can continue to exist at least as kind of backup reservoirs allowing that the system can be restarted after possible breakdowns.

The extension of the geographical areas and the number of persons that can be affected by unpredictable systemic accidents should alone be sufficient to understand how the presence of these spaces can become indispensable. Unfortunately, there is no control system that can completely secure from complex systems breakdowns. As already mentioned, complex systems dynamics are intrinsically affected by a deep uncertainty that cannot be dealt neither with deterministic, nor with statistical methods.[38] The role played by available information on the status of

[37]This practical knowledge may concern orientation, creation of mental maps, memorization of streets names, etc., but may also concern capabilities related to the employment of possibly available alternative technical systems not relying on electricity consumption (e.g. paper maps, time sheets, railroad switches, hand pumps used to refuel vehicles, etc.).

[38]The application of deterministic methods produces indeed good results only within simple aggregates made of few parts interacting according to very simple mechanisms, whilst statistical methods can be applied to aggregates made of many parts with random and loose interactions. Complex systems are instead aggregates exhibiting organization and made of many and strongly coupled components. On this point, see, for example, Weaver (1948).

complex networks and their energy and matter flows, although still fundamental, becomes therefore sensibly weakened in the framework of whatever policy or strategy that can be designed to increase their security and sustainability. This information is indeed still highly necessary to identify a series of possible evolution patterns, but it will never be sufficient to establish ex-ante the actual evolution pattern, because this pattern is deeply affected by and extremely sensitive to how local interactions change. As a matter of principle, no model, no matter how detailed is the information available on the status of the system under investigation, can allow achieving this end. This conclusion has as a consequence that no underlying blueprint, no predetermined mechanism, no *planned* strategy can completely secure these networks from possible shocks. This is what has to be considered under a theoretical point of view when rules to administer the usage of equipment, resource systems and resource units consumed within complex systems have to be defined to increase their sustainability. Under the point of view of the actual implementation and enforcement of suitable control procedures, the situation is then further worsened by the fact that the extension and the complexity of the systems at stake, together with the presence of different types of economics constraints, typically obliges to delegate the implementation of these procedures to a myriad of different companies and actors, this situation causing that no-one can have the overall view of the status of system.[39]

Besides developing purely technical solutions and improving existing control procedures and risk assessment protocols to prevent situations of energy systems disruption, a good strategy to be further developed to cope with the situation of increased fragility that can be expected from future renewable energy systems is certainly that of learning from accidents and blackout already happened in the past[40] or to perform sociological studies in order to assess which solutions can be reasonably implemented to increase systems resilience and people flexibility and adaptation. These solutions typically range from purely technical ones to solutions generated by sociotechnical capacities of people, linked for example to possibilities of shifting the timing of their activities or moving to places where environmental conditions require less energy, etc.[41]

In relation to the aspects stressed in the last part of this section, it would however be a big mistake to assess the practices that people can autonomously generate to provide for themselves only in terms of the associated possibilities of increasing the resilience of the complex systems they depend on. This type of reductionist approach would indeed not consider, among other things, the presence of a huge diversity potential for increasing the sustainability of human activities existing within communitarian practices developed outside and in alternative to what dictated by these complex systems. Most probably, it is nowadays more than ever

[39]For an accurate description of how these situations have actually occurred see, for example, Tainter and Patzek (2012).

[40]On this point, see, for example Trentmann (2009).

[41]See Chap. 13 of this book for further information.

necessary to acknowledge dignity and important sustainability potentials to alternative collective ways of life which manage to develop and remain outside the paradigm of growth enforced by current huge and complex monetary, energy and information systems. More than that, it is necessary to recognize in the ways of life conducted outside these complex systems an opportunity for recovering a sense of agency and a personal and more sustainable dimension within the relationships established with other people and with our environment in general.

References

Allen, T.F.H., J.A. Tainter, and T.W. Hockstra. 2003. *Supply-Side Sustainability*. Columbia: Columbia University Press. ISBN 0-231-10586-X.

Bateson, G. 1972. *Steps to an Ecology of Mind*. Chicago: The University Chicago Press.

Bateson, G. 1979. *Mind and Nature: A Necessary Unity (Advances in Systems Theory, Complexity and the Human Sciences)*. New York: Hampton Press. ISBN 978-0-553-34581-0.

Boulding, Kenneth E. 1981. *Evolutionary Economics*. Beverly Hills, CA: Sage Publications.

Chaisson, E. J. 2001. *Cosmic Evolution. The Rise of Complexity in Nature*. Harvard University Press.

Diaz-Maurin, F., and M. Giampietro. 2013. Complex Systems and Energy. *Reference Module in Earth Systems and Environmental Sciences*.

Fischer-Kowalski, M., and W. Hüttler. 1998. Society's Metabolism: The Intellectual History of Material Flow Analysis Part II: 1970-1998. *Journal of Industrial Ecology* 2S(Jan.): 107–137.

Georgescu-Roegen, N. 1971. *The Entropy Law and the Economic Process*. Cambridge, MA: Harvard University Press.

Giampietro, M., K. Mayumi, and A.H. Sorman. 2013. *Energy Analysis for a Sustainable Future: Multi-Scale Integrated Analysis of Societal and Ecosystem Metabolism*. London: Routledge.

Glansdorff, P., and I. Prigogine. 1971. *Thermodynamics Theory of Structure. Stability and Fluctuations*. New York.: Wiley.

Goerner, S.J., B. Fath, and D. Fiscus. 2015. *Using Energy Network Science (ENS) to Connect Resilience with the Larger Story of Systemic Health and Development*. Article available at https://journal.emergentpublications.com/article/using-energy-network-science-ens-to-connect-resilience-with-the-larger-story-of-systemic-health-and-development/.

Hannon, B. 1973. The Structure of Ecosystems. *Journal of Theoretical Biology* 41: 535–546.

International Labour Office (ILO). 2013. *Providing Clean Energy and Energy Access Through Cooperatives*. Geneva: Cooperatives Unit (ENT/COOP), Green Jobs Program. ISBN 978-92-2-127528-2.

Jacobs, J. 2000. *The Nature of Economies*. New York: Random House.

Jørgensen, S.E. 1992. *Integration of Ecosystem Theories: A Pattern*. Dordrecht, The Netherlands: Kluwer Academic Publishers.

Labanca, N., and P. Bertoldi. 2013. First Steps Towards a Deeper Understanding of Energy Efficiency Impacts in the Age of Systems. In: *Proceedings ECEEE 2013 Summer Study on Energy Efficiency*, Vol. 4, pp. 2189–2200.

Labanca, N., I. Maschio, and P. Bertoldi. 2015. Evolutions in Energy Conservation Policies in the Time of Renewables. In: *Proceedings ECEEE 2015 Summer Study on Energy Efficiency*, Vol. 4, pp. 2075–2086.

Lambing, J. 2012. Electricity Commons—Toward a New Industrial Society. In *The Wealth of the Commons—A World Beyond Market and State*, ed. D. Bollier, and S. Helfrich, 57. Amherst-Florence: Levellers.

Leontief, W. 1951. *The Structure of the American Economy, 1919-1939*. New York: Oxford University Press.

Lindeman, R.L. 1942. The trophic dynamic aspect of ecology. *Ecology* 23: 399–418.
Lotka, A.J. 1922. Contribution to the Energetics of Evolution. In *Proceedings of National Academy of Sciences*, Vol. 8, pp. 147–151.
Morowitz, H.J. 1979. *Energy Flow in Biology*. Woodbridge, CT.: Ox Bow Press.
Newman, P., T. Beatley, and H. Boyer. 2009. *Resilient Cities: Responding to Peak Oil and Climate Change*. Washington, DC: Island Press.
Nicolis, G., and I. Prigogine. 1977. *Self-Organization in Non-Equilibrium Systems*. New York: Wiley.
Odum, H.T. 2007. *Environment, Power and Society for the 21st Century: The Hierarchy of Energy*. New York: Columbia University Press.
Odum, H.T., and R.C. Pinkerton. 1955. Time's Speed Regulator: The Optimum Efficiency for Maximum Power Output in Physical and Biological Systems. *American Scientist* 43: 331–343.
Ostrom, E. 1990. *Governing the Commons: The Evolution of Institutions for Collective Action*. Cambridge: Cambridge University Press.
Polimeni, J.M., K. Mayumi, M. Giampietro, and B. Alkott. 2009. *The Myth of Resource Efficiency*. Earthscan: The Jevons Paradox.
Prigogine, I. 1961. *Introduction to Thermodynamics of Irreversible Processes*. 2nd revised edition. New York: Interscience Publisher.
RAENG. 2016. *Living without Electricity. One City's Experience of Coping with Loss of Power.* ISBN 978-1-909327-26-9. Report available at http://www.raeng.org.uk/publications/reports/living-without-electricity.
Ruzzenenti, F., and R. Basosi. 2008a. The Rebound Effect: An Evolutionary Perspective. *Ecological Economics* 67 (2008): 526–537.
Ruzzenenti, F., and R. Basosi. 2008b. The Role of the Power/Efficiency Misconception in the Rebound Effect's Size Debate: Does Efficiency Actually Lead to a Power Enhancement? *Energy Policy* 36 (2008): 3626–3632.
Schneider, E.D., and J.J. Kay. 1994. Life as a Manifestation of the Second Law of the Thermodynamics. *Mathemaitical and Computer Modelling* 19: 25–48.
Tainter, J.A., and T.W. Patzek. 2012. *Drilling Down. The Gulf Oil Debacle and Our Energy Dilemma*. Berlin: Springer.
Trentmann, F. 2009. Disruption is Normal: Blackouts, Breakdowns, and the Elasticity of Everyday Life. In *Time Consumption and Everyday Life: Practice, Materiality, and Culture*, ed. Elizabeth Shove, Frank Trentmann, and Richard Wilk. Oxford: Berg.
Weaver, W. 1948. Science and Complexity. *Scientific American* 36: 536–544.

Part II
Complex Systems and Sustainable Energy Transitions

Chapter 3
Complexification in the Energiewende

Franco Ruzzenenti and Brian D. Fath

Abstract The path toward a low-carbon economy takes three main parallel roads: the efficiency of energy conversion, the reduction of energy use and the substitution of fossil-fuels with renewable energy. This chapter will focus mainly on this latter aspect of the problem by analyzing how a transition toward renewable energy can pose a new challenge to economy and governance in terms of complexification of the system. The fate of renewable energy sources (RES) crucially depends on the power sector for electricity is still the main vector for renewable energy. The main features of the ongoing transition toward a renewable energy system are: (1) lower intensity of energy sources; (2) high efficiency of conversion; (3) temporal discontinuity; (4) free access to local and more decentralized energy sources; (5) dramatic change in the economic concept of energy scarcity; (6) new, leading role of the network. Is this process leading to a higher complexification? To answer to this question, we will analyze this energy transition in the light of the concept of complexity and sustainability by looking at the history of economic development and societal change prompted by new energy sources and new form of energy conversions. A particular emphasis will be given to the case study of Germany and recent thrust toward an *energiewende*. Finally, it will be advocated the need for a new *market of power* aimed at decoupling the sites of electricity inlet and outlet overcoming the impending limits of RES energy that curbs their development.

F. Ruzzenenti (✉)
Department of Management and Quantitative Sciences,
Parthenope University of Naples, Naples, Italy
e-mail: ruzzenenti@gmail.com

F. Ruzzenenti
Institute of Sociology, Jagiellonian University, Krakow, Poland

B.D. Fath
Towson University, Baltimore, USA

N. Labanca (ed.), *Complex Systems and Social Practices in Energy Transitions*,
Green Energy and Technology, DOI 10.1007/978-3-319-33753-1_3

Introduction

According to the New Energy Outlook of Bloomberg, by 2040 the world's power-generating capacity mix will have transformed from today's system composed of two-thirds fossil fuels to one with 56% from zero-emission energy sources. Furthermore, over the next 25 years, 60% of the new generating capacity installed and two-thirds of the global investment will be in renewable energy. Are we on the verge of a new energy transition in the electricity sector from fossil fuels to renewable energy sources (RES)? Can RES supply the present energy demand in developed countries, securing the continuity with affordable prices? RES generally present a very high efficiency of conversion. For example, solar panels can have an efficiency of up to 40% comparable to the efficiency of best of the best gas turbines used in thermal power plants, but solar insolation has an average power intensity on land of the order of magnitude of 10^2 W/m^2 in contrast to the order of magnitude of 10^3 W/m^2 displayed by fossil fuel or nuclear power plants. The same can be said of hydropower, which has an even higher efficiency (approximately 90%) but a power intensity down to the order of magnitude (OM) of 10^1 W/m^2, with some peaks for alpine power plants reaching the OM of 10^2 W/m^2, or wind, that has an efficiency of up to 60% but a power intensity down to 10^1 W/m^2. Furthermore, the sun is not always shining, the wind not always blowing or the water flowing. The discontinuity of RES can further scale down the power intensity of a OM (from 10^2 to 10^1 W/m^2 in the case of solar energy, for example), but most importantly can create problems for supplying the baseload demand of electricity and balancing the grid (i.e., constantly matching electricity demand and supply). But RES are also available for free and virtually unlimited, the only costs (almost) being capital costs. This can be an advantage, but also a disadvantage for an economic system that proliferated upon the concept of scarcity and limitation to access. Therefore, the main features of the ongoing transition toward a renewable energy system can be outlined as follows: (1) Lower intensity of energy sources (power per unit of surface or volume); (2) generally high efficiency of conversion; (3) discontinuity and unpredictability in energy sources availability; (4) free access to local and more decentralized energy sources ultimately leading to a new definition of property rights on energy; (5) dramatic change in the economic concept of energy scarcity (to be applied to the energy supply and not to the energy source) to which markets will respond by providing energy services, particularly electric power, which is otherwise scarce and thus profitable; (6) new, leading role of the network due to a diffused rather than concentrated power generation. Is this process leading to a higher complexification? To answer to this question, we will analyze this energy transition in the light of the concept of complexity and sustainability. This chapter focuses on RES in the field of power production avoiding the others application of renewable energy, foremost to the transport sector, which will be tackled in Chap. 4. As it will be clearer in the remaining of this chapter, the power sector is but a marginal sector for the fate of RES, as electricity represents the vital energy carrier for renewable energy. Therefore, the quest for an *energy transition* from fossil fuels to RES must depart from this crucial sector. We will begin with analyzing the case of Germany, which has recently been not only the leading country in RES in terms of investments and growth rate of share in the production mix of the last decades, but also a country where the concept of "Energiewende" has been unequivocally stated in formal documents of the German government and parliament and posited as a target for governance.

The Eneregiewende

In the last years, Germany has gained international attention for its declared aim to leave nuclear and fossil energy behind and move to renewable energy sources (RES) (Morris and Pehnt 2015). German policy-makers in the last 25 years have prompted an aggressive energy policy to sustain the development of RES with the explicit aim of replacing nuclear power in the short-term and coal in the long-term for electricity generation. The motivation is to reduce carbon emissions and improve energy security by coupling economic growth and development to local energy sources. This economic and technological process, and the consequent energy policy, is referred to as the German "Energiewende" (energy transition) (Hake et al. 2015). Green-energy skeptics and climate change deniers, on the one hand, and environmentalists and green-energy stakeholders on the other hand, have been both observing attentively, with dismay or hope, the German locomotive charting its own way through this narrow passage.

From the early 1990s, the share of RES in the German power sector grew seven times, from around 3% to more than 28% in 2014 (IEA 2015) and the employment in the RES sector almost doubled from 2005 to 2011 (Morris and Pehnt 2015). Relative to 1990, Germany reduced its carbon emissions by 27% at the end of 2014, overtaking the target of 21% (for 2012) set by the Kyoto Protocol (Morris and Pehnt 2015). In September 2010, the government of Angela Merkel presented its "Energiekonzep" (energy concept) with very ambitious targets, aiming to reduce greenhouse gas emissions by 40% by 2020 and at least by 80% by 2050 compared to 1990. The proclaimed path to a low-carbon energy future envisaged also means to cut on coal and the exit from nuclear power. On June 6th 2011, pressured by the public opinion, the Bundestag voted, with an overwhelming majority, in favor of a final nuclear phase-out that led to the decision to decommission the seven oldest reactors and to cease all nuclear power generation by 2022 (Hake et al. 2015). The green-supporters all over the world were cheering and praising the German-way to de-carbonization and prosperity. Germany became a global leader and trail-blazer, as the only developed country to have taken, deliberately and coherently, the decision of aiming at a future energy supply entirely based on RES and they were achieving this at an outstanding pace. Their policies demonstrated the effectiveness of decoupling economic growth and carbon emissions. The gloomy skeptics of renewable energy could retreat in their caves: a new eve for energy and economy was on the door. In the view of many, Germany was paving the way in the twenty-first century for a *new industrial revolution* based on RES, similar to what Britain did for the age of coal in the eighteenth century.

For a curious circumstance and repetition of history, it is the old, dirty coal—the same who propelled the first industrial revolution—who is now looming threateningly, casting a dark shadow over the bright and full-sail cruise toward RES. Germany has almost halved its nuclear power in 14 years, but failed to reduce the contribution of lignite to power generation. Lignite, with RES, is still the first energy sources in the electricity mix and since 2011 the consumption of lignite actually grew (The Economist 2015). The coal paradox of Germany in the recent years lies in the twofold effect of the nuclear phasing out: the unpredictable high costs of disinvestment from nuclear power, which drained public expenditures from RES, and the room for power generation that, combined with the low or zero marginal costs of RES, fostered carbon-intensive and cheap energy sources like lignite. In a recent special report on climate change, The *Economist* put forward the case of Germany as an example of the frailty and inconsistency of a univocal policy based on RES. For the *Economist*, Germany, once a leader in the World for renewable energy, is now trapped in a state of impasse caused by the pressure posited by the costs of rapidly dismantling power reactors and the need to fill the void with reliable, continuous, and cheap power (The Economist 2015). The pace of intended decommission is faster than the existing grid can absorb variable-delivered renewable fuels, even if the investments to continue their growth were available. Nuclear energy has a very limited range of power capacity, but once you turn the knob, you know exactly how much power feeds into the grid. On the contrary, the power delivered by RES is virtually free, it costs just a sunny or a windy day, but it is unpredictable and intermittent. A common operational and logistical hurdle is that the grid must always be balanced (electricity cannot be stored on the grid). For a network that has been designed and developed with few, big and under control inlet points to provide power for a vast, heterogeneous and uncontrolled landscape of small outlet points, handling RES can be a problem. Furthermore, the surge of solar and wind power had the paradoxical effect on the market of fostering lignite over gas by pushing down the clearing price of electricity (The Economist 2015). The fuel of RES is free and the energy highly subsidized. Hence green-power producers offer the kWh on the wholesale market for nothing, pushing out the bids from the costly (but efficient, compared to coal) gas-fired power stations. In the cradle of the Energiewende, the mixed effect of free market, high incentives, low oil price, and disinvestment's costs created a fertile environment for a coal backspin. Is this a case of complexification brought about by RES[1]?

[1]The answer to this question brings obviously to the concept of complexity and the criterion for assessing its growth (complexification). This is not the context to address such a complex issue as we can only underline that there is not yet an unanimous definition of complexity, let alone a measure. For the sake of the analysis developed in the present chapter, we will bear to an intuitive concept of complexity that can be briefly outlined as follows: (1) a more complex system has a higher number of components (a nineteenth century coach had a number of components in the OM of 10^1–10^2 parts compared to 10^4 of modern cars); (2) an higher number of interactions among the

Indeed, it is a case of heterogenesis of ends which is often the unintended result of assertiveness in a complex context. With the words of Fernand Braudel, a famous historian: *responses to natural challenges thus continually free humanity from its environment and at the same time subject it to the resultant solutions. We exchange one form of determinism for another* (Braudel 1995). He viewed the history of civilizations primarily as the struggle of mankind against nature aimed at overcoming the constraints on development and, in this clash, a central role has always been played by energy conversion—the driving motive behind all activities. He envisaged four main *industrial revolutions*: (1) coal and steam power; (2) oil and internal combustion; (3) electricity, and (4) in his view, was the rising nuclear power.[2] Note that the first two are fossil based, the third is historically and vastly based on fossil fuels as energy source and not as energy carrier. Theoretically the electricity's means of production is adaptable, adjustable, and open to a multiplicity of options. In this manner, it plays a key role as a transition technology. Nuclear power is also nonfossil (although still nonrenewable) and one that humans still have not tamed for many people's satisfaction. Fernand Braudel wrote his marvelous book in 1963, two decades before Chernobyl and almost three decades before Kyoto. He could not forecast how the struggle with Nature would become so compelling and complex in so far that it had to revert its goals, making the entanglement of economy with nature a new *chrism.* Moreover, he could not predict, if not the dim future of a fading away, the failure of the nuclear age to ascend in the following decades. In the 1960s, very few could have predicted that the atom would have not freed humanity from the burden of physical work and perhaps even today many rest in this certainty. Atoms appeared to be the longed perpetual motion sought by science for centuries and centuries; a vast source of free power, or at least very cheap, awaiting only to be (safely) harnessed. But this did not happen, we now well know. Why? Besides the concerns by nuclear economies of restraining the nuclear proliferation, there was probably a more profound and simple reason

(Footnote 1 continued)

components or/and (3) a different structure (topology) of interactions, like for example, an higher hierarchy (see Chap. 5). In defining the complexity of the system according to the criterion explained above, network theory has grown in relevance in the recent years (Oltvai and Barabási 2002). Indeed, network analysis has proven to be a powerful tool to understand the energy metabolism of ecological systems and to explain their immanent tendency to complexification and accumulation of resources (Fath and Patten, 1999). Energy metabolism will be the central topic of the next chapter, but as we will try to make clear in this chapter, the concept of network is essential to understand the destiny of the energiewende.

[2]"There were four successive Industrial Revolutions, each of the last three building on its predecessor's achievements: that of steam, that of electricity, that of internal combustion and that of nuclear energy. The problem for us is to examine as closely as possible how this series of revolutions began. That means looking at the leading position of Britain between 1780 and 1890. Why was she the first to industrialize? How? And before 1780, what was the general situation of Europe in industrial matters? The Word 'industry', before the eighteenth century—or rather, before the nineteenth—risks evoking a false picture. At the very most, then, there was what may be called 'pre-industry'" (Braudel 1995).

behind: nuclear power (and electricity, which is the vector of nuclear energy) has never been a suitable substitute for internal combustion in transport nor it was, in many cases, a doable and cheap substitution for combustion in domestic heating. Combustion is still the votive altar of our economy and society. Indeed, as cleverly pointed out by Smil, energy transitions have always taken long time to fully develop their course (Smil 1994). The middle age sought the rise of mechanical energy drawn by wind and water, the "very first industrial revolution" according to Braudel,[3] providing an unprecedented leap of power for mankind (Smil 1994). Nevertheless, it was still the age of animate prime movers, i.e., human and animal labor, for the vast majority of society. Currently, in the age of fossil fuels, in most of the World, *animate power and biomass fuels* are the primary sources.[4] Yet, when humans began to utilize combustion for producing mechanical power rather than just for cooking, heating, and lighting, the world transition truly was transformational in the sense intended by Smil. The atmosphere too was transformed, as evidenced by the concentrations of carbon dioxide, as well as unintended production of nitrogen oxides, sulfur oxides, and other detrimental chemical compounds.

In the chronology of the Industrial Revolution, the transition from animate to inanimate power flies on the wings of the discovery of coal. Why was mechanical power, harnessed under combustion, discovered after coal? Nevertheless, the steam engine made coal, which has always existed, a source of mechanical power. Coal has nearly doubled the heat capacity of wood and, most importantly, is not bounded to land use. Nor does it directly compete with forests for timber or pulp. It is much more energy dense, but also much more energy *intense,* in terms of land use per energy unit delivered (Moreno-Cruz and Taylor 2014). In the nineteenth century, Britain produced more than one million tons of iron and, as highlighted by Smil, to produce such an amount of iron with biomass-fueled furnaces, would have required to put one quarter of the territory under coppiced wood (Smil 1994). Coal meant not only power, but also iron, besides energy. The last ingredient to trigger the industrial revolution—and the associated *energy transition,* was a suitable scale of the economy to fully release the newly available power. At that time, only *luxury items* and the factories that were providing nobles and upper classes with them could sustain large-scale production (Braudel 1995). The fate of industry was entangled with that of (national) state and *mercantilism* epitomized this common

[3]"The very first 'industrial revolution', may be said to have occurred in the twelfth century, when wind and water mills spread throughout Europe. But after that, for some seven centuries, there was no major technological innovation. Pre-industry, even in the eighteenth century, had only medieval sources and forms of energy. The power of a water-mill was normally in the region of 5 horsepower; that of a windmill, in windswept regions such as Holland, sometimes exceed 10 horsepower, but its output was intermittent. Without abundant energy resources and powerful machines, industrial life was condemned to semi-immobility, despite a multiplicity of small and often very ingenious technical inventions." (Braudel 1995).

[4]"Millennia of dependence on animate power and biomass fuels came to an end only gradually and the great transition to fossil fuels and fuel-consuming engines had highly country-specific onsets and durations", Smil 1994.

destiny. Mercantilism was the true ideology underpinning the industrial revolution, in which science, technology, and economy all concurred to the prosperity of the nation. According to Landes, it was a very remarkable expression in the domain of economic policy of the rationalism and of the Faustian spirit of domination over Nature (Landes 1969). The quest of resources to sustain growth has always been the crux of this policy, in a fashion similar to the basic behavior of organisms in ecological systems: seeking food is the primal instinct. In more recent times, the doctrine of *Lebensraum* in Germany or that of the *autarchism* in Italy followed the wake of the mercantilistic stake for resources. It is interesting to note that, according to the researchers of the Julich Institute, even the German *energiewende* does not differ from this long standing strand of energy history (Hake et al. 2015). In their view the roots of the *energiewende* are in the energy policy of Germany in response to the oil shock and in their attempt to protect the coal industry—the only truly national energy source, while shedding the basis for an economic growth freed from uranium and oil.[5] Indeed, Germany, perhaps unintentionally, succeeded in their first, original goal, as it is becoming apparent by the flourishing industry of coal. Is this the death knell for RES? No, despite the coal-paradox and the setback for the public policy aimed at sustaining RES, in Germany the share of RES is still growing. Germany is still a fossil-fueled economy (81% of its primary energy supply was from coal, oil, and gas in 2014[6]), but this is due to the historical inertia of the transition explained by (Smil 1994); what matters is that neither the collapse of oil price, nor the flaws of the German Energiewende has reversed the trend of RES. This might be due to the aforementioned inertia of the system or to the fact that the transition is escaping the control capacity of governments and policy. Something has been triggered. It seems that the energy transition is an ongoing process, led by forces of the economy, society, and technology, while policy is lagging behind, sometimes following the wave, sometimes swimming upstream. The last report of the energy division of Bloomberg—probably the most important company of media, data, and financial services in the world, stated this concept clearly:

> By 2040, the world's power-generating capacity mix will have transformed: from today's system composed of two-thirds fossil fuels to one with 56% from zero-emission energy sources. Renewables will command just under 60% of the 9,786 GW of new generating capacity installed over the next 25 years, and two- thirds of the $12.2 trillion of investment. Economics – rather than policy – will increasingly drive the uptake of renewable technologies (Bloomberg 2015).

---

[5] "Economic growth without oil and uranium? First ideas and pilot projects for an energy transition. The idea of an energy transition with a shift away from both fossil and radioactive fuels in Germany was born over 30 years ago and the history of renewable energies in Germany even dates back to the 1970s. As a response to the first oil crisis and inspired by the research projects of the Carter administration in the United States, the R&D program on energy launched by the Ministry of Research and Technology in 1974 spent about DM 10 million for research on renewables" (Hake et al. 2015).

[6] See (IEA 2015).

Intensity and Efficiency of Conversion of RES

Renewable energy sources, notably wind and water, have already paved the way for an unprecedented energy transition that took place, mainly in the pre-industrial Europe and North America (Smil 2008). This transition from animate prime movers to inanimate ones as a source of power was featured by an increasing complexification of the artifacts and the society that took over three centuries to unfold. The freedom of reducing and eventually largely releasing humans from ubiquitous physical labor was channeled into new hierarchies of production and organization in all facets of our living arrangement. Greater complexities emerged from this supply side step function. Although human beings have always exploited solar energy to produce mechanical work by harnessing the power of flowing water and wind—the early examples of elementary devices date back to the Greeks and the ancient civilizations of Persia and Syria—it was only in the European middle-ages that water and wind mills became a pervasive and prominent source of power. England in the eleventh century was dotted by at least 6000 water mills—one for every 350 people. They doubled in number in 200 years, and by the nineteenth century there were 30,000 operating water mills. Between 1300 and 1600, the number of wind mills in the Netherlands grew up to 8000. By the twentieth century, there were around 30,000 wind mills in the countries boarding the North Sea for an installed capacity of 100 MW (Smil 2008). Compared to the power of medieval or roman wind mills, with an output in the order of magnitude of 10^2–10^3 W, the Dutch wind mills of the late seventeenth century deployed a power of up to 30 kW and a transmission efficiency of around 60–70% (Smil 2008). In the same time, power of single water plants scaled up, from 3.5 W and an efficiency lower than 10% for the old vertical axis water mills and around to 50% for the horizontal axis ones, to a power above 50 kW, achieving peaks of 200 kW in some gigantic plants and an average efficiency of 60–70% in the twentieth century (Smil 2008). Along with power and efficiency, the complexity of the devices increased, from vertical to horizontal axes up to the modern screw and external vertical wheels in water mills that maximized the torque. The complexification occurred during the pre-industrial and post-middle age concerned also the way of converting animate power. The introduction of pulleys and levers brought the weight a human could lift up to 150–160 kg and the power output from a 100 W of a single worker to 700–800 W of eight workers in a treadwheel or above 1 kW with a couple of animals. The standardization and implementation of these unwieldy first generation contraptions demonstrate the incredible allure these new technologies provided. For example, in the late eighteenth century Colonial America, the harnessing of water power fed the textiles industrial boom, and the fledgling Federal Government Commissioned arms manufacturers first in Springfield, Massachusetts and then in Harpers Ferry, West Virginia due to their location to reliable hydropower. The entire towns became alive with whirling axles and leather pulleys that each factory tapped into (see Fig. 3.1). Considering that factories such as this already existed, the transition from hydro to fossil-based steam was an easy adjustment when that became available one half century later.

Fig. 3.1 Gun smithing equipment on display at Harpers Ferry National Historical Park (https://en.wikipedia.org/wiki/Harpers_Ferry_Armory#/media/File:Armory_at_Harpers_Ferry,_WV_IMG_4683.JPG)

Most remarkably, the complexification induced by the energy transition regarded the society itself through the advent of the first, pre-industrial manufacturing industries powered by water.[7] Nunn and Qian estimated that the introduction of the potato from the New World was responsible for a population growth of 26% in Europe between the eighteenth and nineteenth century and an increase in the urban share (of population inhabiting cities compared to that in rural areas) by 0.36% for every 1% of hectare dedicated to this cultivation (Nunn and Qian 2011). The complexification of society was sustained by an increasing per capita energy budget, which grew from the hunter-gatherer societies, relying on passive[8] solar energy, with 10–20 GJ per year available, to the agrarian societies, with up to 60–80 GJ a year by actively using solar energy. The scale-revolution actually came with the use of fossil fuel for generating mechanical work, which gave society an astonishing energy budget of 220–350 GJ per capita per year (Schlör et al. 2012).

[7]"Some medieval societies began to rely on inanimate prime movers for a number of demanding tasks including grain milling, oil pressing, wood sawing, powering of furnace bellows and forge hammers, and the mechanization of manufacturing processes ranging from wire pulling to tile glazing" (Smil 1994).

[8]For "passive use" of solar energy we refer to the energy embodied in autotrophos and heterotrophos which convert directly solar energy into phytomass or indirectly into biomass. For "active use" of solar energy we refer to the solar energy which is directly harnessed by humans in the form of potential energy of water or kinetic energy of wind.

It is now important noting that this wonderful escalation in efficiency and power output was nonetheless limited in the *intensity,* that is, in the *power per unit of surface* (of the land or air claimed by the conversion device). Wind power ranged in the order of magnitude between 10^0 and 10^1 W/m^2 and only water could get up to 10^2, which is still one or two orders of magnitude less than modern thermal power plants whose power output ranges between 10^3 and 10^4 W/m^2. Globally, the escalation was also vastly unequal in its distribution and access.

As previously highlighted, humankind has mainly relied on animate power as a prime mover for centuries and in slave societies this was freely available, subject to land surface availability and productivity to feed it. However, interestingly, despite the affluence of the free slave labor, no ancient society has ever progressed toward a scale of mass manufacturing such as that witnessed during the industrial revolution. Perhaps this was due to the anatomical limitations of the human body, which could be only partially overcome by means of ingenious mechanical aids. For example, an average modern western family can consume up to 30 kW of installed power in its appliances and in its two cars. This raises the power that must be available to allow this consumption up to 200–250 kW, which is an equivalent in slave labor force of almost 3000 slaves or 400 draft horses (Smil 1994). Indeed, everyone can agree that it is difficult, and hastily inconvenient, to fit 400 draft horses in an average modern western house. Likewise, the power of an average motorcycle is of about 30 kW, which, if it had to be delivered by means of photovoltaic cells, would need almost 200 m^2 of these cells; an extent that can be hardly placed on a ship. Yet, a mere gallon of gasoline contains more than 30 kWh of energy, which means that, with an (optimistic) efficiency of about 40%, it enables me to ride my motorbike at *maximum speed* for half an hour, *even by night.* Efficiency is not the only metric to judge RES. Photovoltaic cells, with an efficiency ranging from 20% to almost 40% is one order of magnitude greater than the efficiency of photosynthesis that ranges from an optimal efficiency of 5–6% to an average efficiency of about only 1–2%; hydroelectric power plants can have an efficiency up to 90%, which is by far higher than the attainable efficiency from the newest combined cycle gas turbines, with a peak of 60%. It is also worth noting that RES, as a recent study has unmistakably and surprisingly shown for the UK electricity sector, can score better than fossil fuels in terms of *energy return on investment* (EROI) to produce electricity (Raugei and Leccisi 2016). EROI accounts for the amount of energy that, directly and indirectly, must be consumed in order to generate one Joule of electricity. It is an important criterion for selecting energy sources in a highly complex society, with a non-modest need of exosomatic energy. However, features of energy sources other than efficiency or EROI are relevant for economic purposes. These features are the energy intensity, as we have explained above, and *time disposal,* as it will be explained in the following section.

Complexity of RES: Discontinuity, Diffusion, and (Im)Predictability of RES

Stanley Jevons was fully aware of the advantage of fossil fuels' time disposal compared to renewables in relation to the cycle of capital. He argued, for example, that what made steam vessels more economical was neither fuel efficiency (wind power is more efficient) nor unit costs (wind vessels are almost costless), but it is instead the availability and certainty of coal disposal as an energy source which had an incomparable positive impact on the capital return cycle.[9] It was the *regularity and rapidity* of steam vessels that rendered them more advantageous. Similarly, in the field of biological systems, evolution has awarded those systems that are faster in tapping the available free energy.[10] This is the so-called "maximum power principle,"first envisaged by Alfred Lotka and later developed by H.T. Odum (Odum 1995). The nexus between power, efficiency and complexity will be treated more in depth in Chap. 5. What is worth noting is that biological evolution and economic development have followed the same paths or, we may say, *paradigms,* by increasing the rate of energy intake per unit of surface, or volume, across their systems. This was achieved by, or perhaps as a result of, an ever growing complexity. It is thus difficult to foresee a future for RES where the system will be able to *reduce its energy intensity* (power per unit of surface) and sustain the level of complexity so far achieved. Power will become a limiting factor to feed the complexity of the system and at the same time complexity will also increase to deliver more power to the system: this is the conundrum of RES that involves significantly the role of the network. This issue will be tackled in one of the following sections.

It is also worth remembering that when we talk about power, we mean a rate of energy flow and the discontinuity of RES inevitably reduces this rate. Nevertheless, the problem of the discontinuity and unpredictability of RES goes beyond the two above mentioned aspects, i.e., the impact on the capital return cycle and the lower power output. It probably has to do also with the twentieth century concept of modernity, and its undying myth of control, which innervates the last vision of progress, so seriously threatened by the impending environmental (and economic) crisis. It is not difficult to accept for a farmer that when it is raining, it is time to rest. It can be otherwise very difficult for an engineer to relax while watching the wind turbines idle. However, the torment of one engineer is not to be the trouble of an economy.

[9]"The regularity and rapidity of a steam vessel render it an economical mode of conveyance even for a heavy freight like coal. The first cost of a steam collier is greater than for sailing colliers of equal tonnage. But then capital invested in the steam vessel is many times as efficient as in the sailing vessel" (Jevons 1965).

[10]"The maximum power principle can be stated: During self-organization, system designs develop and prevail that maximize power intake, energy transformation, and those uses that reinforce production and efficiency" (Odum 1995).

Will the growth of RES aggravate or ameliorate the impact of discontinuity over the need to supply the baseload energy demand? Or, in other words, the question is: will more intermittent power inlet points increase the volatility of the system and its unpredictability? This is a hard question to answer. We must acknowledge that, for example, if the output of a single RES plant fluctuates greatly, the fluctuations in the total output from a large number of geographically distributed RES plants, in different environmental conditions, will be much smaller and more predictable. These generation facilities follow the law of *large numbers*. Several studies have shown that RES, if properly implemented, can provide baseload power for Europe in the foreseeable future.[11] With an accurate mix of energy storage, demand side management and infrastructure enhancement, it is thus in principle possible to feed the European energy system only with RES, guaranteeing the continuity of supply. Hence, discontinuity is not a technical problem, it is rather and *economical and cultural* problem, as explained above.

To conclude, it is important to understand that these two prominent features of fossil fuels, i.e., the higher energy intensity and the time disposal, created the conditions for the complexity leap witnessed in the last two centuries, by shaping the spatial and time dimension of economy and society. The problem of sustainability of RES, from the viewpoint of economic and social sustainability, must thus critically address these two main issues.

The Market of RES: Property Rights and Free Access

Another important viewpoint on the conditions that fostered the Industrial Revolution, is the evolution of property rights, foremost in the eighteenth century England. Landes, among others, has placed the emphasis on the role of the *Enclosure Acts* on the process of capital accumulation that was a fundamental condition for the development of the first industries (Landes 1969). Enclosure was the legal process in England of enclosing a number of small, common landholdings to create one larger privately owned farm. Enclosures were twofold important: (1) they increased the land productivity and (2) created the middle-class that inspirited the Industrial Revolution. As previously highlighted, mercantilism was the ideological expression of this tight bound with land and it is not by chance that the birth of the classical economic theory, with influential scholars of the like of David Ricardo, focused primarily on the marginal productivity of land. The rise of coal and its fabulous applications was disruptive with this view of prosperity: wealth was no longer tied to land, but to the power of mechanical work. The role of marginal productivity was thereby restored in a new frame that elevated production factors, whose remuneration was based on the output of productivity and the *scarcity* of the factor. An unlimited resource, freely available, is not valuable in the

[11] https://ec.europa.eu/energy/sites/ener/files/documents/2012_energy_roadmap_2050_en_0.pdf

context of this economic activity and coal, indeed, was far from unlimited, though abundant and cheap. Coal, like oil, is a finite source that comes from the ground. It is thus possible to clearly assign property rights on it (like for the land once is *enclosed).* It goes along with this that renewables are difficult to price and exploit according to the logic above. The revolution of RES is double: on the one hand they reconnect the link with the land (surface) and on the other hand they sever the link with the property rights. The energy source itself is free and cannot be otherwise. With the impending energy transition, energy utilities are confronting with the same problem that the music and film industries have just faced with the advent of the Internet and the free share of movies and music. It was the technology, with the means of mass reproduction and the physical supports of art artifacts that created the condition for the application of property rights. Records were a *scarce* resource for the reproducibility was restrained to industry. Now, with the Internet, reproduction is free and unlimited and the industry must resort on new laws and enforcement to apply property rights. Likewise, all energy systems, from the network to the market, were conceived for a large, manifold, and uncontrolled demand and a limited, supervised number of suppliers retaining the control of the energy source. The impelling issue of the property rights mirrors that of the balance and regulation of the grid, which was developed in a context of a few large inlet points, almost perfectly under control and a vast and variegate pit of outlet points. The grid must be constantly balanced, and failing in doing this is worth a big loss of tension. The electricity control center of the network is facing the unknown frontier of having to handle a large portion of installed electric power capacity that will be intermittent and uncontrolled. These are practical and technical problems that are entangled with the economic ones. When facing them, we should bear in mind that in the new context of market liberalization, the private and public actors were once wedded in a single, public company which used to oversee all the production chain, from the production to the transmission and the dispatch. Perhaps, this is why big utilities cannot help but hide a certain disbelief, if not nuisance, to RES. Despite it is still very profitable for utilities selling highly subsidized electricity, it is unsettling the prospect of making long-term business plans in a scenario where the revenue depends only on the capital costs and is highly affected by the unpredictability of future energy prices. This is not to mention the burden of the legacy of the traditional plants (some of them, like the nuclear power, of unknown dismantling costs) and the capital invested and frozen in them[12]. However, the traditional way of producing electricity is not free from hazard. Since the *synchronization* of all the energy commodities with the price of oil, even the price of elasticity is bound to the uncertain and capricious fluctuations; but utilities have the possibility to hedge against them either by indexing the retail prices or by relying on the market with the

[12]This is already becoming compelling and paradigmatic, in many countries where gas turbine plants are paid by the government to stay idle. Though apparently appalling and disturbing, this is the result of the ambiguity between the private and public sectors in the field of electricity. Indeed it is true that some of those power plants were once paid by the government and they thus must be a concern not just of the utilities, but of the collectivity altogether.

related financial tools. However, when in the wholesale market the marginal price lowers to *zero* because of the overflow of RES, even these possibilities are dramatically cut off (Ruzzenenti 2015).

We are thus hereby contending that the problem of the governance of the energy transition toward RES, in the context of market economy, is also a problem of setting a favorable environment for a new frame of property rights suitable to the technical constraints posed by renewable energy. In other words, the policy-maker must aid the market to adapt to this new context and quest for the dimension where the *new scarcity lies.* What is scarce for the nexus between RES and society? As we hoped to have explained above, it is *power* and *intensity.* Rather than the present overall supply—*quantus*—the energy in a given time, in a certain place—*hic et nunc*, and for an adequate amount—*aliquanto*, will be the scarce source of the future.

The Governance of RES: The Role of the Network

The shift from the paradigm of *quantus* to that of the *hic, nunc et aliquanto* envisages a new and radically different definition of the role of the network, i.e., the *electricity grid.* It is noteworthy that the network makes the RES a truly competitive and affordable source of energy compared to fossil fuels. In the past, the technical and physical limits of mechanically connecting renewable power generation, like water or wind mills, condemned RES to a marginal role when fossil fuels were scaling up the intensity of power in factories and transport means, but it was only with the onset of the electrical network that the intensity of power could scale up even in daily domestic usage decreasing drudgeries of human labor.

We have already outlined the new technical and managerial challenges that the network will undergo in the next years, but it is its role altogether that will change. The grid now must fulfill constantly the need of consumers, households or firms, pervasively and permanently. While meeting this need, the grid tries also to minimize the losses, for example, by imposing a cap on absorption; by making costumers pay for the so-called *reactive energy*; by differentiating the tariffs to promote consumption in off-peak hours and, for few large users, by negotiating load break off. Nevertheless, these are just embryonal examples of demand side management. In real terms, what drives the current logic of the network is just a matter of *amount.* In the future, rather than the *total,* what will matter will be *the parcel, the peak, and the place* and this is why we will face a tremendous leap in the topological complexity of the networks and in the ongoing dynamics. All facilitated by increased metering and dissemination of electricity but highly reliant on the flexibility and performance of inchoate smart grids. This complexity will be further increased by the fact that what we consider now *exogenous* variables will become *endogenous* ones and who now is a *customer* will also become a *producer* or a *dealer.* The case of power storage technology is an emblematical example. In a recent article, the Guardian forecasted that 2016 will be the turning point for the energy storage technologies (Carrington 2016). Carrington maintains that *industry figures [are] predicting a breakthrough year for a technology not only seen as vital to the large-scale rollout of renewable energy, but also offering the prospect of lowering customers' energy bills.* What is mostly fascinating and promising of small-scale storage technologies is that they are able to accomplish that kind of *paradigm shift* formerly envisaged. Innovate UK, a government agency,[13] is studying the use of electric car batteries as a smart storage network. According to Mark Thompson, there could be 4 GW capacity, something like four medium size nuclear power plants, across the 300,000 electric cars projected to be on UK roads by 2025 (Carrington 2016). Notably, cars are parked for 95% of the time and, although a cleverer integration between information technology (IT) and transports could lead to a more efficient use of vehicles—such are the cases of car-sharing or ride-sharing—converting stationary private vehicles into power reservoir could save billions of pounds, pendant a suitable tariff scheme. Indeed, this new role of the network is prone to accomplishing a new governance system aimed at pricing power rather than energy. By now, both self-production and storage are hindered by the fact that only energy is priced, even if highly subsidized. The paradox of high incentives is that there is little convenience to store (or self-consume) the energy rather than input this energy in the grid. By doing this, however, the grid is transformed in a storage of power,[14] whereby to draw when needed and the costs of the supply of power is shifted toward the collectivity. A good example of an enhanced tariff scheme came from the third Italian solar energy program,[15] where self-consumption (the consumption of the self-produced energy) was subsidized per kWh as much as one-half the energy produced and sold to the grid. Therefore, the sum of the incentives for the self-consumed energy plus the energy not taken from the grid made self-consumption more profitable than selling energy to the grid. Unfortunately, this scheme was recently dropped. Perhaps, the problem is that even this latter scheme, though representing a progress compared to the past, still failed to acknowledge that power should be priced rather than (or together with) energy. It is noteworthy that if we start pricing power, then the costs of RES would probably change altogether. The evidence that power is already becoming a scarce resource was provided by an investigation made by Althesys, an energy consulting company, a few years ago. They demonstrated that Italy, because of photovoltaics, saved in 2013 about 1.4 billion Euro in *peak shaving* (Gualzeri 2013).

[13]https://www.gov.uk/government/organisations/innovate-uk.

[14]Although, in technical sense grid can't store power since it must be used "instantaneously". Here, storage, is meant in an economical sense: the user/producer exchanges of energy are deferred.

[15]http://www.renewableenergyworld.com/articles/print/volume-14/issue-3/solar-energy/italy-overhauls-its-pv-incentives.html.

That said, the goal of an outbreak of a *market of power*[16] requires unprecedented technical and regulatory conditions that only a visionary and far-sighted policy can create.[17] The preconditions are the network, but the regulation thereof has to dramatically change. A double-pronged change in the spatial and temporal dimension of the grid regulation is required. The first requirement is a decoupling of the inlet from the outlet points for every single user, physical or juridical person. That means that one could produce energy in one place and simultaneously use the same energy somewhere else (thanks to different meters). Second, the *work* (energy) and the *power* must also be decoupled within pricing and tariff schemes. That is to say, one should be able to sell/buy the energy produced/consumed in a given time interval, but also the power produced/consumed in a certain time and place. Now, both in the retail market and the subsidy schemes only energy is priced, but in the day-ahead market price power is already being traded. This possibility should be extended to all users and not only to market operators. In order to achieve this *double-decoupling,* however, there is urgency both to connect IT technologies with electronic meters and to develop a suitable legal environment.

Policy Indications

The transition from fossil fuels to RES in the European electricity sector is already an ongoing process that seemingly will not be reverted by low oil prices or by the backwardness of politics. Those who maintain that fossil fuels will remain the primary energy source for power generation neglect the fact that every energy transition historically took decades to fully unfold. In the nineteenth century animate power, wind, and water were still the dominant energy sources, in spite of the emerging role of coal. Hydrocarbons, witch underpinned the fastest energy transition in history, surpassed coal as a primary energy source only in the late 1970s, after more than 50 years (Smil 1994). Interestingly, or perhaps not surprisingly, hydrocarbons became dominant with the second wave of globalization, when transport turned into a major energy demanding sector of economy and most of the available free energy was converted into prime movers, rather than heating

[16]It is worth noting that in English, power and electricity are often used as synonyms and many refer to "power market" or "electricity market" indifferently. Indeed, here, with the notion of "market of power" we are not making any reference to the existing, so-called, wholesale electricity market and its articulations into the day-head or intra-day markets. Here, for market of power, we mean a market of energy at a certain time, in a certain place, linking a producer and an user dispersed in the network. Nevertheless, it is interesting that these financial markets somehow present the same embryonic form of interaction though limited to operators and restricted by technical and economic conditions.

[17]Indeed, there are different options to tackle the problem of intermittency and low intensity of RES other than an open free market of power. For example this issue could be tackled with a more sophisticated pricing system by a public regulator. Nevertheless, the market option is consistent with the current EU policy aimed at fostering power market liberation and integration.

(Ayres et al. 2003). On the contrary, nuclear power, after more than 50 years has never become a leading source of power, even during the ongoing *electrification of society.* The transition toward nuclear energy simply never occurred, probably because of the real costs of this energy. This simple evidence should educate us enough about the real fate of nuclear and the opportunity to rely on this source to de-carbonize the economy.

The present analysis eluded the issues concerning the environmental impact of RES for it was not in the scope of this chapter. Indeed RES are generally not unblemished and may present many environmental hazards, but the real issue at stake concerns climate change and global warming and the transition to RES must therefore be pursued at the fastest possible pace. Other mitigation measures, like carbon sequestration or *climate engineering,* in the view of the authors, are still too uncertain and impracticable to be relied on in the due time. Nevertheless, history taught us that energy transitions are slow. It is therefore a policy duty to accelerate the pace of the ongoing transition, for the sake of future generations.

We will outline below some of the critical issues concerning the foreseeable and desirable transition to RES, in the light of the concept of complexity and the goals of sustainability:

(a) to be environmentally sustainable, the transition to RES must be as fast and pervasive as possible, without compromising the social and economical sustainability of the economic and social systems.
(b) To be economically and socially sustainable, the transition to RES must maintain the present level of complexity of the system, both in terms of the complexity and scale of the production structure, of artifacts and of urban development (see Chaps. 4 and 5).
(c) To sustain this level of complexity of society and economy, the complexity of the energy system must increase: these will concern the design, the governance, and the structure of the network(s). The electricity network, but also the communication networks (information flows) and the transport network (mass and people flows). This determines a situation, which is commonplace, insidious, and paradoxical throughout history, that to maintain complexity at one scale (of society and economy), complexity at another scale (of energy systems) must increase. For example, the complexity of modern production chains would not have been possible if electricity would not have initially permitted the cumulative and complex connection of mechanical work (formerly attainable only with pulleys, mills or threads) and, more recently, with the new frontier of *networks of machines,* a connection at very large distances.
(d) The network(s) will have to undergo a leap of complexity that can be described in the form of a transition from the paradigm of *quantus* to that of the *hic, nunc et aliquanto:* this means the decoupling of energy and power for the market.
(e) This transition if inadequately tackled can cause a dramatic increase in electricity demand, posing a critical limit to the onset of RES. This increase in electricity demand can have two main causes: an increase caused by the shift from energy sources (for example, from gas/coal to electric pumps for heating

or from gasoline to electric cars); an increase in energy demand brought about by automatization and labor-saving technologies, which substitute endosomatic energy with exosomatic energy mostly represented by electricity.

(f) The foreseeable rebound effect following the introduction of electric or hybrid cars can further increase the demand for electricity. At present prices of electricity, the fuel economy of electric cars is one order of magnitude lower than internal combustion (this topic will feature Chap. 7).

Specifically, the present or forthcoming impediments to the completion of a transition to RES that are immanent to the present system and affect the current economy and governance schemes are the following:

1. On the short run: the power glut hat RES can generate can cause the inactivity of traditional power plants. This depends on the fact that RES can generate periods of oversupply (because of discontinuity) making problematic the necessary continuous production of traditional plants, both for economic reasons, the electricity wholesale market being flooded in this case by the electricity at zero-price from the overproducing RES, and technical reasons, the grid being unbalanced due to a the lack of adequate storing and buffering systems.
2. On the long run: the rate at which new renewable power can be installed to replace fossil fuel generated power, (this means that we will still need fossil energy for the foreseeable future).
3. The double bottleneck of the two networks constituted by the electric grid and the transport network: the present electric grid is still devised for a few, big inlet points (big power plants) and unsuitable for a complex governance and management of all the involved factors (storage, demand management, self-production, transmission, etc.). Moreover the transport sector is still closely bound to hydrocarbons.
4. The present (modern) concept of control in the face of the increasing complexification of network and unpredictability of RES.
5. The new concept of scarcity (related to power rather than to energy source availability), and consequently the new property structure brought about by RES. Power, rather than energy, will become the valuable item of the market.

Finally, we would like to highlight the fact that two networks are crucial to RES: the electricity grid, as is obvious for any energy source that relies mainly on electricity as a vector; and the transport network, transport being arguably to date the only sector where hydrocarbons are not completely replaceable. In the next chapter, we will explain how the transport sector crucially contributed to build the complexity of our economy by interconnecting productive sites globally. It is noteworthy that the transition from a Fordian productive structure, where the production's chain is uni-located, to a post-Fordian one, where the production's chain is pluri located, entailed a change in the spatial symmetry of the system, underlining the complexity leap featuring the second wave of globalization (Ruzzenenti et al. 2010).The notion that transport networks play a fundamental role to sustain and develop the complexity of the system is not new in science

(West et al. 1997). Therefore, the policy-maker in the quest for removing the obstacles to RES must address the critical issue of the transport sector, in the light of the paradigm of networks. This is to say that rather than addressing the energy source of prime movers, i.e., switching from internal combustion to electricity, it is more important to approach the problem in a systemic way. Flooding the market with incentives for electric cars is for example less important than creating the technical and institutional conditions for a market of power. As previously described, this means essentially to decouple inlet and outlet points or, in other words, the electric meter where energy or power is sold and the electric meter where electricity and power are bought. It is also recommendable, in the long run, to set the target of decoupling the physical from the legal inlet/outlet points. In other words, the grid should allow the purchase/sale of energy/power anywhere on the condition that there exists a free and suitable electric meter (and inlet/outlet point) to enter the grid. This ambitious goal could only be achieved by means of IT and a further thrust in the development of the grid. All the aforementioned changes would constitute a breakthrough for the triumph of RES, not just in the transports sector, but also in the household, service, and industrial sectors. But the question is: what has to be done in order to achieve this goal?

1. Strengthen the physical infrastructure of the network (and possibly, maintain the public control of it, in the light of the crucial role this will play in the coming future).
2. Foster the switch from hydrocarbons to electricity (even in the heating sector): electricity is the carrier for RES and combustion must be reduced, foremost in densely inhabited areas.
3. Implement the needed measures to enhance the grid from dull grid to smart grid: energy storage, transmission infrastructure, information technology applied to the grid, decentralization of control, synchronization of demand and supply, demand management.
4. Develop a suitable regulatory and fiscal framework to create a market of power separated from that of energy (something similar to the wholesale electricity market).
5. Split the inlet point from the outlet point in the grid: individual production and consumption must be spatially (and temporally) decoupled.
6. Split the physical from the legal contact point to the grid: the user must be different from the node of the grid.

As foreseen by Bloomberg, the predicament of RES lies in the politics rather than in the market. However, by implementing these measures, the policy-maker would most probably enable the market to accelerate dramatically the pace of the transition to RES, outpacing the foreseeable return of coal, breaking the resistance represented by hydrocarbons and counterbalancing the demand increase due to the rebound effect.

References

Ayres, R.B., L.W. Ayres, and B. Warr. 2003. Exergy, power and work in the US economy, 1900–1998. *Energy* 28:219–273.

Bloomberg. 2015. New energy outlook (NEO). Available at: http://www.bloomberg.com/company/new-energy-outlook/.

Braudel, F. 1995. *A history of civilizations*. Harmondsworth, Middlesex, England: Penguin Books.

Carrington, D. 2016. From liquid air to super capacitors, energy storage is finally poised for a breakthrough. *The Guardian*, February 4, 2016.

Fath, B.D., and B.C. Patten. 1999. Review of the foundations of network environ analysis. *Ecosystems* 2 (2):167–179.

Gualzeri, V. 2013. Rinnovabili, uno studio assolve i costi: "Alla fine ci guadagniamo quasi 50 miliardi". *La Repubblica*, April 18, 2013.

Hake, J., W. Fischer, S. Venghaus, and C. Weckenbrock. 2015. The German Energiewende—History and status quo. *Energy*, Part 3 92: 532–546.

IEA. 2015. Energy statistics of OECD countries. http://data.iea.org.

Jevons, S. 1965. *The coal question—An inquiry concerning the progress of the nation, and the probable exhaustion of our coal-mines*. Augustus M. Kelley Publisher: New York, 1965. First edition: 1865.

Landes, D.S. 1969. *The unbound Prometheus: Technological change and industrial development in Western Europe from 1750 to the present*. Cambridge University Press.

Moreno-Cruz, J., and M.S. Taylor. 2014. *A spatial approach to energy economics: Theory, measurement and empirics*. CESifo Working Paper Series 4845, CESifo Group Munich.

Morris, C., and M. Pehnt. 2015. *Energy transition: The German Energiewende*. Heinrich Böll Stiftung. http://energytransition.de/.

Nunn, N., and N. Qian. 2011. The Potato's contribution to population and urbanization: Evidence from a historical experiment. *Quarterly Journal of Economics* 126:593–650.

Odum, H.T. 1995. Self-organization and maximum empower. In *Maximum power: The ideas and applications of H.T. Odum*, ed. Hall, C.A.S. Colorado: Colorado University Press.

Oltvai, Z.N., and A.L. Barabási. 2002. Life's complexity pyramid. *Science* 298 (5594):763–764.

Raugei, M., and E. Leccisi. 2016. A comprehensive assessment of the energy performance of the full range of electricity generation technologies deployed in the United Kingdom. *Energy Policy* 90:46–59.

Ruzzenenti, F. 2015. *Series of papers on changes in the relationship between the financial and the real sector and the present financial and economic crisis: The energy market*. FESSUD Working Paper Series No 105. Available at: http://fessud.eu/working-paper/.

Ruzzenenti, F., D. Garlaschelli, and R. Basosi. 2010. Complex networks and symmetry II: Reciprocity and evolution of world trade. *Symmetry* 2 (3):1710–1744.

Schlör, H., W. Fischer, and J.F. Hake. 2012. The meaning of energy systems for the genesis of the concept of sustainable development. *Applied Energy* 97:192–200.

Smil, V. 1994. *Energy in world history*. Boulder: Westview.

Smil, V. 2008. *Energy in nature and society: General energetics of complex systems*. MIT press.

The Economist. 2015. When the wind blows, November 28, vol. 417, no. 8966.

West, G.B., J.H. Brown, and B.J. Enquist. 1997. A general model for the origin of allometric scaling laws in biology. *Science* 276 (5309):122–126.

Chapter 4
Present Energy Metabolism and the Future of Renewables

Franco Ruzzenenti and Brian D. Fath

Abstract Metabolism refers to the process of energy and material flows required to sustain the structure of an organism, ecosystem, or socioeconomic system (such as an urban area). The study of energy metabolism of an economy is insightful on both a local scale (city, region, or country) and on a global scale (world economy). A key feature contributing to the complexity of socioecologic systems is feedback, manifest in the presence of *cycles*. Material cycles in ecological systems are closed: mass is conserved throughout all cyclic paths. Furthermore, the incoming solar energy is maximally dissipated throughout cycles. Ecological systems have developed intricate couplings in order to reduce or eliminate energy or material waste, in juxtaposition to economic systems. What makes then an economy so *inefficient* compared to nature? On a local scale, the study of metabolism indicates that cities or countries are not a self-sustaining systems: they draw materials, energy, and information from the surrounding ecological and economic environment. Cyclic metabolic paths in the world economy are typically strictly (anti)-correlated to oil price. As showed in this chapter, the percentage of cycled material in trade was negatively correlated to oil price; this anti(correlation) scoring from 85 to 62% between 1960 and 2011. This shows that world metabolism is remarkably connected to the price of oil. In the long run, world metabolism is correlated to oil price because of the architecture of trading relationships. With low oil prices, the productive chain tends to unfold across countries, whereas with high oil prices the productive chain tends to shrink. Constraints and impediments to the complete success of renewable energy sources (RES) over fossil fuels are therefore based on certain factors which can be determined from a metabolic analysis of the economy: (1) energy source intensity, (2) the nonfungibility of oil in the transport sector, and

F. Ruzzenenti (✉)
Department of Management and Quantitative Sciences,
Parthenope University of Naples, Naples, Italy
e-mail: ruzzenenti@gmail.com

F. Ruzzenenti
Institute of Sociology, Jagiellonian University, Krakow, Poland

B.D. Fath
Towson University, Towson, USA

N. Labanca (ed.), *Complex Systems and Social Practices in Energy Transitions*,
Green Energy and Technology, DOI 10.1007/978-3-319-33753-1_4

(3) scale of production. Each factor raises particular questions which will be answered in this chapter. For example: Is the scale of the present economy/society (cities, countries, or world) strictly dependent on the intensity of fossil fuels? Can these scales of processes be sustained with energy sources at a lower intensity? What is the appropriate feedback between the scale of ecosystem services and scale of governance? Is circular economy attainable at the scale of the present global economy? These questions will be addressed in the light of energy metabolism.

Introduction: The Study of Energy Metabolism Concepts and Methods

The simple to ask, yet difficult to answer, question "What is life?" provides a good starting point to better understand complex adaptive systems.[1] Basic textbooks in Biology—the science of life—have trouble answering this question,[2] but at core they identify life with the reproduction of three main functions: metabolism, repair, and replication. Wilson et al. (1978) state, "Metabolism is the most obvious hallmark of life" (p. 6) as it is the release of chemical energy that allows for repair, replication, and other processes to continue maintenance of the individual. Fiscus et al. (2012) found it useful to distinguish between discrete life and sustained life, the former being the typical study of biology textbooks: an individual and the internal relations that keep it functioning. An individual is alive but it is not sustained—it will perish if severed from its environment even for a short time. The latter, sustained life, refers to the complex interrelations that exist allowing an ecosystem to persist. Life is sustained at the scale of the ecosystem, not at the individual (Keller and Botkin 2008), yet this basic fact is not reinforced in our understanding of life. In both discrete life and sustained life, metabolism occurs to release the energy needed to support the activities. In this thermodynamic perspective of complex system behavior, the first step is the acquisition of the energy resources and incorporation of it across the system boundary. For living systems, this begins with the conversion of solar energy to organic compounds during photosynthesis in the chloroplasts of plant cells. This primary production is then

---

[1]A complex adaptive system (CAS) is a complex macroscopic collection of connected components organized in order to adapt to the changing environment. With the words of Holland: CAS are systems that have numerous components, often called agents that interact and adapt or learn (Holland 2006).

[2]Classic textbooks by Wallace et al. (1981) and Raven and Johnson (1989) do not include a straightforward definition, nor do they have "Life" listed in the glossary. A book by Wilson et al. (1978) has this line, "All living creatures must metabolize, grow and reproduce, protect themselves and their offspring, and evolve in response to long-term changes in their environment." (p. 6).

utilized by the entire ecosystem including primary producers, primary and secondary consumers,[3] and decomposers through a complex network of trophic interactions. This represents a transformation from exosomatic to endosomatic energy, the internalization of "environmental" resources.

A similar perspective, incorporating the energy and material flow through human systems, has been found useful and has developed into the field of urban metabolism. This approach was first used by Wolman (1965) who used input–output analysis to study flows in cities. Current examples of urban metabolism study the energy, carbon, nutrient, and water flows in cities (Baker et al. 2001; Zhang et al. 2012). Some research provides a complete accounting framework to identify the sources and recipients of each flow (Zhang et al. 2010, 2016). Other studies focus on the similarities and difference between urban systems and natural ecosystems (Villarroel et al. 2012). Urban metabolism studies can use the energy and material flow to investigate the trophic relations or the relationship types in the system. Concerning the trophic relations, ecological systems typically display a pyramid structure in that they are supported by a large amount (of biomass or number of species and biomass) of primary producers with a diminishing number of first level consumers, second level consumers, etc. (Elton 1927). This is believed to be a healthy overall structure because the energy requirements are in sufficient supply at the base to support the smaller populations of consumers. Research has shown that in urban system, these pyramids are often inverted, highlighting the fact that energy resources supporting the city are not in abundance locally (Zhang et al. 2011). This may in part be a boundary issue, in that cities by design are separated from, yet linked to, a surrounding countryside for resource support such as food, water, energy delivery, and waste reception. In contrast, ecosystems are bound to local constraints thus relying on present conditions such that all activities occur in one space with less obvious spatial specialization. An open discussion remains whether it is better to mimic ecosystems in this manner and thus promote more activities such as urban farming and onsite waste incineration or to maintain the specializations and dependencies. It depends on the scale at which a system should be considered as sustainable or not. Our view is that cities by design are not sustainable systems in a local sense and it is not efficient to make them so. A city's purpose is as the human niche and it can provide residences, jobs, eateries, shops, services, social settings, etc. most efficiently by concentrating these functions within some areas whilst reserving other areas for food, energy, timber, and mineral production as well as watershed and waste management. Many urban metabolism studies have shown that urban regions, in contrast to ecological systems, display inverted or fusiform pyramids structures with fewer mutualistic relations (e.g., Zhang et al. 2010, 2012).

[3]Producers are photosynthetic organisms, first consumers are species that feed directly from producers (such as herbivores) and second consumers are predators.

Global Scale Metabolism

The metabolism of living systems is the process of breaking down the intake of organic matter and building up of the components of cells such as proteins and nucleic acids for the sake of the organism's development and sustenance. In order to occur, this process needs a source and a sink of both energy and raw material.[4] On a global scale, such as the level of organization of an ecosystem, this process is not linear, but cyclical: there is no waste of material. As previously discussed, the parallel with an economy on the local scale of firms, cities, or countries, is straightforward, although in this latter case the process is linear rather than circular (the economy taps into the natural sources without restoring them). The analogy with the metabolism of the economy on a global scale is much less evident, and possibly, less understood. This is because many believe that the country-economy is the optimal dimension of the production system and thus look at international trade as a mere process by which specialization of national economies exchange their peculiar endeavors, either technological or natural. Nevertheless, the largest share of trade in the world economy involves the means of production (Miroudot et al. 2009). In the modern economy, countries import the means of production and export goods or intermediate goods that are further processed somewhere else. This means that nowadays production chains are global rather than national (Baldwin 2006). Goods at every step of the global value chain (GVC) embody the value-added augmented in every step of production. This notion delivers a thoroughly different picture of the global economy, more similar to a system, rather than a market. This *World System* is hierarchical and deeply interconnected (Prell et al. 2014). Like in ecosystems, where matter flows cyclically along food webs, matter in the World System runs through cyclic paths alongside the global value chains. In order to detect the path of conserved value added through the global economy (the global value chains), we need to assess the probability of a certain amount of money, embodied in a good, moving across countries in the world trade network along a cyclic path. One way to do so is by means of a cycling index, a measure based on Markov chain theory.

A cycling index was first developed to assess the share of matter that is recycled throughout food chains in an ecological network, from the primary producers (photosynthesis) to the top predators and detritus feeders (Finn 1976, 1980). Likewise, we can think of value as matter in food chains and look by means of cycling how this is conserved throughout the stages of production internationally,

[4]It is worth noting that at the scale of biosphere only a source (short wave radiation) and a sink (long wave radiation) of energy is needed, being matter almost completely recycled (almost, because you still have a sequestration and de-sequestration of undecomposed biomass, which can become coal and hydrocarbons and varies according to climatic factors).

where the sink and the source of value are national economies. The world trade network is an open system: matter is not conserved through every stage of the international production chain because goods are produced and consumed at every network's node, and every country is a source of raw material and a sink of waste. However, the analogy, methodological and semantic, builds on the conservation principle that applies, respectively, to matter in the case of ecosystems and to value in the case of the economic World System.[5]

Figure 4.1 shows the cycling index Γ,[6] which accounts for the share of cyclical value over the noncyclical value (closed over open loops of the value) for the world trade network, between the 1960 and 2010, along paths of increasing length, from two steps to infinite steps (Picciolo et al. 2017).

The cyclical value in trade amounts, statistically, to one-fourth–one-third, depending on the period, of the total value traded, signaling, as previously stated, that world trade operates as a reticulated system rather than a market, where the value added is exchanged between partners and, supposedly, does not return to the seller.[7] In other words, this picture tells us that, statistically, one-third of the value in trade will sooner or later return to the country of origin along direct and indirect paths. Figure 4.1 shows also that this share is not stationary, meaning that it is not a stable, immutable feature of the system. It actually varies greatly in time, but it did not steadily increase, as one might expect given the process of globalization and market integration. This is seen as a process that is universal and irreversible. Nevertheless, the cycling index reveals that the share of cyclical value in trade dwindled twice in the recent history, after the 1970s and the mid-1990s. Therefore,

[5]In ecological networks, when we want to assess the amount of mass that is conveyed through one species (prey) to the other (predator) at every step of the food chain, from primary producers (grass) to the last predators (and decomposers), we cannot tag every atom of the organism and map every passage. We can only weigh the body mass of organisms through the food chain. If we know that the species A feeds 50% on the species B and 50% on the species C, then we know that the atoms of the species A have 0.5 probability of coming from B and 0.5 of coming from C. We can do this for all the species of the food chain and represent this by a continuous, steady food flow. If in the previous example the species C feeds on the species E for 50%, the species A has 0.25 probabilities of having atoms from species E, even if it does not directly feed on species E. Upon this, we can calculate the probabilities of an atom to go from one species to the other through all the possible direct and indirect paths. This is referred to as transition matrix, and in the transition matrix, we can calculate the share of atoms that make a cycle, i.e., that start from species A and come back to species A along all the possible paths (i.e., not only directly via the species B and C, but also indirectly along the species E). Now, suppose we are not talking about atoms, but value of a product. If, for example, Italy sells cars to USA, where the engines of the Italian cars are produced, the share of value of car relative to engine is cyclical with USA. Suppose now that the USA buys iron from China and that Italy sells cars to China. Even if Italy does not buy iron directly from China, the share of the value of iron in the engine of the car is cyclical.

[6]The formal definition of this index is complex. For a detailed definition see Picciolo et al. (2017).

[7]It is also worth noting, that by measuring the share of embodied value in traded goods, we also provide an indication of the embodied emissions in producing them. This issue concerning the increasing share of embodied emissions in traded goods have prompted many scientists to switch from a production-based to a consumption-based accounting of global emissions.

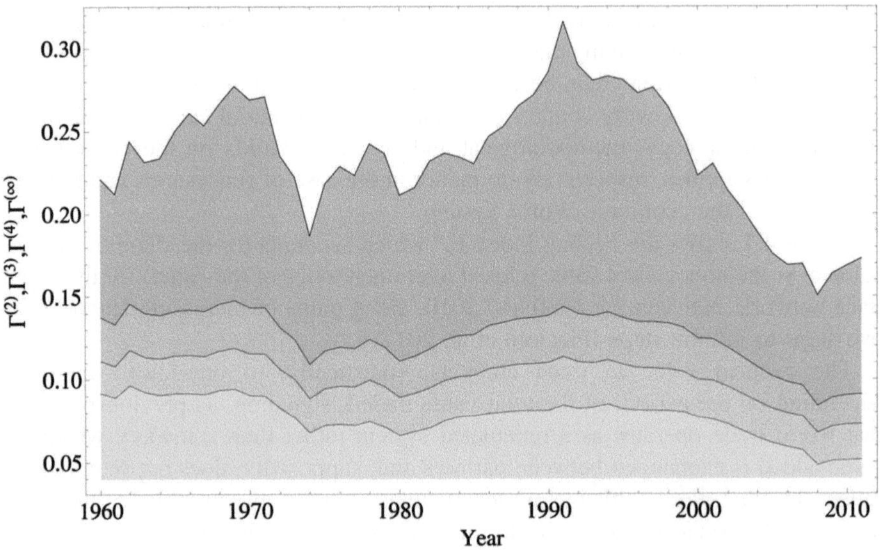

Fig. 4.1 The trend of the cyclical flow index as a share of global trade of the World Trade Web calculated from 1960 to 2011. *From lighter to darker grey*, cyclical index at different path-length: Γ^((2)) two steps, Γ^((3)) three steps, Γ^((4)) four steps, and Γ^((∞)) infinite steps of the path of the value. Adapted from: Picciolo et al., 2017

a question arises: what drives the cycling index trend? Figure 4.2 shows the cycling index (normalized on the maximum value), along different path lengths, compared to the price of the crude oil (Brent).

The correlation between cycling and oil price ranges between −62 and −85% (from infinite-steps cycling to two-steps cycling). This is a remarkable, striking correlation that is unmatched by any former analysis addressing the correlation between crude oil price and any macro-economical indicator (Picciolo et al. 2017), like inflation, industrial production of gross domestic product. Indeed, there is general understanding that oil price drives the national economy and the balance of payments, but, as it will be clearer later on in this chapter, it is difficult to explain why the global value chain is correlated to oil price. Remarkably, in fact, this tight correlation unveils the way productive sites worldwide connect and thereby that the international division of labor is shaped by oil price. This analysis reveals that, metaphorically, but also empirically, the World metabolism is strictly linked to the swings of oil price.

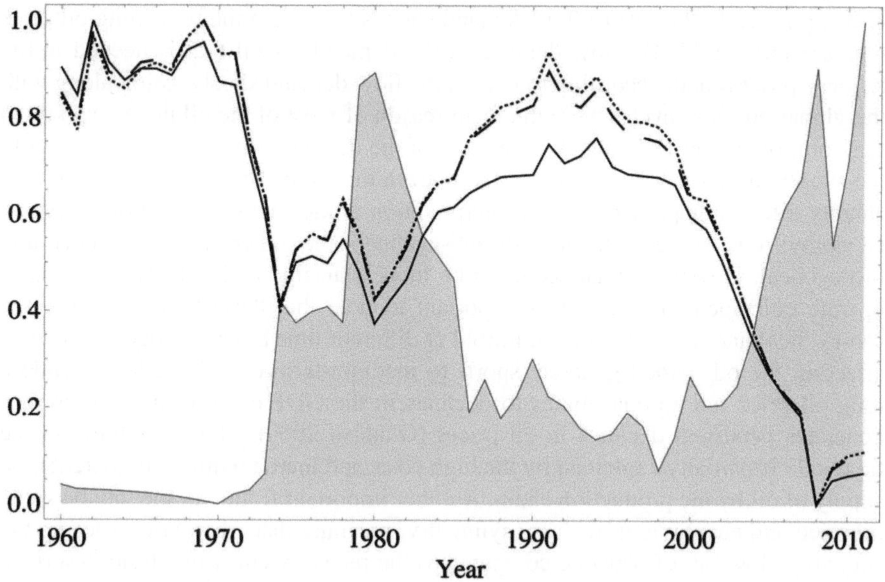

Fig. 4.2 The normalized crude oil price (*black solid line* with subtended *grey area*) is shown. The trend of the normalized value of Γ^((2)) (*solid black line*), Γ^((3)) (*dashed line*), and Γ^((4)) (*dotted line*) are also shown. The linear normalization is done based on the maximum and minimum value reached by each quantity. The final scaled values lie between 0 and 1. Adapted from: Picciolo et al., 2017

Oil and Economy: A Tale of a Happy, Arranged Marriage

In the aftermath of the oil crises of the 1970s, the relationship between oil price and economy became a hot topic for the public debate and the scientific milieu. In 1983, James Hamilton published an influential article showing that an oil price increase had preceded all but one recession (that in 1960) in the United States since the end of World War II (Hamilton 1983). Since then, a large empirical literature has looked into the connection between oil prices and real economic growth, and many studies have found a significant negative correlation (Donald and Leiby 1996; Allsopp and Fattouh 2011). For instance, oil price shocks in the period 1948–1986 would have a cumulative impact on USA growth rate between −5 and −7% and for OECD countries between −2.4 and −10.8% (Donald and Leiby 1996). But what causes oil prices' upswings? Hamilton's research focused on exogenous factors like oil supply shocks. The OPEC embargo, the Iranian revolution, or the Iraq wars were headliners and considered credible sources of disruption of oil supply by the market operators, politics and public opinion. Although seldom have such crises actually ever led to a shortage of supply, the general perception was otherwise, and it cherished a prolific literature focusing on supply driven shocks. Nevertheless, as demonstrated by Barsky and Kilian, supply shocks only explain at most 25% of the observed oil price increase in 1973–1974 (Barsky and Kilian 2004). More recently

Kilian pointed to the role of flow demand shocks to explain the remaining oil price increase (Kilian 2014). Flow demand is the demand for oil that is needed in the refinery process and, according to Kilian, the flow demand shocks going along with the global business cycles were the main reason of most of the oil price's spikes so far, comprising the 1973 shock. The view of the demand-driven shocks opposed to the supply driven shocks is a now a leitmotif in this field. This dichotomy casts two sharply different approaches to the interpretation of the nexus between oil price and economy: one is exogenous and anti-cyclical, the other is endogenous and pro-cyclical. However, what seems to be more plausible is that both mechanisms operate coincidentally. It is also important to note that these two dynamics, the pro-cyclical and the anti-cyclical, unfold at different time scales. Exogenous events affecting the oil market generate short- to medium-term oil price effects, while a long oil price run-up contributes to declines in the GDP components, this in turn generates persistent declines in oil prices (Oladosu 2009). This dichotomy in the timescale is partially explained by the high costs and inertia required to increase the supply of oil by the production chain. Another important feature of the link between oil price and economic growth is asymmetry, meaning that when price goes up the economy slows at a faster rate compared to the recovery effect that follows a price decrease (Cologni and Manera 2009). There are also many lag patterns to be considered.[8] The correlation between oil price (variation) and economy also depends strongly on what kind of oil price-definition we are looking at.[9] Finally, the last question addressed by the literature on the issue was: did the correlation between oil price and economy change over time or is it a stable feature of an economy? For example, while the first oil shock delivered clear information about the nexus, the extent of effect after the second oil shock is more disputed (Donald and Leiby 1996; Donald et al. 2004). There is now a general consensus on the notion that the relationship between oil price and the economy did not cease,[10] but it has become more elusive, perhaps because the transmission mechanism behind it is still little understood. The fundamental problem is that the direct cost of energy is

[8]Almost all empirical studies have found the largest impacts of oil prices on economic growth in the third and fourth quarters lag, with further negative effects in even later quarters (Donald et al. 2004).

[9]Naccache showed that the highest negative correlation results by defining price variation in terms of second derivative, that is, as price accelerations (Naccache 2010).

[10]According to Hooker there is no evidence of correlation between 1973 and 1994; for Mork and Alvarez-Ramirez, it weakened, but persisted while for Hamilton it is still statistically significant, albeit in a nonlinear analytical form (Hooker 1996; Mork 1989; Alvarez-Ramirez et al. 2010; Hamilton 2010). Did this tie weaken with time or disappear since the time of oil shocks? Recent studies, with more refined statistical tools and price specifications, have accomplished in restoring a stable relationship between the oil price and the economic activity beyond the 1986, the negative oil shock, when oil prices reached a minimum (Hamilton 2010; Naccache 2010; Papapetrou 2001; Oladosu 2009; Cologni and Manera 2009). According to these studies, there is evidence of a negative correlation, in some cases, between oil price and economy up to 70%, depending on the economic indicator (like industrial production or inflation) or the wavelet component of economic output.

too small a part of GDP (between 4 and 6% for OECD countries) to explain a productivity slowdown between 2 and 10%, depending on the countries and observed in the decades between the 1960s and the 2000s (Barsky and Kilian 2004). So the observed significant response to oil price increases cannot be attributed to the direct effects of decreased energy use within the production function of industries (supply shock).

We have herein presented a brief summary of the economic literature on the nexus between oil and economy. This is of course neither an exhaustive, nor a detailed account of all the scientific endeavor made to understand why oil price is so entangled to most of the economic variables. This overview was restrained to the discipline of economics by virtue of the fact that price is a category of economics. It is worth noting that for the present state of the economy, there is nothing like oil price to epitomize the metabolic activity of the economic system, as clearly and quantitatively depicted in the previous paragraph. Indeed, this relationship is complex and to sum-up, the synopsis of the tight bond is the following:

1. it is mutual: spikes cause recession (anti-cyclical), but sluggish demand causes prices reduction (pro-cyclical) and vice versa (fast economic growth pushes up oil prices)
2. it is asymmetrical: a price increase has a higher impact on the economy than a price reduction
3. it is temporal, depending on both the time scope and lag patterns: sudden increases have higher impact compared to steady increase (but demand shocks driven by business cycles have a longer and slower dynamic), one year lag shows the highest correlation
4. it is sensitive to the price specification: the best price specifications are composite refiner acquisition cost expressed in national currency, and the best analytical forms are net variations or price accelerations (second derivative)
5. it may be periodic: it appears to have weakened with time (but this is still controversial and not fully supported by empirical evidence)

As it is clear from the description above, oil price is linked to the economic activity in two ways: (1) it raises or decreases when the business cycle is going upward or downward, and (2) it stimulates or depress economic growth when it is high or low. Even news nowadays tell us contradictorily that low oil prices are expected to foster growth while informing us that financial markets are alarmed by the plunging oil price, depressing stock and commodity prices, sending some oil companies into bankruptcy and signaling stagnant demand. Should we cheer for low oil prices or should we be warned? This is somehow puzzling if we look at this phenomenon as a mechanism. Indeed, it presents a circular causality and a counteractive causality (both positive and negative feedback loops). This means that the causal relationship works in both directions: oil price is a signal and a control, a cause and effect, and output and an input of the economic system. It is bewildering if we look at it as a linear mechanical system, rather than as a circular complex system. This phenomenology resembles very much what in cybernetics is referred to as a homeostatic system (or a self-regulating system), like the mechanism presiding in living systems that regulate temperature, balance between acidity/alkalinity or blood glucose with insulin and glucagon. Indeed, having at the same time a pro-cyclical and an anti-cyclical effect, oil price works like the negative feedback control system of hormones, where the negative feedback is triggered by the overproduction of an "effect" of the hormone. This picture is consistent with the results of the analysis of the metabolic paths of world economy presented in this chapter, where the cyclical share of value of trade is correlated to oil price in a complex, circular manner. If we perform a causality test, then it is not possible to claim whether oil price causes the cycles of value or vice versa (Picciolo et al. 2017). However, there is a major difference between the nexus linking business cycle to oil price with the case of the cycles of value. In this latter case, it is not a scalar quantity, like GDP, inflation or employment rate to be affected by oil, but a *structural property of the system*. It is the international division of labor and the structure of production that are surprisingly, and enigmatically, entangled with crude oil price. Enlightened by these results, it is convenient to address the question of the transmission mechanism underpinning the link between oil price and economy, bearing in mind that, given its phenomenology, it should probably look like a complex organic system rather than a mechanical one.

Oil and Economy: The Transmission Mechanism

The search for the routes by which oil price shocks work their way through the economy has mainly addressed microeconomic mechanisms: (1) employment leakage from energy intensive sectors affects aggregate unemployment; (2) investments slowdown in the uncertain climate caused by oil price shocks; and some aggregate channels: (1) demand adjustments and following reallocation of production; (2) tight monetary policy to fight inflation (Donald and Leiby 1996; Donald et al. 2004; Barsky and Kilian 2004). Most of the microeconomic research on the mechanisms by which oil price shocks operate has focused on either product or labor markets. More recently, research addressed the re-allocative effects of oil price shocks in capital markets, showing that stock prices reflect oil shocks through their expected negative effects on cash flows (Donald et al. 2004). Indeed, stock markets have shown to be correlated with oil price, pointing that positive oil shocks have caused negative returns to most oil importer countries (Park and Ratti 2008). Goods and financial trade make up the balance of payments, which is ultimately influenced by oil price and, like stock markets, in a mirror image between oil importer and exporter countries. For example, many have viewed the existence of large current account imbalances between large economies as a possible cause of the recent financial crisis and there is growing evidence that these are correlated to oil prices worldwide, thus pointing to a possible further transmission mechanism (Rebucci and Spatafora 2006; Carrasco and Serrano 2014).

More broadly, we must acknowledge that oil for the present economy works as a sort of *international monetary system* and it represents the material equivalent of the international currency, the dollar, like gold for the former *gold standard.* Indeed, the current monetary system is practically based on the role of U.S. dollars (the so-called, *petrol-dollars*) as the international currency for oil (the first commodity in international trade). This system raised after the dim of the *dollar-standard*, when the value of U.S. dollar was locked to that of gold, and took off in the mid-1970s when OPEC agreed to sell oil only in dollars and still holds, given that more than 90% of oil is traded in dollars. Therefore, the convertibility of the dollar is now with oil, rather than gold. The role of oil in the international monetary system has thus a deep impact also on finance, globally and not just because the dollar is the main currency in financial flows. The reason is that commodity prices (and financial derivates on them) are strictly correlated to the price of oil on global financial markets (Tang and Wei 2010). This is the so-called "commodity bubble" that had so many detrimental consequences on the price of food, for example, and has one cause on the financialization of energy and commodities markets (Ruzzenenti 2015). A second reason, perhaps, lays down to the role of oil as a feedstock for many industries worldwide. About 10% of the oil in the world is transformed into plastic and various kinds of polymers. Oil in the form of materials and chemicals (petrochemicals) has profoundly changed the shape of our world, from asphalt, to naphtha, from PVC to polypropylene and even the pharmaceutical industry, which uses olefins, has grown exponential since the oil age. Oil is the *apeiron* of the present economy and society; yet, by many scholars it is viewed only as an energy source. A recent striking example of the impact of oil price on the real and *material* economy has been provided by the recycling industry in U.S. In an article of *The New York Times*, it is noted that when oil and other commodity prices were high, companies, cities, and counties were all able to make money through recycling. Now, the falling prices of oil are bringing down the prices of other commodities, including paper, aluminum, and copper, thus dramatically reducing the margins for recyclers around the country, which are now facing bankruptcy in the worst case, or charging local administrations for waste disposal in the best case (Gellesfeb 2016).

We have given an account of some of the possible contexts or *gears* that could frame the *transmission mechanism.* Some of them have been scrutinized by economists, some other mostly overlooked. However, despite the amount of research so far delivered, the actual mechanism is still obscure[11]. The problem is that this stack of theories are tentative as well as inadequate to explain the nexus between oil price and global metabolism. It is difficult to explain the above illustrated relationship with the herein envisaged transmission mechanisms because these are either spatially undifferentiated, like those that can be traced to the business cycle, or pertinent only to direct mutual relationships, when discerning between countries, like imbalance. But the cycling index tell us that it is the global

[11]With the words of Jones, "Development of policies to deal with oil price shocks, other than broad monetary and fiscal policies and holding strategic crude oil stocks, if any satisfactory ones are to be found, awaits firmer and more detailed understanding of the mechanisms by which those shocks work their impacts" (Donald et al. 2004).

architecture of trade, which means both direct and indirect trading relationships, that is correlated to oil (Picciolo et al. 2017).

Interestingly, researchers in the field of oil studies have never focused on the transport sector, which remains nearly totally reliant on refined products from crude oil, to investigate the transmission mechanism. Notably, 80% of oil is used in the transport sector. Not surprisingly, the link between oil price and transport has been an object of investigation in the field of international economics. In the post-war period, world trade grew at a faster pace than world GDP. According to Hummels (2007), this remarkably high rate was propelled by a dramatic decline in air freight costs and in a less pronounced, but steady reduction in ocean shipping costs. Other scholars also found evidence of the declining importance of distance for trade and showed that this was contingent on the oscillation of oil prices (Coe et al. 2007; Anderson and Wincoop 2004; Feenstra and Hanson 1997). Perhaps, the notion that trade grew amid globalization and that this was sustained by declining transport costs is not a surprise. What is more surprising is that intermediate and capital goods, in the last decades, grew faster than final products and now account for the largest part of trade in the world (Miroudot et al. 2009). While, in the aftermath of the Second World War, the integration of a products' market was the main effect of the first wave of globalization; in the second wave of globalization that began in the mid-1980s, the displacement of production was primarily affected (Baldwin 2006). This process lead to the fragmentation of production and to a higher vertical integration of productive sites internationally (Hummels et al. 2001). Despite acknowledging the crucial role played by the transport sector, most scholars focused their research on other factors in order to explain the slicing up of the value chain (the process of adding up value across every step of production), like the pursuit of cheap labor or more favorable fiscal/environmental legislations (Levinson and Taylor 2008). Amador and Cabral (2014) recently suggested that the strong increase of trade associated with the development of the global value chains (GVCs) in the 1990s coincides with a period of low oil prices, admitting though that there is little empirical evidence linking these two factors.

Perhaps, the lack of empirical evidence lays both in the measure and in the model employed to detect global value chains. Indeed, the cycling index previously introduced provides a clear evidence of this tight and durable relationship, although it does not establish a clear causation in the process, nor enable us to identify those causes. But the transport sector seems to be the full-fledged candidate to explain this nexus. First of all, because this crucial sector is bound to nonfungible, oil-derivate fuels; second, because the transport costs make international outsourcing more or less profitable (IEA 2009). Interestingly, if we look at the trend of the cycling index it is possible to spot two major phases that are connected to two well-documented breakthroughs in the efficiency of transport means. The first phase underlies a trend of increasing, short-range (less than 5000 km) cycling, from the early 1980s until the mid-1990s, sustained by a dramatic enhancement of the efficiency in the road freight transport sector (Ruzzenenti and Basosi 2009). This phase is featured by an integration of factors within regional markets. The second phase, from mid-1990s until the global crisis of 2008, shows a decreasing global cycling (over all distances) amid an increasing long-distance cycling and the emerging role of China as a hub. This latter phase, that points to a shift of the production chain overseas, follows a significant increase in the efficiency of the international air and cargo shipping (Hummels 2007; IEA 2009).

In a more general perspective, these results indicate that the production structure could be approached as an energy system, constrained by energy efficiency in the transport sector, whereby its metabolism is wedded to the price of oil. This view of the economic system is akin the paradigm expressed in the work of scholars like Ayres or Kummell, who approached growth as an outcome of energy efficiency and the resulting amount of available work (Kummel et al. 2002; Ayres et al. 2003). It also builds on the fundamental advances in the study of allometric scaling, aimed at explaining the structure and size of many biological processes as the result of general features of efficient transportation networks (West et al. 1997; Banavar et al. 1999; Brown et al. 2004). If efficiency in the transport of organic fluids can explain the structure of living systems, then why should not the efficiency in transport means help us in understanding the structure of global economy and its metabolic system? But if this is the case, we must admit that oil has to be regarded as a fundamental feature of the global metabolism and, in the perspective of the de-carbonization of economy, a major issue that need to be addressed in all its scope and extent.

Interactions Between Scales of Ecosystems, Economics, and Governance

In the light of energy metabolism, governance needs to address the issue of scale at which processes, natural and anthropic, unbundle and consider that there is a tension with the spatial scale at which sustainable solutions should be applied. Ecosystems contain cycles at different scales that function together to produce a local sustainable outcome. For example, the energy balance occurs at the scale of the solar system, receiving energy from the sun and passing energy back to space. The water budget is a planetary scale with evaporation, transpiration, transportation, and precipitation moving water steadily and readily around the entire globe. Within days, an individual water molecule could move from the Amazon Basin to Central Europe to the Indian Ocean to East Coast of Tasmania. Mineral nutrient cycles, at an ecological temporal scale,[12] are influenced most by the local geology, climate, and biota. Availability of nitrogen, phosphorus, potassium, sodium, calcium, and other essential elements occurs at this local scale within an ecosystem. One exception is the oxygen-carbon cycle driven by photosynthesis and respiration (and more recently fossil fuel consumption and deforestation), which are truly global

[12]The temporal scales of ecosystems vary greatlly, from geological, with the order of magnitude of 10^5 years to that of population growth rates, with the OM 10^0–10^1 years.

cycles with global consequences. The result is that we have systems nested within other systems creating intricate cross-scale, couplings that promote overall homeostasis and resilience of ecosystems.

With economic and governance systems, there is a discrepancy between the scales they were designed and that at which they operate. Economics has always been about local purpose but also about nonlocal trade. Locally, a town or region must meet its basic needs, these require local activities. This type of sustenance actions typically has a long term, intergenerational impact in that failure to accomplish them would leave the region impoverished. The economic scale expands when considering trade and growth that comes with that trade. Economic growth was believed to be facilitated by, even dependent upon, exports. Classical theories hold that value added to imports is what makes exports profitable, entailing growth. Jane Jacobs astutely pointed out that imports need not to occur only over space but also over time. In other words, regions are endowed with natural resources that have accumulated over time and human industry finds ways to harvest and utilize those resources and stretch them into higher value commodities for trade. This natural capital is an "unearned" gift that activates further development. "Every settlement starts with at least one useful resources, maybe several, already in place as a gift from nature. It's an inheritance from Earth's past development and expansion". The role of nature's gifts has long been overlooked as a vital ingredient to human wealth and well-being, but has recently been under considered investigation in the context of ecosystem services. The objective is through proper valuation of nature's services to entice better conservation. However, a trade-based growth paradigm puts pressure on rapid utilization of the local natural capital. The core problem of sustainable development is when we run through this endowment faster—much faster—than it can be replenished. The relevant point here is that economies start locally but extend to regional trading partners and in the modern world are globally interconnected. The ubiquitous shipping, rail, trucking, air transport, and exchange that underpin the twenty-first century economy eventually, perhaps inevitably, rapidly demand and metabolize vast amounts of energy (primarily oil as stated above) and matter to maintain our twenty-first century economy and lifestyle.

The third leg of the stool, governance also operates at multiple scales. Again, historically, most problems arose and were settled at local scales. However, the global impact of many current problems (e.g., climate change, ozone depletion, ocean acidification, human population) implies that local solutions are not sufficient. Countries can individually take extensive action, say to reduce greenhouse gas emissions, yet may have little or no effect on the problem. On the other hand, some countries that take aggressive action might not be the main beneficiaries. Therefore, it makes sense in these cases to have international agreements and treaties. Due to the lack of a world government, these treaties succeed or fail based on the willingness of individual nations—peer pressure can only do so much. Interestingly, the greater the interconnections between countries, one would think the greater the leverage to impose common interests. However, it appears to have reached a point of too big to fail in that each piece checks the other in a game that no one can blink. For example, in the current oil price fluctuations, countries such as Saudi Arabia and Russia that rely heavily on oil revenue are hurting financially. But, rather than Saudi Arabia putting pressure on OPEC to limit production—thus raising prices—they would rather try to use the low prices to drive out more expensive production in the U.S. and Canada, and at the same time to punish Russia for backing the current Syrian president. In other words, it is a mix of national, political, and economic self-interest that blocks aggressive action. The point is that the scale of governance is not in line with the scale of the global energy and material metabolism and is unlikely to be so any time soon.

Policy Indications

As we have seen, economic growth is correlated to oil price and, beyond that, the unemployment rate and the industrial production are all correlated to oil price. Most remarkably, the international division of labor and the development of global value chains are also correlated to oil price. Not just real economy, but finance too is correlated to oil price, from stock markets, to currency rate exchanges, markets of commodities and finally, the balance of payments, globally. We have also seen that this correlation hints to a complex nexus that works in both directions, running from oil price to economy and from economy to oil price (homeostasis) and on different time scales. Besides that, the fate of RES themselves is linked to the price of oil and that of the recycling industry as well.

The reason why the metabolism of present economy is so deeply affected by oil price is many fold:

- oil is still the irreplaceable energy source in transportation;
- the international monetary system is based on the petrol dollar-standard;
- oil is the largest commodity traded and one of the most important assets in financial markets;
- oil is the main feedstock for chemicals and materials;
- oil is crucial to the current global geopolitics (some of the most powerful countries control either the oilfields, or oil price, or its currency).

Our economy, at any temporal and spatial scale, has been shaped by oil more than any other energy source. This is probably the main lesson we can get by observing the energy metabolism of the economy, in order to individuate existing constraints and impediments to energy demand adaptability and to a massive shift to renewable energies. The *Oil Age* has increased both: (1) the scale of the economy and (2) the energy metabolism of countries and cities. Oil has the highest energy density among fossil and nonfossil energy sources in terms of specific heat (calorific value) per unit of volume.[13] It is also easy to store, transport, and most importantly, to use. Oil and oil derivatives (together with natural gas) have spurted the cycles of value across countries, by integrating markets. Furthermore, the energy yield of the primary source is not bound to the land surface, like primary production (photosynthesis) for ecosystems or like the RES. Therefore, the energy pyramids of urban systems are often inverted, highlighting the fact that energy resources supporting the city are not in abundance locally.

The global economy and the externalization of production have further extended this metabolic paradigm to national economies: the most developed countries consume, directly and indirectly, more resources then their territory can supply. Oil has had a threefold role in the onset of this metabolic paradigm by: (1) decoupling primary energy production from land (like any fossil fuel); (2) increasing the energy density of both energy production and use; (3) dramatically potentiating the transport sector, which molded the integration of the production structure globally.

The fact that the primary energy production is not bound any longer to land has had also an impact on the evolution of capital cycles and the frequency of production, ultimately putting an increasing pressure toward the rapid utilization of local natural capital.

There are two main conclusions from this analysis that can be drawn for the sake of energy policy and the governance of the energy transition from fossil fuels toward RES:

(a) The scale of the governance system must be adequate to the scale of the (metabolic) process, economic, and ecological.
(b) Oil is not just as any other energy source and not even just the most important energy source: it is the *Ding an sich* of present economy that shapes its energy metabolism.

From these two main guiding lines descend further implications for effective and sound policy aimed at sustaining the transition to RES in the light of the present energy metabolism:

1. The study of metabolism tells us that the scale of governance (control function) must be adequate to the scale of cycles in the economy and in the biosphere. Only trans-national governance can cope with the scale of the present economy and with the pressure posed on the environment by it. This is obviously a lesson of paramount importance for Europe, if not vital.
2. The scale of economy increased not only because of the integration of production globally, but also because of the integration of financial markets and the gigantic growth of international corporations and major financial operators, which are now more endowed than the economy of many countries. In Chap. 5, it is shown that there are some international corporations that consume more

[13]The issue of the constrain on the path toward a law carbon economy posed by the energy intensity of fossil fuels compared to RES is addressed in Chap. 3.

energy than most of the countries of the world. This means that the scale of the *country* is no longer adequate not only for governance, but also for understanding the processes.

3. Crude oil is the major resource responsible for the anthropocentric metabolic paradigm, but the ways oil shaped the present economy and society are still very little understood. Oil should not be regarded just as any other source of fossil energy on the route toward de-carbonization: an atom of carbon from oil is different from an atom of carbon from coal or gas. Rather than generally addressing the de-carbonization of economy, we must address the *de-oilification* of economy: more emphasis on the specific role of oil.
4. Oil it is still not fungible in the transport sector: implications go beyond the single sector and has a deep impact on the structure of production and the international division of labor, let alone the environment. In order to exit the oil area, we must particularly address the transport sector. This issue was tackled in depth in Chap. 3 of this book.
5. The transition to RES could lead to a disentanglement of local and global metabolism from oil price. This could have predictable positive effects on the governance of local metabolic process, like the implementation of a circular economy, but also on that of monetary and financial process on national and international scales.
6. Nevertheless, dismantling of the oil-economy could also have negative effects on the problem of governance at a global scale: oil price is a *control variable* for the global metabolism which a future, desirable global governance could use to regulate the system (homeostatic function).
7. The transition to RES means also a *return to land* as a source of primary energy production. The impact of this transition on geopolitics is unpredictable, foremost in the light of the present geopolitics that is shaped by oil, not only in terms of location of oil fields, but also in terms of oil price (supply disruptions or gluts) and sales' currency.
8. The European Union is devoting a significant effort to achieve the ambitious goal of the transition from a linear economy, where processes, from feedstock to waste, are featured by open loops, to a circular economy, where processes of matter's transformation are described by closed loops, like in ecosystems. This is evidence in the rise of fields such as Industrial Ecology, Ecological Engineering, and Biomimicry. The impact of oil price on the metabolic paths of economy is detrimental. Low oil prices, as we have seen, tend to increase the share of cyclical value in the world economy, meaning that global value chains will develop greater in amount and length. Is this good for the sake of circular economy? Paradoxically it is not: international outsourcing makes it more difficult to implement any policy aimed at controlling, from cradle to grave, the lifecycle of goods. Within an international context it is, for example, troublesome to enforce a law aimed at making the producer responsible for the recycling of the material. Low oil prices also have a negative impact on the economic feasibility of recycling by reducing the commodities' prices. In

conclusion, the goal of circular economy is doomed by low oil prices and is seriously hampered by the scale of the present economy.

A foreseeable and desirable, if not necessary, massive shift to renewable energies is hampered by the entanglement of the present economy with oil. Energy and environmental policy should thus address not just the issue of reducing the carbon load of the economy, but more specifically and more emphatically, its hydrocarbon content. Beyond the positive, direct effects on the environment and on energy security, severing the link between crude oil and economy would have beneficial indirect effects in terms of fiscal and monetary policy for national and local governance. *De-oilificating* the energy sector, starting from the transport sector, would have an unpredictable consequence on the economy, at any scale, but also on the global financial system. Indeed, it would downsize the scale of energy metabolism by partially restoring the link with land: even in a globalized economy the ultimate source of energy would be bounded to surface. A further envisaged consequence would be that of liberating the energy commodities and commodities price from the tight link with oil price. These two processes would have a twofold positive consequence on the energy demand adaptability, by fine-tuning the scale of energy metabolism to that of governance and by returning to national governance fiscal and monetary tools in the field of energy market and beyond that. In any case, human civilization will continue to run on the metabolism of its energy and material resources. Bringing these resources closer to home will instill greater knowledge and stewardship of their use and management.

References

Allsopp, C., and B. Fattouh. 2011. Oil and international energy. *Oxford Review of Economic Policy* 27 (1): 1–32.

Alvarez-Ramirez, J., J. Alvarez, and R. Solis. 2010. Crude oil market efficiency and modeling: Insights from the multiscaling autocorrelation pattern. *Energy Economics* 32.

Amador, J & S. Cabral. 2014. *Global Value Chains: Surveying Drivers, Measures and Impacts* (European Central Bank). Technical Report No. 1739.

Anderson, J.E., and E. van Wincoop. 2004. Trade costs. *Journal of Economic Literature* 42: 691–751.

Ayres, R.U., L.W. Ayres, and B. Warr. 2003. Exergy, power and work in the US economy, 1900–1998. *Energy* 28 (3): 219–273.

Baker, L.A., D. Hope, Y. Xu, J. Edmonds, and L. Lauver. 2001. Nitrogen balance for the central arizona-phoenix (CAP) ecosystem. *Ecosystems* 4: 582–602.

Baldwin R. 2006. *Globalisation: The Great Unbundling(s).* Technical Report, Economic Council of Finland.

Banavar, J.R., A. Maritan, and A. Rinaldo. 1999. Size and form in efficient transportation networks. *Nature* 399: 130–132.

Barsky, R. and L. Kilian. 2004. *Oil and the Macroeconomy Since the 1970s*. Working paper 10855, NBER.

Brown, J.H., J.F. Gillooly, A.P. Allen, V.M. Savage, and G.B. West. 2004. Toward a metabolic theory of ecology. *Ecology* 85 (7): 1771–1789.

Carrasco, C.A., and F. Serrano. 2014. *Global and European Imbalances: A Critical Review*. Technical Report, available at http://www.fessud.eu/wp-content/uploads/2013/04/Global-and-European-Imbalances-a-critical-review-Working-Paper-42.pdf.

Coe, D.T., A. Subramanian, and N.T. Tamirisa. 2007. The missing globalization puzzle: Evidence of the declining importance of distance. *IMF Staff Papers* 54 (1): 34–58.

Cologni, A., and M. Manera. 2009. The asymmetric effects of oil shocks on output growth: A Markov–Switching analysis for the G-7 countries. *Economic Modelling* 26 (1).

Donald, W.J., and P.N. Leiby. 1996. *The Macroeconomic Impacts of Oil Price Shocks: a Review of Literature and Issues*. US Department of Energy, January 1996.

Donald, W.J., P.N. Leiby, and I.K. Paik. 2004. Oil price shocks and the macroeconomy: What has been learned since 1996. *The Energy Journal* 25 (2): 1–32.

Elton, C. 1927. *Animal Ecology*. London: Sudgwick & Jackson.

Feenstra, R.C., and G. Hanson. (1997). Productivity *Measurement and the Impact of Trade and Technology on Wages: Estimates for the U.S., 1972-1990* (National Bureau of Economic Research). Working Paper 6052.

Finn, J.T. 1976. Measures of ecosystem structure and function derived from analysis of flows. *Journal of Theoretical Biology* 56: 363–380.

Finn, J.T. 1980. Flow analysis of models of the Hubbard Brook ecosystem. *Ecology* 61: 562–571.

Fiscus, D., B.D. Fath, and S. Goerner. 2012. The tri-modal nature of life with implications for actualizing human-environmental sustainability. *Emergence, Complexity and Organization* 14 (3): 44–88.

Gellesfeb, D. 2016. Skid in Oil Prices Pulls the Recycling Industry Down With It. *ENERGY & ENVIRONMENT The New York Times*.

Hamilton, J.D. 1983. Oil and the macroeconomy since World War II. *Journal of Political Economy* 91: 228248.

Hamilton, J.D. 2010. *Nonlinearities and the Macroeconomic Effects of Oil Prices*. NBER Working Paper No. 16186, JEL No. E32,Q43.

Holland, J.H. 2006. Studying complex adaptive systems. *Journal of Systems Science and Complexity* 19 (1): 1–8.

Hooker, M.A. 1996. What happened to the oil price-macroeconomy relationship? *Journal of Monetary Economics* 38: 195–213.

Hummels, D. 2007. Transportation costs and international trade in the second era of globalization. *Journal of Economic Perspectives* 21 (3): 131–154.

Hummels, D., J. Ishii, and K. Yi. 2001. The nature and growth of vertical specialization in world trade. *Journal of International Economics* 54 (1): 75–96.

IEA. 2009. Transport, E. CO_2: Moving towards sustainability. *International Energy Agency* 44.

Keller, E.A., and D.B. Botkin. 2008. *Essential Environmental Science*. ISBN 9780471704119.

Kilian, L. 2014. Oil price shocks: Causes and consequences. *Annual Review of Resource Economics* 133–154.

Kummel, R., Henn J. Julian, and D. Lindenberger. 2002. Capital, labor, energy and creativity: modeling innovation diffusion. *Structural Change and Economic Dynamics* 13 (2002): 415–433.

Levinson, A., and M.S. Taylor. 2008. Unmasking the pollution haven effect*. *International Economic Review* 49: 223–254.

Miroudot, S., R. Lanz, and A. Ragoussis. 2009. *Trade in Intermediate Goods and Services*. OECD Trade Policy Papers, No. 93, OECD Publishing.

Mork, K.A. 1989. Oil and the macroeconomy when prices go up and down: An extension of Hamilton's results. *Journal of Political Economy* 97: 740–744.

Naccache, T. 2010. Slow oil shocks and the "weakening of the oil price–macroeconomy relationship". *Energy Policy* 38 (5): 2340–2345.

Oladosu, G. 2009. Identifying the oil price–macroeconomy relationship: An empirical mode decomposition analysis of US data. *Energy Policy* 37 (12): 5417–5426.

Papapetrou, E. 2001. Oil price shocks, stock market, economic activity and employment in Greece. *Energy Economics* 23 (5).

Park, J., and R.A. Ratti. 2008. Oil prices and stock markets in the U.S. and 13 European countries. *Energy Economics* 30: 2587–2608.

Picciolo, F., A. Papandreou, K. Hubacek, and F. Ruzzenenti. 2017. How crude oil prices shape the global division of labor. *Applied Energy* 189: 753–761.

Prell, C., K. Feng, L. Sun, M. Geores, and K. Hubacek. 2014. The economic gains and environmental losses of us consumption: A world-systems and input-output approach. *Social Forces* 93: 405–428.

Raven, P.H., and G.B. Johnson. 1989. *Biology*, 2nd ed. St Louis, Missouri: Times Mirror/Mosby College Publishing.

Rebucci, A., and N. Spatafora. 2006. *Oil Prices and Global Imbalances. International Monetary Fund.* Technical Report, available at https://www.imf.org/external/pubs/ft/weo/2006/01/pdf/c2.pdf.

Ruzzenenti, F. 2015. *Series of Papers on Changes in the Relationship Between the Financial and the real Sector and the Present Financial and Economic Crisis: The Energy Market.* FESSUD Working Paper Series No 105. Available at: http://fessud.eu/working-paper/.

Ruzzenenti, F., and R. Basosi. 2009. Evaluation of the energy efficiency evolution in the European road freight transport sector. *Energy Policy* 37 (10): 4079–4085.

Tang, K., and X. Wei. 2010. *VIndex Investment and Financialization of CommoditiesV.* NBER Working Paper w16385 (2010).

Villarroel Walker, R., and M.B. Beck. 2012. Understanding the metabolism of urban–rural ecosystems. *Urban Ecosystem* 15: 809–848.

Wallace, R.A., J.L. King, and G.P. Sanders. 1981. *Biology: The science of Life*. Santa Monica, California: Goodyear Publishing Company Inc.

West, G.B., J.H. Brown, and B.J. Enquist. 1997. A general model for the origin of allometric scaling laws in biology. *Science* 276 (5309): 122–126.

Wilson, E.O., T. Eisner, R.B. Winslow, R.E. Dickerson, R.L. Metzenberg, R.D. O'Brien, M. Susman, and W.E. Boggs. 1978. *Life on Earth*, 2nd ed. Sunderland, Massachusetts: Sinauer Associates, Inc.

Wolman, A. 1965. The metabolism of cities. *Scientific American* 213 (3): 179–190.

Zhang, Y., Z.F. Yang, and B.D. Fath. 2010. Ecological network analysis of an urban water metabolic system: Model development, and a case study for Beijing. *Science of the Total Environment* 408: 4702–4711.

Zhang, Y., S. Li, B.D. Fath, Z. Yang, and N. Yang. 2011. Analysis of an urban energy metabolic system: Comparison of simple and complex model results. *Ecological Modelling* 223: 14–19.

Zhang, Y., H. Liu, Y. Li, Z. Yang, S. Li, and N. Yang. 2012. Ecological network analysis of China's societal metabolism. *Journal of Environmental Management* 93: 254–263.

Zhang, Y., H. Lu, B.D. Fath, and H. Zheng. 2016. Modelling urban nitrogen metabolic processes based on ecological network analysis: A case of study in Beijing, China. *Ecological Modelling* 337: 29–38.

Chapter 5
Hierarchies, Power and the Problem of Governing Complex Systems

Franco Ruzzenenti

Abstract The concept of hierarchy is central to thermodynamics. Energy processes can be evaluated in terms of entropy content and the higher the entropy the lower they are positioned in the hierarchy of irreversibility. Hence, a Joule of heat at 500 K has a higher *quality* that the same amount of heat at 400 K. Introducing irreversibility into the Carnot machinery—the intellectual device by which we have historically developed the concept of *efficiency*, leads to the concept of maximum power output at suboptimal efficiency level. Introducing irreversibility—the hierarchal criterion for thermodynamics, means that time becomes a binding variable in thermal machines. Interestingly and perhaps not surprisingly, hierarchy is also a key concept of complexity. Along the line of an increasing hierarchical complexity, economic progress and evolution have been rewarding larger organizations or organisms throughout sentient or accidental selection. From microbes to whales, from villages to nations, from family firms to international corporations, the scaling up of the system has been achieved at the expenses of a growing complexity and hierarchy. To sustain the increasing complexity, processes have been increasing their power capacity thorough evolution and economic history. Is this intriguing parallel important to understand the fate of renewable energy? In this chapter I will try to expand upon the ideas of hierarchical scaling and power maximization to the problem of governing RES, with insights from finite-time thermodynamics, algometric scaling and complex science.

F. Ruzzenenti (✉)
Department of Management and Quantitative Sciences, Parthenope University of Naples, Naples, Italy
e-mail: ruzzenenti@gmail.com

F. Ruzzenenti
Institute of Sociology, Jagiellonian University, Krakow, Poland

N. Labanca (ed.), *Complex Systems and Social Practices in Energy Transitions*, Green Energy and Technology, DOI 10.1007/978-3-319-33753-1_5

Introduction

Hierarchy is generally considered a salient property of complexity. What distinguishes a primitive tribe from a contemporary nation is the number of activities covered by and the number of levels constituting the hierarchical organization. Likewise, in food chains, the larger the number of branching levels, the higher the complexity of the underlying ecosystem. We tend to visualize hierarchy as a branching tree. What is hierarchy when connections are intricate, dense and mutual like in economic systems? To what extent does hierarchy relate to the scale of the system, to its boundaries or to its intrinsic dynamics? These questions will be addressed in the light of complex networks theory, thermodynamics and growth theory to highlight how hierarchies can be considered as a dissipative structure evolving under the pressure of an increasing flow of energy (power capacity). The first part of this chapter will illustrate the concept of hierarchy in thermodynamics and how this is related to the ladder of entropy (irreversibility): from more abstract and reversible systems to more real and irreversible ones. It will be shown how the introduction of a higher degree of irreversibility in the Carnot machinery, i.e. a thermodynamic system operating under ideal conditions of reversibility, leads to infer that thermodynamic machines need to operate at suboptimal efficiency in order to maximize their power output. The efficiency level at maximum power output is the real, observed efficiency of thermic machines. The second part will show how hierarchy (of the sustaining network) is important to understand how the power capacity of natural organisms (i.e. their metabolic rate) is a driver of the their evolution and induces a scaling up in their size: from prokaryotic cells to the biggest mammals. Lastly, it will be shown how this paradigm linking hierarchical complexity throughout scaling with the power maximization is also present in the evolution of anthropic systems: it is the increasing power density of energy processes (extensively described in Chap. 3) that enabled the scaling up of firms and corporations and the hierarchical complexity displayed by the modern productive structure. Finally, some considerations for policymaking will be drawn.

Efficiency at Maximum Power Output: An Insight from Finite-Time Thermodynamics

The first approach to the concept of *hierarchy* for sophomores of physics, chemistry or engineering, is the *ladder of entropy* in thermodynamics. What they are taught is that in spontaneous processes (e.g. during heat exchanges from hot to cold sources) this ladder can only be descended (i.e. entropy can only increase). When this ladder is ascended within a given system, the associated entropy decrease must be overcompensated by an increase in the entropy of the surrounding environment. Entropy is a state function and thus it depends only on the current equilibrium state of the system, regardless of the path done to achieve it. Even the potential energy in mechanics or the eigenvalues in quantum mechanics define a state of the system independently from the path insofar as they are scalars and like any scalar, they provide a hierarchical metrics. They tell you which state has more energy, but unlike entropy, they cannot tell you what *happened first*. This is less evident if we compare the potential energy of two states of a falling body: the lower, the later. What if we compare the potential energy of two states of a planet in its orbit. Can we say what happened first? Entropy was a groundbreaking concept in the history of science, as much as in the career of every student, as it introduced the idea of *irreversibility* and upon this, that of spontaneous direction of processes. Therefore, it is just surprising, if not confounding, that thermodynamics was born when Sadi Carnot conceptualized a purely abstract engine meant to extract kinetic energy from heat by *reversible* operations (Smil 2008).

Lotka was among the first scholars who enucleated this contradiction for real processes (Lotka 1956). He noticed that the actual performance of real heat engines *always fall short-and usually far short-of the theoretical maximum attainable under ideal conditions of reversibility.* In his own words: *the first service rendered by the laws of thermodynamics is thus a negative one, to save us from vain efforts to achieve the impossible. They tell us what we cannot do; they give us no guarantee as to what we can do, in this matter of engine efficiency.*[1]

This is not to say that the Carnot engine was of no use at all. Indeed, it gave us a theoretical ground to understand the maximum work attainable from two temperature reservoirs. A limit achievable only if processes are reversible, but from a theoretical as well as a practical viewpoint, only models of *irreversible* heat engines enable us to foresee the system evolution (Apertet et al. 2012). It is the presence of dissipative elements (irreversibility) that ensure us on the causality of processes; hence, in the Carnot engine, like in the planetary orbits, there is no *arrow of time*. It is also worth noting that, in order to assume reversibility in the Carnot engine, processes need to occur at an infinite slow pace, that is, *at equilibrium*. This is the well-known condition e.g. for an isothermal process, that can occur only if the system can slowly adjust its temperature with the heat reservoir, transferring heat *without temperature gradient*. It follows that, for the sake of the maximal theoretical efficiency, Carnot posits an incredibly implausible condition: that the power output of the system is (close to) *zero*. In the 1950s, physicists and engineers faced the need to amend this unrealistic condition in the attempt of theorizing the best efficiency attainable from a nuclear power plant working at the maximum power. Novikov (1957) was the first who formally defined the efficiency of an heat machine working at the maximum power output, paving the way for the onset of a new branch of thermodynamics, the *finite-time thermodynamics*, which found a complete formalization by Curzon and Ahlbron in the 1970s, when it emerged as a new theory and a generalization of the thermodynamics of irreversible processes (Curzon and Ahlborn 1975). The Cuzorn-Ahlborn model it is still an abstract

[1]After Lotka, Odum and Pinkerton, in 1955, formalized for the first time this concept by proposing the optimal effcicieny at maximum power output (Odum and Pinkerton 1955).

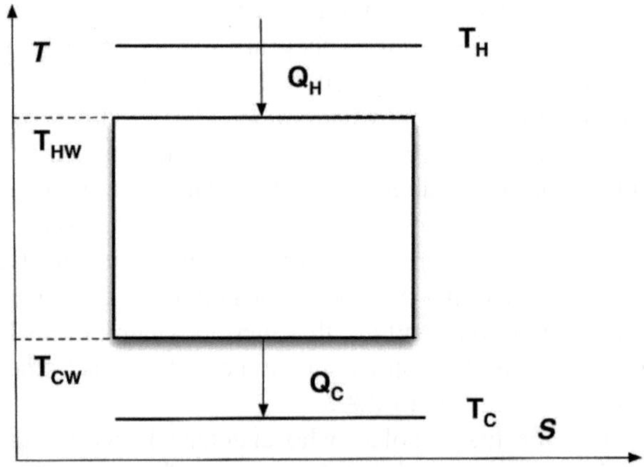

Fig. 5.1 This figure shows the modified Carnot cycle on a entropy/temperature diagram. T_h and T_c are the temperatures of the hot and cold reservoirs and T_{hw} and T_{cw} are the two extreme temperatures of the working fluid. The symmetrical coupling of the Cuzorn-Ahlborn model implies that the gradient between T_h and T_{hw} is the same as between T_{cw} and T_c. The more we increase the two gradients—in order to increase the speed of heat transfer, the closer the extreme temperatures of the working substance. Ultimately, the two isothermal stages take place with no change in the temperature of the working substance. Heat flows directly from the hot source to the cold sink and no work is done. Hence the power output is zero and the engine has zero efficiency as well. Adapted from: Basosi and Ruzzenenti (2010)

machinery (*endoreverisble*, meaning that its internal processes occur without dissipation of heat), but it introduces irreversibility—and thereby *time*, in the heat transfer with the environment. The rate of heat transfer is proportional to the thermal gradient between the heat reservoir and the thermal working fluid, which is assumed to be symmetrical (meaning that the heat flows from the machinery to the hot and to the cold source occur at the same speed, see Fig. 5.1).

Therefore, if we increase this gradient we increase the speed of the cycle but we also decrease the efficiency,[2] which is proportional to the *internal* thermal gradient set by the difference between the maximal and minimal temperature of the working fluid (see Fig. 5.1).

The introduction of *time* in two machinery's stages enables us to formally address the question of *power*, which is the rate of work output and thereby, by means of a new control variable, opens the gates to a new conceptual framework: what is the efficiency of a machine that maximizes the rate of heat exchange and thus the power?

The efficiency (CA efficiency in the following) for a thermal machine operating at the maximum power output between two heat sources is:

[2]Interestingly, this principle, the strong coupling of antithetical forces, is common to the Hatwood machine Odum and Pinkerton used to introduce the maximum power principle in 1955 by means of a purely mechanical device (Odum and Pinkerton 1955).

$$\eta_{CA} = 1 - \sqrt{\frac{T_c}{T_h}}$$

This efficiency, compared to the Carnot efficiency ($1 - (T_c/T_h)$), describes very well the real efficiency of many thermal power plants (Chen et al. 2001). Unfortunately, the CA efficiency still bears a certain—and significant, degree of abstraction when it assumes at symmetric dissipation, i.e. that the heat transfer between the working fluid and the two heat reservoirs occurs at the same rate. A second important limit of the CA efficiency is that it only applies to thermal machines. Recently, it has been shown that both assumptions can be relaxed for further generalizing the model by assuming strong coupling in generic forces[3] and by differentiating the rate of heat transfer between the working fluid and the hot and cold reservoir respectively (Van den Broeck 2005; Esposito et al. 2010). Within this generalized framework, knowing that the Carnot efficiency, is:

$$\eta_C = 1 - \frac{T_c}{T_h}$$

it can be shown that the efficiency at maximum power output ranges between:

$$\frac{\eta_C}{2} \leq \eta^* \leq \frac{\eta_C}{(2 - \eta_C)}$$

The two ends of the inequalities correspond to two extreme operating regimes; the upper bound obtained by optimizing[4] with respect to the temperature of the hot reservoir, the lower bound with respect to the temperature of the cold reservoir. In Fig. 5.2 the efficiency at maximum power output is plotted as a function of η_C for some real power plants.

Although not displayed, all observations lie below the ideal line $y = x$ (which means that they all relate to a real efficiency lower than the Carnot efficiency) and almost all are in the range provided by the theory, with some deviations though. Interestingly, thermal power plants lie closer to the lower limit, meaning that cooling is optimized, whereas nuclear and geothermal power plants are positioned around the upper bound, meaning that the heat transfer from the hot source is maximized. Perhaps this is not surprising given the higher temperatures of

[3]In the case of CA machinery, the two coupled forces are the heat injection and rejection. Another example of heat engine operating in a regime of strong coupling is a couple of thermoelectric generators (Apertet et al. 2012).

[4]To optimize the power in a thermal machine we can either increment the speed at which the heat is transferred to the working fluid from the hot reservoir (the combustion) or from the working fluid to the cold reservoir (the environment). A very intuitive example is that of cars: when we introduce a turbocharger, we increase the heat transfer speed (the phase of heat addiction at constant volume of the cylinder) by increasing the pressure at the same volume ratio (the piston's size) and when we are introducing a cooling system, like a water or air cooling device, we are increasing the heat rejection by diminishing the temperature of the machine.

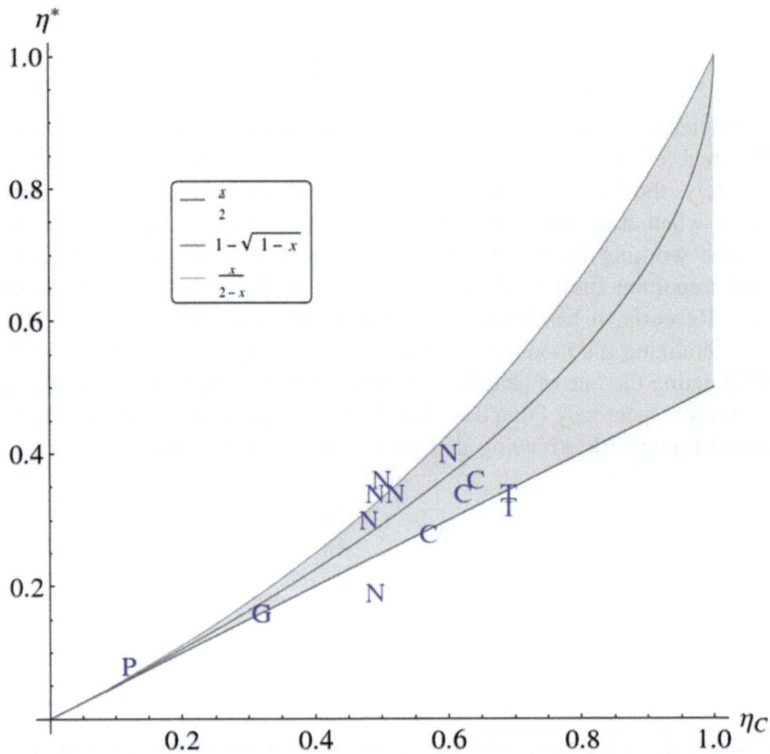

Fig. 5.2 Efficiency at maximum power output versus Carnot efficiency for several power plants and photosynthesis. The first number in the following couples of numbers is the Carnot efficiency reported on the *x* axis, whilst the second is the observed efficiency reported on the *y* axis (therefore, the maximum theoretical efficiency lays on the $Y = X$). Data points refer to either nuclear power plants (*N*), coal power plants (*C*), gas turbines (*T*), geothermal power plants (*G*) and photosynthesis (*P*): (1) Almaraz II (Nuclear, Spain) 0.52–0.34; (2) Sizewell B (Nuclear, UK) 0.5–0.36; (3) Cofrentes (Nuclear, Spain) 0.49–0.34; (4) Heysham (Nuclear, UK) 0.60–0.40; (5) West Thurrock (Coal, UK) 0.64–0.36; (6) CANDU (Nuclear, Canada) 0.48–0.30; (7) Larderello (Geothermal, Italy) 0.32–0.16; (8) Calder Hall (Nuclear, UK) 0.49–0.19; (9) (Steam Coal/Mercury,USA) 0.62–0.34; (10) (Steam Coal, UK 0.57–0.28; (11) (Gas Turbine, Switzerland) 0.69–0.32; (12) (Gas Turbine, France) 0.69–0.34; (13) Photosynthesis 0.12–0.08. Author elaboration on: Espostito et al. (2010), Smil (2008)

combustion in thermal plants (see Fig. 5.2) and the same working fluid (vapor). What is important is that both strategies[5] (those maximizing the heat transfer from the cold reservoir and those from the hot reservoir) aim at *maximizing power at expenses of efficiency*. This is clear if one observers that all the data points lie under

[5]It should be noted that we are hereby referring to a broader concept of "strategy" that concerns not only the operational conditions of thermal machines but also their design. Indeed a car running at the speed of a bicycle would be much more energy-efficient, but a car is conceived to run at one order of magnitude faster.

the level of maximum theoretical Carnot efficiency (the $y = x$ line). Predicting suboptimal efficiency amid maximum power output is a remarkable result of the model and finds empirical evidence in real processes. However, it bears one important and unexplained question, *why?*

Forces' coupling generates, roughly speaking, a form of trade-off or, more formally, a control variable whose profile is represented by a convex function with at least a point of maximum (Ruzzenenti and Basosi 2008). Optimization is thus theoretically encrypted in the model *ab origine*, which might be considered a strength or a weakness of the theory and has indeed moved some scholars to exacerbated critiques contending that finite-time thermodynamics is a mere intellectual exercise without any substantial relation with reality (Gyftopoulos 1999, 2002).

Perhaps, we cannot but glimpse in the acrimony with respect to finite-time thermodynamics the same positivistic refusal to any approach or theory undermining the fulgent myth of prosperity based on energy efficiency. The same refusal that is addressed in Chap. 7 describing a wide spread attitude towards rebound effect. Nevertheless, the aforementioned question is still lingering: why do we have to expect thermodynamic machines to maximize power rather than efficiency?

It is difficult, in economics for example, to envisage whether economic optimization would lead to maximize efficiency (reducing costs) or maximizing output (increasing revenue) and in what conditions one strategy would prevail over the other. Curzon and Ahlborn, aware of this epistemic leak, advanced the hypothesis that more complex machineries would turn on power maximization in order to pay back the higher costs of the *apparatus* (Curzon and Ahlborn 1975). This view is consistent with the economic theory regarding *economy of scale*, which states that when capital costs are overwhelming with respect to variable costs the firm tends to increase the rate of production in order to maximize depreciation. In this view, time becomes the binding variable and the more complex and bigger is the structure the larger the output flow for unit of time required to payback the machinery.

Truly, this phenomenon is a common feature of human made systems and natural systems and it thus bears a more general explanation (see Chap. 7). For example, it has been studied that the maximal theoretical efficiency of photosynthesis would range between 12 and 13%[6] in optimal conditions but in reality, at best, it reaches a level of 8–9% (4–5% when some resources are scarce). The reason is that photosynthesis maximizes growth rates and hence improves the chances of early survival and competitive maturation, with irreversible losses at expense of

[6]Theoretical efficiency for plants is calculated comparing the energy carried by the photons with the energy converted in ATP by the photosynthesizer apparatus (the light absorption by pigments in disk-like thylakoid membranes inside chloroplasts in specialized leaf cells). Real efficiency compares the solar energy hitting the surface with the growth rate of the energy (calories) embodied in the phytomass, thus considering any energy loss, from plant's respiration to inefficiency in the related cycles (Calvin cycle, etc.). For an extensive description see the Chapter dedicated to Photosynthesis in Smil (2008).

efficiency[7] (Smil 2008). In this view, similarly to the ideas of Lotka, natural selection awards power maximization rather than efficiency optimization when resources are copious. In the light of power maximization as a result of competition, can we establish a nexus between power capacity and hierarchical complexity? This question will be addressed in the remaining part of the chapter. Another remarkable example of suboptimal efficiency at maximum power output is provided by cancer cells. In tumors the pyruvate formed from glycolysis is converted to lactic acid and energy is produced anaerobically, delivering only 2 ATPs molecules instead of the 36 that could be yielded by a complete oxidation of a glucose molecule (Coller 2014). Why is a less efficient catabolic pathway—glycolysis instead of oxidative phosphorylation, induced in tumor cells? One accredited hypothesis is that glycolysis is part of a metabolic profile "that channels glucose among the available pathways in a way that facilitates rapid proliferation and growth" (Coller 2014).

Hierarchy: A Salient Property of Complexity

Indeed, hierarchy is not only peculiar to thermodynamics. The concept of hierarchy is also familiar to the field of social and natural sciences. The word itself has an ancient, sacred origin, descending from the merging of *hieros* "sacred" and *arkhia* "rule" (Verdier 2006). For centuries—in the European civilizations, this concept has strictly pertained to the cosmogony or the classification of celestial spheres, whence semantically evolved towards a register for the description of the ecclesiastical state, and lately the society in general (Verdier 2006). In the field of natural sciences, and prominently in ecology, the concept of hierarchy is pivotal. It is a common knowledge that living systems are structured in a set of successive nesting levels of organization: genome and protein, cell, tissue, organism, community, population, ecosystem and lastly biosphere. Today, not only evolutionists, but biologists and ecologists question and debate about the mechanisms that led to the actual *nested hierarchy of organized entities and trophic networks* (Pavé 2006). Despite being still a vague and embryonal field of science, complexity theory is infused in all his variegate branches by the concept of hierarchy which is a pillar of this theory as much as it is a pillar of thermodynamics. This is not just a semantical

[7]"Actual short-term increments of new phytomass are at best 50% or, more likely, just 33% of these rates. The top seasonal or annual additions are between 20 and 25% of the ideal rates, and long-term, large-scale averages are merely 10% and all the way down to just 2% of the best hypothetical performance. The two main reasons for these disparities are the respiration costs and the inevitable losses that go with rapid rates of photosynthetic reactions. In order to conserve as much light as possible during the limited hours of intensive insolation, the rates must be quite fast, but this rapidity results in two kinds of considerable inefficiencies. Unless the plant's enzymes can keep up with the radiation flux coming into the excited pigments, the absorbed energy will be reradiated as heat. Utilization must be immediate because the chlorophyll molecules cannot store sunlight. Only at very low light intensities, when radiation would be the only factor limiting the rate of the terrestrial photosynthesis, is there such a perfect match" (Smil 2008).

coincidence, rather one of the most fascinating and intriguing conjunction that Nature in its mysterious schemes has delivered. With the words of Schenider and Kay: *when highly ordered complex systems emerge, they develop and grow at the expense of increasing the disorder at higher levels in the system's hierarchy* (Schneider and Kay 1994). Hence, hierarchy is much more than just a recursive concept of complex theory; it is a salient property of complexity. We can go further and contend—as a golden rule of cybernetics states, that a system grows in complexity when it grows in hierarchical levels, given other properties equal (von Bertalanffy 1968). However, it should be noted that introducing a higher hierarchical level is not equivalent to *centralizing the information processing*. This distinction is important because in a system designed by a command-and-control strategy, such as most of those governed by technology, hierarchy is generally intended as a centralization of the information-elaborating process. In Nature, though, higher hierarchy is not always higher centralization and, most importantly, higher centralization does not always mean higher complexity. For example, it is rather difficult to claim, on the coarse observation of the cybernetics of the homeostatic (regulatory) process, whether ectotherms are less or more hierarchical complex than endotherms—both having a decentralized regulation of body temperature. The brain and the central nervous system are actually more involved in ectotherms' than in endotherms' regulation system. In the former case the animal needs external heat sources to be searched and reached by senses and motion (central nervous system), whereas in the latter case the heat is produced by the internal metabolic activity (mitochondria), and in some cases, like for sweat glands, the sympathetic, autonomous system is involved (for sweating or shivering, for example). Nevertheless, we know that endothermia came after ectothermia in evolution (Mesozoic) and was crucial for the development of mammal species diversity. On the contrary, eukaryotic cells are manifestly more complex than prokaryotic cells (due to the presence of organelles and an internal transport structure, i.e. a larger number of interacting components) while displaying a centralized versus decentralized DNA encoding system. Hence, what is akin to both regulatory systems and serves us as unequivocal criterion for defining complexity, is the expression of the hierarchy as a scale property rather than a centralized/decentralized information processing system. In the case of endotherms, the regulation system lays at the scale of organs as much as DNA processing occurs at the scale of organelles in eukaryotic cells: a lower hierarchical scale compared to that of the organism. On the contrary, for both ectotherms and prokaryotic cells, the scale of the regulatory process is invariant. According to this view, what qualifies the degree of complexity in the light of hierarchy is the *scaling* of information processing rather than its internal orientation (central/peripheral). In Nature, the hierarchical complexity that transpires in the nesting and the scaling of phenomena is its metrics.

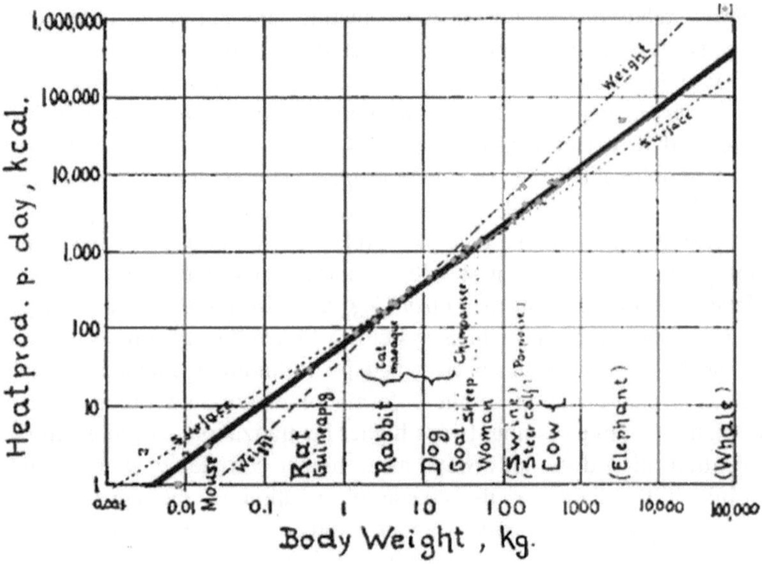

Fig. 5.3 The original graph, hand-drawn by Max Kleiber, representing the body size, in kg, versus the metabolic rate, in kcal for various species of mammals; *Source* Kleiber (1947)

Hierarchy and Scale in Natural Systems

The typical graphical representation of hierarchy is a spanning tree and, notably, a spanning tree is formally a *network*. Interestingly—and perhaps not surprisingly, network theory coupled with fluid dynamics has been successfully applied to one of the oldest riddle in biological sciences: Kleiber's allometric laws (West et al. 1997). Allometry is the study of the relationship of body size to shape and the most famous law of allometry, which dates back to 1932, it is Kleiber's power law which states that animal's basal metabolic rate is ¾ power law function of the animal's body mass (a straight line with a ¾ angular coefficient on a logarithmic scale). This universal scaling has been puzzling scientists for decades. The scaling factor might change, but it remains always lower than 1, so indicating that the scaling of metabolism is generally decreasing with size, or the intercept with respect to group of species (unicellular, invertebrates, vertebrates, etc.) or species of mammals (see Fig. 5.3), and it is strikingly ubiquitous in Nature (West 2006).

Furthermore, according to some scholars, albeit this it still debated and by some circumscribed to organisms that present vascular systems, the scaling factor is typically a multiple of ¼, the so called *quarter-power law* (Glazier 2014). It has been proposed by West et al. that these universal quarter powers have their origin in general properties of the various hierarchical, fractal-like branching transport network that sustains life at several scales (West et al. 1997; West 2006). The fractal-like branching network is the resource distribution, volume-filling network where energy dissipated is minimized, given the constrains placed by hydrodynamics (West et al. 1999). According to West, not only the metabolic rate to the body size ratio is determined by hierarchical networks, but also the time scale of several basic biological processes—like for example the development time, depend on the combination of variables of M $^{1/4}$ and some function of the temperature (Gillooly et al. 2002). The scaling with the temperature is determined by the temperature dependence of the energy producing biochemical reactions within mitochondria which are governed by the classic Boltzman factor $e^{E/kT}$, typical of thermodynamics and statistical mechanics and explaining the population density of energy levels.

The combination of the temperature scaling with the quarter power scaling, *represents the joint effects of the scaling of the production of energy at the "microscopic" intramitochondrial level and transport* (network, n.d.r.) *constraints at the "macroscopic" whole-body level* (West 2006).

In conclusion, hierarchical complexity of (transport) networks is the strategy evolution adopted to overcome the physical constraints placed by fluid dynamics or thermodynamics in order to increase the scale of organisms.[8] This is not to say that bigger organisms are a goal of evolution, but that a greater energy flux (metabolic rate, for example) founds its evolutionary pathway through hierarchical branching and scaling. This is the main lesson Nature gives us about the nexus between power capacity and hierarchy.[9] Bearing on what we have been describing so far for Nature, what is the parallel for human made systems?

[8]"Quarter-power scaling laws are perhaps as universal and as uniquely biological as the biochemical pathways of metabolism, the structure and function of the genetic code, and the process of natural selection. The vast majority of organisms exhibit scaling exponents very close to 3/4 for metabolic rate and to 1/4 for internal times and distances. These are the maximal and minimal values, respectively, for the effective surface area and linear dimensions for a volume-filling fractal-like network. On the one hand, this is testimony to the power of natural selection, which has exploited variations on this fractal theme to produce the incredible variety of biological form and function. On the other hand, it is testimony to the severe geometric and physical constraints on metabolic processes, which have dictated that all of these organisms obey a common set of quarter-power scaling laws. Fractal geometry has literally given life an added dimension" (West et al. 1999).

[9]We tend to think of hierarchy as a designed process, the result of a sentient subject. How can be hierarchy the outcome of evolution and the result of a spontaneous process? The interesting topic of hierarchy genealogy goes beyond the scope of the present analysis, it is so complex and vast that would probably require a chapter for its own. It is the opinion of the author that an investigation on the process of hierarchy creation should expand upon the concept of symmetry breaking. Geoffrey West himself hinted to the fact that the hierarchical branching in elementary particles derives from a symmetry breaking (West 2006). The first who seemingly first envisaged the nexus between hierarchy and symmetry was Gregory Bateson, who suggested that the information for symmetry breaking may be embodied in physical or chemical gradients (Bateson 1972). The etiological bond between symmetry breaking and spatial gradients has been central to a former paper by Ruzzenenti and Basosi titled *Complexity change and space symmetry rupture* (Ruzzenenti and Basosi 2009).

Hierarchy and Scale in Anthropic Systems

Biological evolution, as we have seen, has awarded the scaling up of size and power capacity throughout hierarchical branching. In the same fashion, as it was shown in Chap. 3, economic evolution has pursued power enhancement leading to increasing societal complexity which has been historically sustained by an ever growing energy density of the featuring forms of energy. Not only cities scaled up, but also the energy metabolism of households (and countries). Nevertheless, on the line of this paragon, it is difficult to envisage how and where to apply the concept of hierarchical growth to the structure of cities or countries, let alone households. Are modern megalopolis or states more hierarchically complex than ancient ones? The very contrary seems to be true if you look back at the intricate hierarchical structure of feudal system or the cast system of the precolonial India and compare them with the ubiquitous and unanimous social class of the modern society's *consumers*. Consumerism, more than communism, has had the unintended effect of abolishing social hierarchy, sacrificed on the altar of mass production. The electrification and the digitalization of economy is further flattening the horizon of societal structure alongside the barriers-demolishing pathways of an expanding communication network. It is interesting, for example, to notice how Italy is presently being affected by a backspin of young illiteracy and how this phenomenon is less related to the social and economic conditions of the parents than to the education and the interests of their sons (Save the Children 2016).

There is one domain of the anthropic sphere where the growth in scale has proceeded side-by-side with a hierarchical growth: capital. More specifically this is happened in case of: (1) the size and the hierarchical structure of economic organizations, and in case of (2) the complexity in terms of number of components and structure of artefacts. Owing to a need of concision, I further develop the former aspect of scaling growth in the light of increasing power capacity.

The biggest companies in the World have been ranked by several journals or institutions in terms of profit, revenue,[10] market value,[11] number of employees[12] or even environmental performance,[13] but, to the knowledge of the author, no ranking in terms of *energy use* has ever been attempted. Yet, after a brief investigation, it turned out that the company Arcelor-Mittal, the largest iron producer in the World (responsible for almost 50% of the global production) consumes 2.6 billions of GJ per year, which is more than the primary energy consumption of Poland and almost half of that of Italy, just for producing iron (see Table 5.1). Wal-Mart is the largest private corporation in the world, both in terms of employees and gross revenue, and it consumes more than 30 millions of GJ of electricity per year. The Department of

---

[10]http://fortune.com/global500/.

[11]http://www.forbes.com/sites/liyanchen/2015/05/06/the-worlds-largest-companies/#513ecbeb4fe5.

[12]https://en.wikipedia.org/wiki/List_of_largest_employers.

[13]http://www.newsweek.com/green-2015/top-green-companies-world-2015.

Defense of the USA is the single largest employer of the World (even larger than the Chinese Popular Army) with more than 3 millions of employees and a federal budged that surpasses the gross revenue of Wal-Mart and any other multinational corporation. The Department of Defense consumes 0.7 billions of GJ, which is more than Delta airlines, the largest air company in the World (0.5 billions of GJ)

Table 5.1 The scale of economy: energy use, number of employees and revenue of a sample of major transnational corporations

	Gross revenue (billions $)	Employees	Energy (GJ)	
Wal-Mart	485.70	2,200,000	3.02E+007	Electricity (estimation)
Arcelor-Mittal	79.28	222,000	2.16E+009	Production of steel—2013
U.S. Department of Defense	585.00	3,200,000	7.70E+008	Primary energy consumption—2014
Delta airlines	37.77	80,000	5.01E+008	Primary energy consumption—2014
Fed-Ex	47.50	300,000	1.85E+008	Fuel use for shipping—2014
Google	74.5	61,400	1.58E+007	Primary energy
Coca Cola	46.00	129,200	6.33E+007	Primary energy consumption—2014
	GDP (billions $)	Population	Energy (GJ)	
USA	17,419	309,349,689	9.60E+010	Primary energy consumption—2013
Italy	2174	60,795,612	6.75E+009	Primary energy consumption—2013
Poland	508	38,483,957	1.92E+009	Primary energy consumption—2013
Kenya	65	38,610,097	1.89E+008	Primary energy consumption—2013

Sources

- Arceror-Mittal Corporate Sustainability Report, available at: http://corporate.arcelormittal.com/sustainability (Retrieved: 2/4/2016)
- Delta Corporate Responsibility Report 2014, available at: http://www.delta.com/content/www/en_US/about-delta/corporate-responsibility.html (Retrieved: 2/4/2016)
- FedEx Annual report 2014, available at: http://annualreport.van.fedex.com/2014/ (Retrieved: 2/4/2016)
- Walmart's Approach to Renewable Energy, available at: http://cdn.corporate.walmart.com/eb/80/4c32210b44ccbae634ddedd18a27/walmarts-approach-to-renewable-energy.pdf (Retrieved: 2/4/2016)
- Coca-Cola Climate Protection Report, available at: http://www.coca-colacompany.com/stories/position-statement-on-climate-protection/ (Retrieved: 2/4/2016)
- U.S. Energy Information Administration, 2016. Annual Energy Outlook 2015. This publication is available at: www.eia.gov/forecasts/aeo (Retrieved: 2/4/2016) http://www.nytimes.com/2011/09/09/technology/google-details-and-defends-its-use-of-electricity.html)

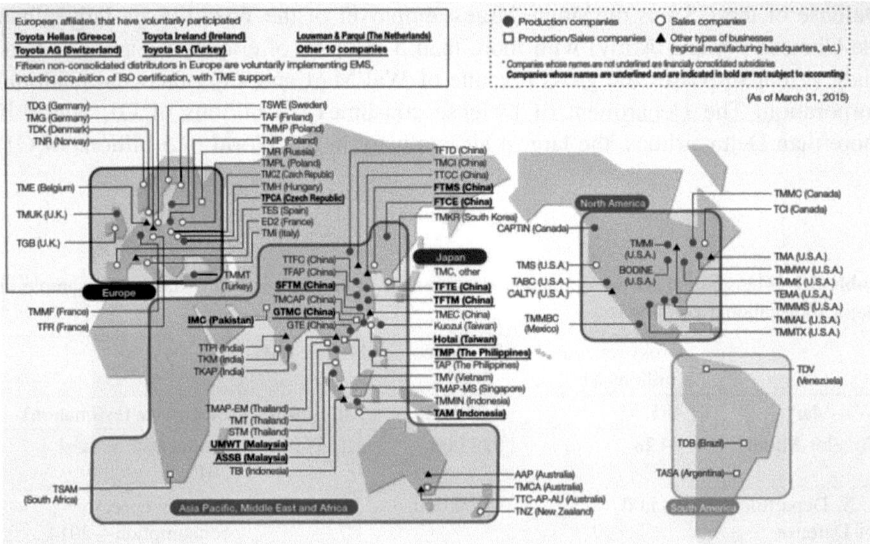

Fig. 5.4 Main companies subject to the environmental management system of Toyota Corporation. Courtesy of: Toyota 2016

and four times more than the primary energy consumption of Kenya (see Table 5.1).

Such an astonishing growth in size has paired with an increase in hierarchical complexity. Rather than on the scale or the company internal structure, this growth in hierarchical complexification has developed on the spatial distribution of the value chain's nodes on a global and international scale (Rajan and Wulf 2006; Cantwell and Janne 1999; Picciolo et al. 2017). That is, the hierarchical complexity of the productive space evolve in the external space rather than in the organization internal space. For example, Toyota has (1) 164 subsidiaries which are financially consolidated and under the direct control of Toyota Motor Corporation (TMC); (2) 51 major production companies and overseas distributors that are not subject to consolidated accounting; (3) one organization from other types of businesses; (4) 340 subsidiaries that are financially consolidated and under the indirect control of TMC (managed via consolidated subsidiaries). The scope and the complexity of the network is depicted in Fig. 5.4 (Toyota 2016).

It is evident that such a complex and giant system could evolve only by relying upon the increasing power capacity of the underlying physical and information transport networks. A network of such a scale needs a fast information processing —at a time scale comparable of that of decision making—and a fast (and energy efficient) matter processing—at a time scale comparable of that of the production process.

Conclusions and Policy Indications

Finite-time thermodynamics shows that real processes operate at a suboptimal efficiency in order to maximize power output when time becomes a binding factor. Competition for survival in the realm of Nature and in the domain of economy set time as the general force of evolution and development. Hence, it is a persistent and ubiquitous pressure on the existing energy gradients that enhances the rate of energy throughput—the power capacity. The creation of hierarchies is functional to the power increase in complex systems to sustain the scaling up of processes and, at the same time, hierarchical nesting in complex systems increases throughout scales. The creation of the hierarchies throughout scales of complex transport networks is the universal *dissipative strategy* whereby energy gradients are degraded.

The gigantism and complexity achieved by international corporations are thus, by the view points of thermodynamics and cybernetics, consistent, if not consequential, to the energy path of evolution in Nature. Although, paradoxically, it is this gigantism and the ever-growing metabolism that are placing, among other factors, a serious threat to the stability of natural cycles and to the health of human being (see Chap. 4). In other words, the evolution of corporations and, more broadly, of the productive structure is theoretically coherent with the evolutionary pathway, but practically irreconcilable and, on the long run, possibly noxious for human being.

The gigantism of corporations and the globalization of production are a new and possibly underestimated predicament for governance and for the achievement of either energy conservation, or decarbonization, or renewable energy supply policy targets. Can RES with their low energy density and supply discontinuity sustain the achieved scale of production and the energy intensity and range of the transport network that this scale of production implies? Can we reconcile the scale of the closed cycles of value (see Chap. 4), on the one hand, and the open cycles of energy and matter, on the other, that a global system of production exhibits with the scale of the cycles of Nature (sometimes smaller, sometimes larger) and society (local by definition)? Is *circular economy* attainable within a framework of global production and with corporations that process more energy and matter than many countries' economy? Lastly, do we have an adequate governance for challenging corporations or network of corporations (subsidiaries, etc.) that are scattered in several states and endowed with a gross revenue superior in amount to many government budgets (see Table 5.1).

In Chap. 3 we have extensively tackled the issue of the critical factors of RES in the light of the increasing energy demand and energy density of society and economy. The problem of energy metabolism of countries and global economy with respect to natural cycles has been tackled in Chap. 4. Here, I would like to raise the attention on the specific and often overlooked problem of the metabolism of giant, global corporations within the framework of the impending energy and environmental crisis, or, with a more optimistic outlook, in relation to the foreseeable and desirable energy transition to a low-carbon economy.

Now, it is obviously very difficult, if not impossible, to assess how much energy is claimed by corporations globally for their metabolic functions. National energy accounting and surveys do not cover such a category and the energy used up by firms is embodied in the Industry and Tertiary (services) sectors mainly and in the Transport sector (freights) partially; but in a broader perspective, such as the one raised by the targets of circular economy, the energy and mass metabolism of corporations range over the all spectrum of sectors, from the extraction of minerals, to the final products, both in terms of use and disposal.

Therefore, what we need in the field of energy studies in order to fully understand and practically address the scale and the dynamics of change described above, is a shift of perspective—if not a change of paradigm, from the analysis' scope of consumers/sectors/cities/countries to that of corporations. This is not to say that we have to abandon the canonical scheme based on the concept that human beings are the logical type (elementary unit) that in all their several forms of aggregation, from individual consumer, to sectors (in the form of working force), from cities to countries, are the agents of change; but that we must add another logical type as an agent of change: firms.

An interesting perspective is raised by Life Cycle Analysis that, being a methodology with a growing consensus among scholars, but also practitioners and policy makers, deploys the kind of synoptic paradigm centred onto firms' metabolism that we are here advocating. The inventory analysis, from the cradle to the grave, epitomizes perfectly the quest for an analysis' scope placing the firm at the fulcrum of change, with the status of elementary unit of it. For LCA this change of perspective is just necessary to reconstruct the real energy and mass load required to produce a good or delivery a certain service. Can we expect governments to embrace the same perspective in relation to the energy and environmental issues by adopting, for example, measures aimed at: evaluating the carbon and environmental load over the entire (and global) cycle of production, preventing infringement dislocation and enforcing externalities internalization at any scale? National and transnational governance is now facing a similar problem with the raising and unbounded role of global finance, notably off-shore finance, in affecting real economy, as many are calling for a global governance of the phenomenon.

Indeed, a first step in this direction would be to develop an accounting systems based on firms and corporations. The only information we have about firms' metabolism is voluntary provided and deemed to the environmental commitment of the firm's stakeholders or to the targets that some local (national) legislation has imposed, until the next organizational reshuffling and units reallocation. In conclusion, the problem is very similar to the one prospected in Chap. 3 in relation to inlet/outlet points in the electricity network: given the present scale of production, the legal entity of firms based on country–economy is not any longer a suitable logical type for enacting any energy/environmental policy aimed at positively involving firms and corporations (the principal agents of change as explained above) and the their essential partners in their effective governance for a transition towards a low carbon economy. Therefore, either we raise the scale of governance or we find a different way to do it. One alternative way could be that of enhancing transparency (of information), knowledge (of processes) and awareness of the consequences of production among the public. Indeed, ultimately firms and corporations need consumers' consensus to prosper. We are their substrate.

A second major indication for energy policy we can draw from finite-time thermodynamics is that power maximization, rather than energy efficiency optimal condition, shape the energy path of economy and society. Hence, any energy conservation policy aiming at reducing energy demand by just imposing efficiency mandates or promoting energy efficiency enhancement should carefully look at the unintended effects that power maximization can brought about when a more efficient technology is used—or misused, under the pressure of time or for the opposite sake of increasing the output rather than reducing the input. In general, the dimension of time in assessing both behavioural responses by the consumers and systemic adaptations by the producers to a more efficient process need to be considered in designing policy measures and in constructing energy scenarios for the policy makers. In a fully connected and fast developing world, time is becoming a new dimension (or perhaps it is just the same old one, but newly framed), of power and governance needs to adequately face this rising domain of the economy and society. In Chap. 3 the new dimension of time for the ongoing transition towards RES—focusing particularly on the rising role of the network, as a paradigm and as an infrastructure, has been addressed by explaining how this represents a new challenge for institutional science and governance. In Chap. 7 we will focus on the implications of power maximization for energy conservation policy by tackling the rebound effect as a pivotal case of analysis.

References

Apertet, Y., H. Ouerdane, C. Goupil, and P. Lecoeur. 2012. Efficiency at maximum power of thermally coupled heat engines. *Physical Review E* 85 (4): 041144.

Basosi, R., and F. Ruzzenenti. 2010. Energy growth, complexity and efficiency, energy efficiency. In *InTech*, ed. Jenny Palm. doi:10.5772/9827. Available from: http://www.intechopen.com/books/energy-efficiency/energy-growth-complexity-and-efficiency.

Bateson, G. 1972. Steps to an Ecology of Mind. San Francisco, USA: Chandler Pub. Co.

Cantwell, J., and O. Janne. 1999. Technological globalisation and innovative centres: The role of corporate technological leadership and locational hierarchy. *Research Policy* 28 (2): 119–144.

Chen, J., Z. Yan, G. Lin, and B. Andresen. 2001. On the Curzon-Ahlborn efficiency and its connection with the efficiencies of real heat engines. *Energy Conversion and Management* 42 (2): 173–181.

Curzon, F., and B. Ahlborn. 1975. Efficiency of a Carnot engine at maximum power output. *American Journal of Physics* 43: 22.

Coller, H.A. 2014. Is cancer a metabolic disease? *The American Journal of Pathology* 184 (1): 4–17.

Esposito, M., R. Kawai, K. Lindenberg, and C. Van den Broeck. 2010. Efficiency at maximum power of low-dissipation Carnot engines. *Physical Review Letters* 105 (15): 150603.

Gillooly, J.F., E.L. Charnov, G.B. West, V.M. Savage, and J.H. Brown. 2002. Effects of size and temperature on developmental time. *Nature* 417 (6884): 70–73.

Glazier, D.S. 2014. Metabolic scaling in complex living systems. *Systems* 2 (4): 451–540.

Gyftopoulos, E.P. 1999. Infinite time (reversible) versus finite time (irreversible) thermodynamics: A misconceived distinction. *Energy* 24 (12): 1035–1039.
Gyftopoulos, E.P. 2002. On the Curzon-Ahlborn efficiency and its lack of connection to power producing processes. *Energy Conversion and Management* 43 (5): 609–615.
Kleiber, M. 1947. Body size and metabolic rate. *Physiological Reviews* 27: 511–541.
Lotka, A.L. 1956. *Elements of Physical Biology*. New York: Dover Publications Inc. (First publication: Elements of physical biology. The Williams and Wilkins Co., Inc. 1924).
Novikov, I.I. 1957. The efficiency of atomic power stations. *AtomnayaEnergiya* 3 (3): 409.
Odum, H.T., and R.C. Pinkerton. 1955. Time's speed regulator: The optimum efficiency for maximum power output in physical and biological systems. *American Scientist* 43 (2): 331–343.
Picciolo, F., A. Papandreou, K. Hubacek, and F Ruzzenenti. 2017. How crude oil prices shape the global division of labour. *Applied Energy* 189: 753–761.
Pavé, A. 2006. Hierarchical organization of biological and ecological systems. In *Hierarchy in Natural*.
Rajan, R.G., and J. Wulf. 2006. The flattening firm: Evidence from panel data on the changing nature of corporate hierarchies. *The Review of Economics and Statistics* 88 (4): 759–773; Social sciences. In *Methods Series*, 39–70. Springer.
Ruzzenenti, F., and R. Basosi. 2008. The role of the power/efficiency misconception in the rebound effect's size debate: Does efficiency actually lead to a power enhancement? *Energy Policy* 36 (9): 3626–3632.
Ruzzenenti, F., and R. Basosi. 2009. Complexity change and space symmetry rupture. *Ecological Modelling* 220 (16): 1880–1885.
Save the Children. 2016. Illuminiamo il futuro 2030 - Obiettivi per liberare i bambini dalla Povertà Educativa. Report available at: http://www.savethechildren.it/IT/Tool/Pubblicazioni/Related?id_object=274&id_category=40. Accessed on 04/06/2016.
Schneider, E.D., and J.J. Kay. 1994. Life as a manifestation of the second law of thermodynamics. *Mathematical and Computer Modelling* 19 (6): 25–48.
Smil, V. 2008. *Energy in Nature and Society, General Energetics of Complex Systems*. Cambridge, Massachusetts and London, England: The MIT Press.
Toyota. 2016. Sustainability report 2015, available at: http://www.toyota-global.com/sustainability/report/sr/. Accessed on 04/06/2016.
Van den Broeck, C. 2005. Thermodynamic efficiency at maximum power. *Physical Review Letters* 95 (19): 190602.
Verdier, N. 2006. Hierarchy: A short history of a word in western thought. In *Hierarchy in Natural and Social Sciences*, 13–37. Netherlands: Springer.
von Bertalanffy, L. 1968. *General System Theory: Foundations, Development, Applications*. New York, NY, USA: George Braziller.
West, G.B., J.H. Brown, and B.J. Enquist. 1997. A general model for the origin of allometric scaling laws in biology. *Science* 276 (5309): 122–126.
West, Geoffrey B., James H. Brown, and Brian J. Enquist. 1999. The fourth dimension of life: Fractal geometry and allometric scaling of organisms. *Science* 284 (5420): 1677–1679.
West, G.B. 2006. Size, scale and the boat race; Conceptions, connections and misconceptions. In *Hierarchy in Natural and Social Sciences*, 71–80. Netherlands: Springer.

Chapter 6
Polycentric Governance Approaches for a Low-Carbon Transition: The Roles of Community-Based Energy Initiatives in Enhancing the Resilience of Future Energy Systems

Thomas Bauwens

Abstract An understanding of the resilience of energy systems is critical in order to tackle forthcoming challenges. This chapter proposes that the polycentric governance perspective, developed by Vincent and Elinor Ostrom, may be highly relevant in formulating policies to enhance the resilience of future energy systems. Polycentric governance systems involve the coexistence of many self-organized centers of decision-making at multiple levels that are formally independent of each other, but operate under an overarching set of rules. Given this polycentric approach, this chapter studies the roles of community-based energy initiatives and, in particular, of renewable energy cooperatives, in enhancing the institutional resilience of energy systems. In this perspective, the chapter identifies three major socio-institutional obstacles, which undermine this resilience capacity: the collective action problem arising from the diffusion of sustainable energy technologies and practices, the lack of public trust in established energy actors and the existence of strong vested interests in favor of the status quo. Then, it shows why the development of community-based energy initiatives and renewable energy cooperatives may offer effective responses to these obstacles, relying on many empirical illustrations. More specifically, it is argued that community-based energy initiatives present institutional features encouraging the activation of social norms and a high trust capital, therefore enabling them to offer effective solutions to avoid free riding and enhance trust in energy institutions and organizations. The creation of federated polycentric structures may also offer a partial response to the existence of vested interests in favor of the status quo. Finally, some recommendations for policy-makers are derived from this analysis.

T. Bauwens (✉)
École Polytechnique Fédérale de Lausanne. EPFL ENAC IIE HERUS,
GR C1 482 (Bâtiment GR), Station 2, 1015 Lausanne, Switzerland
e-mail: thomas.bauwens@epfl.ch

N. Labanca (ed.), *Complex Systems and Social Practices in Energy Transitions*,
Green Energy and Technology, DOI 10.1007/978-3-319-33753-1_6

Introduction

Energy systems are constituted of technological, social, and ecological components and processes that interact with each other in a complex fashion. Acknowledging this complexity is crucial to coping with the challenges that energy systems currently present. The analyses of the political economist Elinor Ostrom and her collaborators may be very insightful in this endeavor. Although Ostrom's approach comes within the scope of rational choice theories, she was conscious of the implications of complexity theory and incorporated some of its concepts, such as nonlinearity, self-organization and feedback loops into her own work (Morçöl 2014). Based on her contributions, energy systems can be conceptualized as complex "social–ecological systems" (SESs), i.e., systems of interdependent biophysical and nonhuman biological units interacting with social and institutional components.[1] SES scholars have been particularly interested in the determinants that reinforce or hinder the "resilience" of such systems. Resilience of an SES refers, essentially, to its capacity to retain the same system characteristics despite changes in the behavior of its component parts or the surrounding environment (Carlson and Doyle 2002; Walker et al. 2004).

An understanding of the resilience of energy systems is critical in order to tackle forthcoming challenges. In particular, climate change and the depletion of fossil energy sources require massive transformations of our models of energy production and consumption. In response, the necessity of a transition to low-carbon sources is increasingly acknowledged. This transition will most likely imply the displacement of fossil fuels by various renewable, intermittent and distributed energy technologies along with energy demand reduction, and will be highly disruptive for established energy actors. The ability of future energy systems to implement this transition process will depend, therefore, on their resilience, understood as their capacity to deploy these renewable energy (RE) technologies fast enough so as to maintain their essential functions, such as the provision of energy services. However, several socio-institutional barriers can severely undermine this resilience. In this chapter, three barriers are discussed: the collective action problem, arising in the diffusion of more sustainable energy technologies and practices; the lack of trust from the public in established energy actors; and the existence of strong vested interests within the energy industry in favor of the status quo. Regarding the first obstacle, averting climate change is an action of global and public interest, because everyone benefits from a reduction of greenhouse gas emissions even if not contributing to it. This problem can therefore be configured in terms of a collective action problem (Sandler 2004). According to the conventional theory of collective action, rational actors pursuing their own interest will indeed not participate in

[1]Institutional components are constituted by the formal and informal rules shaping and structuring the interactions between people within collective settings (families, local communities, markets, business organizations, etc.) (Ostrom 2005).

collective efforts because they have incentives to free ride on the constructive behavior of others (Olson 1965; Hardin 1968).

It is proposed in this chapter that the polycentric governance perspective, developed by Vincent and Elinor Ostrom, may be highly relevant in formulating policies to enhance the resilience of future energy systems. Polycentric systems involve the coexistence of many self-organized centers of decision-making at multiple levels that are formally independent of each other, but operate under an overarching set of rules (Ostrom et al. 1961). Given this polycentric framework, this chapter focuses on the roles of community-based energy (CBE) initiatives and, in particular, of RE cooperatives, in enhancing the institutional resilience of energy systems. CBE initiatives are formal or informal citizen-led initiatives which propose collaborative solutions on a local basis to facilitate the development of sustainable energy technologies (Walker and Devine-Wright 2008; Bauwens et al. 2016). As a specific form of CBE initiative, the cooperative model enables citizens to collectively own and manage RE systems at the local level (Huybrechts and Mertens 2014). CBE initiatives are therefore inspired by a self-organization principle, which is at the core of polycentric systems.

This is organized as follows: the first section presents Elinor Ostrom's approach to complexity and the main building blocks of her work, including the Social–Ecological System framework. The second section explores the concept of resilience of SESs and specifically emphasizes the need for institutional adaptability and polycentrism. The third section applies these theoretical lenses to energy systems. As a first step, the section shows how energy systems can be considered as social–ecological systems and describes some essential features of the historical model of present energy systems, i.e., a model based on centralized extraction and conversion of fossil energies. Secondly, after discussing the need for a transition to low-carbon sources, this section outlines three major barriers to this transition. The fourth section describes why the development of CBE initiatives and RE cooperatives may offer effective responses to these obstacles, and, finally, the last section provides some main conclusions and recommendations for policymakers.

Elinor Ostrom and Complex Thinking

There are many different notions and measures of complexity (Mitchell 2009). One way to define the complexity of a system is to characterize it in terms of its degree of hierarchy or level of organization (Simon 1962; McShea 2001). This approach sees complex systems as composed of multiple nested subsystems. Each subsystem possesses unique emergent properties, which depend on its own constitutive elements, but appear only when these elements are integrated. In this sense, each subsystem constitutes a whole, but is also, at the same time, a part of a larger structure. This nested, hierarchical structure is common to physical, biological and social complex systems (von Bertalanffy 1969). As an example in the biological context, the human body is composed of cells which are organized into tissues, which themselves constitute organs, which are parts of physiological systems. In the social context, individuals are part of families, which organize into villages or tribes, which organize into larger groupings, etc.

Elinor Ostrom's conception of complexity finds its roots in this hierarchical vision. Ostrom focused primarily on institutions, which are the rules, the norms, the dos and don'ts that structure all kinds of social interactions. She built on the concept of "holon" to describe and analyze the multilevel nature of complex institutional systems (Ostrom 2005). A holon refers to what, being a whole in one situation, is simultaneously itself just a part of another larger system (Koestler 1973). In this hierarchical approach to complexity, these part-whole units are the fundamental constituents of any complex system. This holon property also holds for institutional systems: rules affecting one situation are themselves parts of a larger system of rules designed by individuals interacting at a higher level of decision-making. Ostrom and her collaborators used this idea to develop the "Institutional Analysis and Development" (IAD) framework to help structured thinking about the elements that influence decision-making in social situations and, in particular, collective action situations and social dilemmas.[2] This framework offers a nested set of variables, which can be used to investigate human interactions and outcomes in a wide variety of settings. It has notably been used as a basis for developing a theory of common pool resource management and for studying decentralized natural resource policies (Yandle and Dewees 2003; Clement 2010).

The IAD framework has been especially useful for studying social dilemmas at a microscale, but pays little attention to the broader social, institutional and physical environment, including demographic and market pressures. For this reason, a common criticism addressed to the IAD framework is that actors have often been presented as "independent of the larger historical and social context in which [they] operate" (Clement 2010: 131; see also Agrawal 2001). In order to include the influence of broader contextual variables, Ostrom and her colleagues designed the "Social–Ecological System" (SES) framework (Ostrom 2009; McGinnis and Ostrom 2014). A SES encompasses "interaction between, on one hand, a society's cultural and institutional arrangements, and, on the other hand, its physical environment" (Aligica and Tarko 2014: 55). Indeed, human beings transform the physical environment into usable resources (food, raw materials and energy). In addition, SESs reflect the hierarchical approach to complexity described above, as they are composed of multiple subsystems, whiles being embedded in multiple larger systems (Anderies et al. 2004). The elementary unit of this framework is constituted by an "action situation," in which multiple actors interact with each other under the influence of different contextual variables. These interactions produce outcomes, which are linked to contextual variables through feedback paths (Fig. 6.1).

[2]A social dilemma is a situation in which an individual profits from selfishness unless everyone chooses the selfish alternative, in which case the whole group loses (see https://en.wikipedia.org/wiki/Social_dilemma for further information).

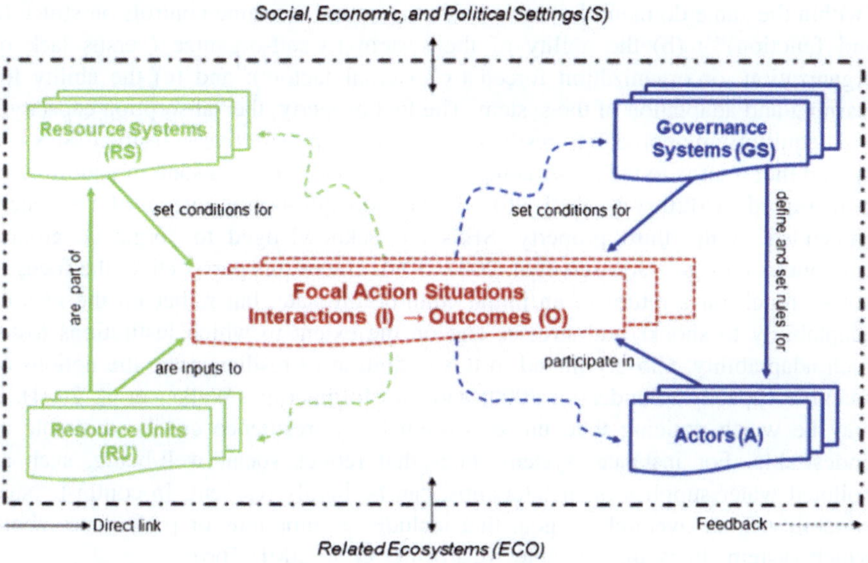

Fig. 6.1 Graphical representation of the social–ecological framework. *Source* McGinnis and Ostrom (2014)

Contextual variables relate to four core interacting subsystems: Resource Systems, Resource Units, Governance Systems and Actors. Resource Systems designate the biological/technological systems from which Resource Units are extracted. These Resource Units can then be consumed, used as inputs in a production process or exchanged for other goods and services. Governance Systems include "the prevailing sets of processes or institutions through which the rules shaping the behavior of the [actors] are set and revised" (McGinnis 2011: 181). Actors are individuals or collective entities who participate in relevant action situations and are defined by some shared attribute(s). Social, Economic, and Political Settings and Related Ecosystems represent respectively the broader social and ecological contexts that may influence the focal SES exogenously.

Ostrom's Approach to Resilience of Complex Social–Ecological Systems

Most discussions about resilience currently unfold within the SES framework just presented. Social–ecological resilience encompasses three properties (Carpenter et al. 2001): (a) the amount of disturbance a system can absorb while remaining "within the same domain of attraction (that is, retain the same controls on structure and function)"[3]; (b) the ability of the system to self-organize (versus lack of organization, or organization forced by external factors); and (c) the ability for learning and adaptation of the system. The first property, the "absorption capacity," is an equilibrium approach to resilience. Under this perspective, a resilient SES is a system that returns to the pre-existing situation, conceived as a state of equilibrium, if disrupted (Holling et al. 1995). In the nonequilibrium vision of resilience, embedded in the third property, SESs are acknowledged to "organize around continuous change" (Janssen et al. 2007: 309). Under this perspective, the focus is not so much on a return to an initial equilibrium state, but rather on the idea of adaptability to shocks and stresses and on the extent to which institutions foster such adaptability. Closely linked to this definition of resilience are the notions of adaptive capacity (Gunderson 2000) and transformability (Walker et al. 2004). It may be worth noticing that, unlike sustainability, resilience can be desirable or undesirable. For instance, system states that reduce social well-being, such as polluted water supplies or dictatorships, can be highly resilient. In contrast, "sustainability is an overarching goal that includes assumptions or preferences about which system states are desirable" (Carpenter et al. 2001: 766).

Janssen et al. (2007) showed that interactions between biophysical and social elements of an SES are bidirectional: pressures on ecological systems can affect the social–economic configuration and, conversely, threats to social–economic institutions can impact the biophysical environment. Social–institutional components and, in particular, institutional adaptability are thus crucial to analyzing resilience of an SES. "An institutional arrangement that inhibits innovation or does not secure it at a fast enough rate is an institutional arrangement that undermines resilience" (Aligica and Tarko 2014: 57). On the top of this capacity of institutions to adapt and change, Ostrom stressed the importance of self-organized initiatives by local actors in ensuring resilience of SESs. Indeed, under certain conditions, coordination and rules do not require external drivers or hierarchically superior forces to happen and can emerge from actors' interactions. Relying on Shepsle's (1989) definition of a robust institution, she sought to define the set of general principles that enable the maintenance of self-organization over time. These principles have been demonstrated empirically (Ostrom 1990) and theoretically (Wilson et al. 2013) to promote resilient SESs. These design principles are: (1) clearly defined and generally understood boundaries and institutional roles; (2) effective monitoring against free riding; (3) graduated sanctions against offenders; (4) proportionality between costs and benefits; (5) conflict resolution mechanisms generating outcomes perceived as fair; (6) minimal recognition of rights to organize[4]; (7) effective collective choice arrangements such as consensus, which avoid decisions imposed by some members at the expense of others; and (8) subsidiarity and nested or "polycentric" structures.

---

[3]See Carpenter et al. (2001: 766).

[4]In Wilson et al.'s (2013: 522) words, "groups must have the authority to conduct their own affairs. Externally imposed rules are unlikely to be adapted to local circumstances and violate principle 3".

This last principle, which refers to the concept of "polycentricity", is, according to Aligica and Tarko (2014), the most distinctive feature of the Ostromian approach to resilience. Polycentricity describes the coexistence of many self-organized and autonomous centers of decision making, all operating under an overarching set of shared rules (Ostrom et al. 1961). Polycentric systems involve the existence of local resource governance units nested in larger, general-purpose units located at higher levels of decision-making (often governments, but not necessarily), according to a principle of "Russian nesting dolls". As stated above, self-organization is an essential guiding principle of polycentric systems. Accordingly, a key assumption of polycentric approaches is that governance arrangements are more effective when citizens have the juridical and material capabilities to self-organize multiple governing bodies at diverse scales (Andersson and Ostrom 2008).

Polycentric governance presents several practical advantages over highly centralized systems. First, it enhances the institutional resilience of an SES. By creating redundancy of local centers of decision making, it fosters the conditions for the experimentation and creativity needed to explore novel and potentially superior combinations of rule systems, i.e., for adaptability of rules. This redundancy also mitigates the effects of a governance failure by limiting such effects to one locality, compared to the substantial costs induced by a failure of a centralized unit that covers a large area. Second, polycentric systems may exhibit informational benefits compared to highly centralized systems by encouraging the use of local knowledge to devise rules that are better adapted to each local situation than any general set of rules. Highly complex systems of rules involve many interconnected factors that have to be taken into account, so that "no one, including a scientifically trained, professional staff, can do a complete analysis" (Ostrom 2000: 12). In these conditions, it is often better to rely on the knowledge that local resource users have accumulated, since they are likely to devise rules that are better adapted to their local needs than "one-size-fits-all" rules created at a very centralized level (Irwin 1995; Wynne 1996). A third, related benefit of polycentrism is that it enables feedback on the performance of rules to be captured in a disaggregated way (Ostrom 1999). Fourth, local resource users know each other. They are thus more likely to select trustworthy partners and exclude untrustworthy ones, enhancing the conditions for cooperation and reciprocity between participants (Powers and Thompson 1994; Andersson and Ostrom 2008). Fifth, polycentrism lowers enforcement costs by strengthening local perceptions of the legitimacy of rules, and also by making it easier to fashion rules that can affordably be monitored. Indeed, if local actors are involved in the design of rules and the monitoring of compliance with these rules, they will be apt to craft rules that make infractions highly obvious so that monitoring costs are lower. Further, by creating rules that are seen as legitimate, local actors encourage higher conformance.

While polycentricity is opposed to highly centralized governance, the existence of an overarching set of shared rules implies that it should not be equated to full decentralization either. If the governance regime lacks coordination among self-organized initiatives, it will not operate as a system, but rather as a network of fragmented and unstructured actors who may dissipate their efforts in unproductive directions. This may lead to contradicting actions and poor efficiency (Black 2008).[5] This is why, asides from bottom-up and self-organized initiatives, top-down institutions are crucial in the creation and maintenance of any polycentric governance system to coordinate the activities of the multiple participants, the resolution of conflicts between lower level units and the exchange of information about what has worked well, and may be transferable, from one local setting to others. Mansbridge (2014) identifies at least four roles for these higher level governance units: (1) the threat of imposing a solution if local parties cannot come to a negotiated agreement; (2) the provision of a relatively neutral source of information; (3) the provision of "institutional facilities" to facilitate negotiations; and (4) the monitoring of compliance and sanctioning defection from compliance in the implementation phase after the negotiators have reached agreement.

One example of large-scale successful polycentric structure is the scientific community, as explained by Tarko (2015). This community lacks any central management or a formalized legislative or rule enforcement body and multiple research centers coexist, each with its own somewhat different research agenda and preferred methods of investigation. Yet, the success of the scientific community and the progress of science is the result of an overarching set of shared informal rules, which limits free riding and enables the whole system to work. In environmental contexts, the polycentric perspective has been applied to analyze water (Marshall et al. 2013; Pahl-Wostl and Knieper 2014) and forest resource management (Nagendra and Ostrom 2012). References to the polycentric perspective have been made to a limited degree in previous work on governance of energy systems (Sovacool 2011; Goldthau 2014; Koster and Anderies 2013) but without significant or systematic development.

Resilience of Energy Systems and Low-Carbon Transition

Energy Systems as Social–Ecological Systems

Recent studies argue for framing energy systems as SESs (Hodbod and Adger 2014; Bauwens et al. 2016). Production, distribution and consumption activities within energy systems involve interactions between, on the one hand, ecological processes and technological artefacts and, on the other, social practices and systems

[5]In this perspective, some authors have pointed out various limitations linked with fully decentralized governance systems, among which are: the high cost of self-organization (Meinzen-Dick 2007); the risk of local tyrannies, i.e. the lack of democratic governance or the domination of self-organized systems by local leaders who change rules for their own advantage (Platteau and Gaspart 2003; Platteau 2004); the problem of stagnation, i.e. actors' reluctance to produce new rules and institutions to innovate due to the complexity of the resource system involved; and the risk of conflict among user groups (Alston et al. 1999).

of institutional rules. As such, they can be analyzed through the lenses of the SES framework introduced above. Energy systems consist of "resources that are converted through various means to provide energy services" (Löschel et al. 2009: 391), energy services being defined as the benefits that energy carriers produce for human well-being (e.g., mobility for an automobile, heat for a stove, mechanical energy for air circulation). Energy resources can be renewable and nonrenewable, as well as primary or secondary. Renewable resources include solar radiation and all of its biospheric transformations, such as wind, geothermal heat, moving water or biomass. Their main common characteristic is that they are naturally regenerated over a short period of time. Nonrenewable resources, on the other hand, cannot renew themselves within time frames that are meaningful to humans. They include fossil fuels, coals, hydrocarbons and radioactive minerals. As for the second distinction, primary energy means the energy "embodied" in natural resources and not yet converted into other forms of energy. Secondary energy includes the forms of energy generated by conversion of primary resources, e.g., petroleum products, manufactured solid fuels and gases, or electricity. Crossing these two criteria yields four types of resources (Table 6.1).

It follows that energy systems can be subdivided into two major types of Resource Systems: biophysical Resource Systems, which are natural, and technological Resource Systems, which are human-made. Biophysical Resource Systems refer to the systems from which primary energy resources are extracted and, in the case of renewable resources, through which the levels of the focal resource are regenerated by natural dynamic processes. Nonrenewable resources "are defined in terms of reserves or energy stocks, which can be depleted over time", whereas "renewable energy resources are defined in terms of energy flows (e.g., energy production per year)" (Löschel et al. 2009: 396). For this reason, the distinction between Resource Systems and Resource Units is blurred in the case of renewable resources. Resource Units of biophysical Resource Systems are, for instance, the tons of oil or gas withdrawn from reservoirs, the photons radiated by the sun or the liters of water flowing from water sources. Biophysical Resource System variables encompass the type and abundance of resources, their renewable or nonrenewable nature, their location, etc.

Technological Resource Systems, on the other hand, are defined here as the set of humanly constructed facilities and infrastructures that enable the conversion, transport, distribution and consumption of primary or secondary energy. They can be decomposed into multiple constituents: the generation assets, the "delivery

Table 6.1 Different types of energy resources and fuels

	Renewable	Nonrenewable
Primary	Solar radiation, plant mass, wind, moving water	Coals, crude oil, natural gases, uranium, other minerals
Secondary	Biodiesel, ethanol, processed wood pellets, electricity	Coke, coal gas, refined crude oils, nuclear fuel rods, electricity

Source Sovacool and Dworkin (2014)

mechanisms" and the "prime movers". Prime movers are "the technology that converts primary and secondary fuels into useful and usable energy services" (Sovacool and Dworkin 2014: 38), such as human muscles, steam engines, jet turbines or household electric appliances. Delivery mechanisms are the delivery infrastructures used to connect primary resources to prime movers, including pipelines, tankers, and electric transmission and distribution lines. The Resource Units of technological Resource Systems are the flows of primary or secondary energy circulating in these infrastructures. Technological Resource System variables cover the type of primary resource used, the characteristics of generation assets (their size, their load factor,[6] their distance from the grid, whether they are intermittent or not), delivery mechanisms and prime movers.

In addition to these biophysical and technological characteristics, energy systems are shaped to a considerable extent by Governance Systems and Actors variables. For instance, Moe (2010) remarkably shows that the differences observed in the energy structure of countries cannot solely be explained by their different energy resource endowments. The occurrence and the pace of transitions from one energy source to another, for instance, are also largely determined by the level of political power and influence of established energy actors. In addition, energy systems are shaped to a large extent by households, grassroots actors and civil society (Smith 2012; Stern 2014). By supporting, accepting or opposing changes in larger energy systems, individuals acting as citizens can influence public policies or private organizations' decisions and can contribute to shaping the transformation of energy systems. These reactions sometimes organize into social movements, e.g., public reactions to nuclear power, shale gas extraction infrastructures or wind farms.

The Historical Model of the Energy Supply Industry

It is worth highlighting two aspects of the way energy supply systems have been historically shaped: the dominance of fossil fuels as the main primary energy source and the development of a centralized model of energy supply. Regarding the first aspect, our dependence on fossil fuels is striking. Out of the 157,482 terawatt-hours (TWh) of primary energy generated in 2013 by humankind, fossil fuels provided about 81.4% of this total, while nuclear energy provided 4.8%, and renewable sources provided the remaining 13.8% (10.2% by biomass and waste, 2.4% from hydropower and less than 1.2% from other RE sources) (IEA 2015). Regarding the second aspect, the rise of fossil fuels coincided with the construction of very centralized technological Resource and Governance Systems, which have prevailed until today. At the technological level, the dominant model of energy infrastructure is characterized by large centralized power stations generally located close to sources

[6]The load factor is the percentage of time an asset is operated at full load.

of fossil fuels and remote from load centers, which supply huge grids run by regional or national monopolies. The energy sector has emerged as a vertically and horizontally integrated system with important technical interdependencies. At the governance level, the energy industry was historically institutionalized along with this technological configuration, according to a highly hierarchical and centralized organization under the control of the State. It was vertically integrated, which means that firms operating in the different functions of the energy value chain, i.e. production, network activities and sales, were strongly interconnected through ownership rights, contracts and regulation (Künneke 2008). This was somewhat modified with market liberalization. In practice, however, although the liberalization process significantly changed e.g. the European landscape of the power industry, institutions are still strongly marked by this historical configuration. For instance, according to DTI/Ofgem (2006), the market and regulatory models adopted in the UK at privatization reflected the predominantly centralized model of transmission and distribution. As a result of these technological and institutional evolutions, industrial economies have, to a large extent, become "locked" into fossil-fuel-based, centralized energy systems through a path-dependent process driven by technological and institutional feedbacks and mutual reinforcements—so-called "increasing returns to adoption" in economists' jargon (Unruh 2000). Once a country is "locked in," it faces persistent market and policy failures that reduce the chances for alternative technologies to join the market (Arthur 1994; David 1985).

Furthermore, in this model, actors from the demand-side, i.e. energy consumers, are minimally engaged in energy generation (Eyre 2013). Energy users do not need to know where energy is coming from, how it is produced and transported. Centralized generation "has led to the design and deployment of a range of energy technologies, services and procedures, from meters to bills to regulatory institutions to power stations, that foster minimal public engagement" (Devine-Wright 2007: 68). From a supply-side perspective, it has led "designers, developers and installers of new energy technologies [to] aim to minimize public engagement since this would be assumed to increase the risk of resistance, delay, planning refusal and inefficient or incorrect use of technologies".

The dominance of fossil fuels poses major threats to the ecological and social sustainability of energy systems. Climate change at a global scale and local air pollution associated with greenhouse gas emissions are probably amongst the most alarming ones. Other challenges include energy security in a context of finite fossil sources and price volatility, the geopolitics of energy, universal access to energy services and energy poverty. While there is little consensus about when fossil resources will exhaust, the expected global demand expansion ensures an accelerated decline of current reserves. Furthermore, the world's known remaining oil reserves are concentrated in unstable regions of the world, especially the Middle East. Similarly, the other main conventional energy fuels—coal, natural gas and uranium—are distributed very unevenly. This highly concentrated distribution of oil has caused the transfer of immense wealth from oil-importing countries to oil producers (Sovacool 2012). The concentration of fossil fuels also generates price volatility and interruptions in supply (Sovacool et al. 2014).

These challenges require massive transformations of energy systems and call for a transition toward low-carbon energy systems. In turn, the ability of energy systems to implement this transition process depends on their resilience, understood in this context as their capacity to deploy low-carbon technologies so as to maintain their essential functions.

Institutional Obstacles to a Low-Carbon Transition

There are many constraints to a low-carbon transition, but most existing studies focus on the technical and economic feasibility of alternative energy systems in meeting energy demands at the same time as meeting a carbon reduction target (e.g., Ekins et al. 2013). The remainder of this chapter concentrates however on constraints of a socio-institutional nature. These mostly derive from a lack of support from society, organizations or government agencies or a lack of appropriate institutions to govern change. These challenges are thus more directly related to Governance Systems and Actors variables rather than to Resource Systems and Units. The focus is drawn on three specific challenges that are particularly threatening for institutional resilience of energy systems: the collective action problem arising from the diffusion of sustainable energy technologies and practices, the lack of public trust in established energy actors and the existence of strong vested interests in favor of the status quo.

Regarding the collective action problem, following Samuelson (1954), economic goods are frequently classified into two categories: private goods and public goods. A good is purely private when the producer bears all the costs of production and a single consumer enjoys all the benefits of consumption. A pure public good, in contrast, is characterized by non-rivalry and non-excludability. Non-rivalry means that an individual's consumption of the good does not limit the capacity of others to consume the same good. Non-excludability implies that it is difficult to exclude individuals who have not paid for the good from its consumption. The collective action problem is intimately related to the attribute of non-excludability. More precisely, a person who cannot be excluded from the benefits of a public good will have no incentive to bear a part of the costs of its production and will thus have a strong incentive to behave as a "free rider" (Olson 1965). Collective action problems constitute a threat to the resilience of any SES, because they lead to overharvesting of common resources or to the underprovision of public goods and, eventually, to the collapse of the system.

In the context of energy systems, averting climate change is a global and public interest. Past energy transitions (e.g., from traditional biomass to coal and from coal to oil) have been driven by a large minority of consumers who were willing to pay considerably more for privately accruing services associated with new energy sources or technologies (Fouquet 2010). In contrast, the environmental benefits of the current low-carbon transition are shared by all individuals and thus clearly present characteristics of a public good. It is likely that too few consumers will be willing to pay more for the environmental improvements, although their number is growing (Longo et al. 2008). For instance, free riding has indeed been identified as one of the major barriers to the diffusion of RE technologies. While attitudinal surveys demonstrate high levels of public support for green power products (Batley et al. 2001; Nomura and Akai 2004), the green marketing literature consistently reports a large gap between the number of residential customers willing to pay a premium for them and actual participation rates in green pricing programs (Byrnes et al. 1999; Wiser 1998). The collective action problem has then also been identified as a barrier to sustainable electricity consumption within households (Ohler and Billger 2014).

A second socio-institutional obstacle lies in the lack of trust in traditional energy actors. Trust in institutions can be defined as "believing that a person(s) or organization(s) can be relied upon to accomplish objectives because they are competent and possess values and intentions that are consistent with all or part of the public" (Greenberg 2014: 152).[7] Trust is important for institutional adaptability and resilience because it enhances cooperation and enables shared cognition. That is to say, people feel they can rely on the statements of others without having to go back to first premises to check their validity (Cvetkovich 1999). This is why trust also enables actors to cope with new situations more quickly. In addition, trust appears to be a crucial element as far as energy systems are concerned, mainly because public concerns about risk have intensified in recent years (Slovic 1993). Nuclear power plants and waste management facilities, natural gas plants, fracking, oil refineries, giant hydropower dams and many other issues are examples of areas of public concern as regards energy. Trust is also an important ingredient in the transition to a low-carbon society, because the implementation of decentralized RE installations and smart metering technologies need to be steered by individuals and organizations that are highly trusted and rooted in local communities (Eyre 2013). Trust in actors that are responsible for the development of a technology is critical when it comes to social acceptability of this technology, especially when people know little about it (Jobert et al. 2007; Walker et al. 2010; Huijts et al. 2012). In the wind power context, Eltham et al. (2008) have documented, through the study of public opinions of a local population living near a wind farm, how suspicion of the developers' motives by the public, distrust of the developers and disbelief in the planning system may preclude the success of wind farm projects. Moreover, evidence shows a general lack of trust by the public in traditional energy actors as far as the development of alternative energy is concerned (Mumford and Gray 2010; Greenberg et al. 2012). This lack of trust in conventional energy actors is likely related to the centralized institutional configuration of energy systems described above. Institutions involved in energy (e.g., governments and multinational companies) form part of the expert systems of global politics, commodity markets and

[7]It is worth distinguishing institutional trust, which refers to trust in organizations and institutions managing energy projects, such as public authorities, developers, power utilities and other actors, from interpersonal trust, which describes trust among the members of a community and is closely connected to the notion of social capital (Walker et al. 2010).

large-scale engineering which are not easily accessible to ordinary citizens (Mumford and Gray 2010). The centralized model of energy supply also increases the spatial, social and political distances between actors and, therefore, undermines trust.

A third important hindrance to low-carbon transition is the existence of strong vested interests. Profound shifts in the energy system toward a low-carbon society create "winners" and "losers" and "those that stand to gain or lose out will be at the heart of change debates" (Kuzemko et al. 2016: 101). Generally, incumbent energy actors, including those in the fossil fuels and nuclear industries, and electric utility companies, have a vested interest in preserving the current system. For instance, traditional power utilities are directly damaged by the increase in the proportion of decentralized renewable technologies—especially photovoltaics—forming part of the total installed electricity capacity (Groot 2014). Furthermore, incumbent actors generally have enormous political influence and substantial resources to resist any change that threatens their interests. Actually, established actors do not always seek to resist change intentionally, but as they fight for their own interests (regulations, subsidies, favorable institutional arrangements, etc.), they often do this to the detriment of alternative energy. Vested interests are a threat to the resilience of energy systems because they lead to institutional rigidity. As Olson (1982) shows, an economic sector which becomes economically prosperous also typically acquires political influence and seeks to secure institutional arrangements that are beneficial to itself, but not for society at large. If a society is controlled by vested interests, it loses its ability to adapt and shift the status quo (Moe 2010). Based on a comparison between Japan, China, the United States, Germany, Denmark and Norway, Moe (2015) shows that whether or not renewable energy has been a success is determined by the extent to which countries have been successful in controlling these vested interests and prevented them from unduly influencing energy institutions. In turn, the ability of incumbent actors to be politically influential depends on the historical economic and political importance of the industries they represent (Kuzemko et al. 2016).

The next section explains how community-based energy initiatives in general and energy cooperatives in particular may contribute to overcome these barriers to low-carbon transition and thus enhance the institutional adaptability and resilience of energy systems.

Community-Based Energy Initiatives and Institutional Resilience

Community-based energy (CBE) initiatives are typically characterized by a high degree of citizen agency and involvement in the ownership, management and benefits of projects (Walker and Devine-Wright 2008) and, as such, strongly echo the principle of local self-organized decision-making units characterizing polycentric systems. The RE cooperative model is arguably one of the strongest forms of CBE initiative in Europe. It is not by chance that this model is the only one that is represented at the European level by a federation.[8] RE cooperatives are also often strongly embedded in the international cooperative movement, an international network of cooperatives and advocacy organizations that aim to promote and spread the cooperative principles of solidarity and democratic governance (Birchall 1997). Furthermore, while access to finance during the at-risk stage is acknowledged as a barrier to the development of community energy projects (Nolden 2013), cooperatives are particularly suitable to ensure the financial viability of small-scale projects, through fundraising among community individuals, compared to other models depending on grants or loan schemes.

The cooperative model enables citizens to collectively own and manage RE projects at the local level (Bauwens et al. 2016; Huybrechts and Mertens 2014). Through this model, citizens produce, invest in and, in some cases, consume RE. The following cooperative principles, adopted by the International Cooperative Alliance (ICA) in 1995 (ICA 1995), are generally common to all types of cooperatives around the world: voluntary and open membership, democratic member control (e.g., the "one person-one vote" rule), economic participation by members, autonomy and independence, education, training and information, cooperation among cooperatives, and concern for community. These principles clearly do not uniquely define cooperative structures and in Western Europe there is considerable variety in the legal forms of democratic enterprises (Borzaga and Defourny 2001),[9] but in general these forms share features that embrace the above cooperative principles (Spear 2004). From an economic standpoint, cooperatives present a model of ownership different from conventional business organizations (Hansmann 1996). They are generally owned by their members/users rather than investors, unlike capitalist corporations. Part of the surplus goes to indivisible reserves, which are unavailable for distribution to members, even if a cooperative were to be dissolved. These reserves represent the collective assets of the organization. Another part of the surplus can in principle be divided pro rata among the members according to the volume of transactions (not members' shares) they have conducted with the organization. When the net income is partially allocated as a return on capital shares, such profit distribution is subject to a cap, which suggests that maximization of return on capital may not be a key objective. Finally, as previously mentioned, cooperatives use a democratic governance structure, which involves equal individual voting rights ("one person, one vote") and the absence of barriers to entry for new members. This is another major trait of the cooperative identity, as in other company types the default governance rule is "one share, one vote".

Why would CBE initiatives, and cooperatives in particular, help solve the collective action problem that arises in the diffusion of sustainable technologies and practices? To understand this, it is crucial to acknowledge the importance of local

---

[8]See http://www.rescoop.eu.

[9]See Fici (2013) for a comparative analysis of the legal identity assigned to cooperatives in several European jurisdictions.

actions in mitigating climate change. Many analysts call for an institutional solution at the global level, because global threats such as climate change are believed to require "global solutions", negotiated at the international level (Nordhaus 1994; Stern 2007; Wiener 2007). Solutions to the climate crisis certainly demand efforts at the international level, where most efforts are now being concentrated. Yet, in line with the polycentric governance approach, a global policy is not the only strategy needed and positive actions are required at multiple, smaller scales to start the process of climate change mitigation and secure the efforts made at the global level (Bulkeley and Betsill 2005; Bulkeley and Kern 2006; Ostrom 2010, 2012). Indeed, collective action problems faced by large groups, such as the problem represented by climate change mitigation, are often decomposable into social dilemmas at a smaller scale, some of which are typically surmountable given the existence of social norms and, especially, of pre-existing trust networks (Ostrom 2010; Bauwens and Eyre 2017). Accordingly, several studies have argued that community-based energy initiatives facilitate collective action for climate change mitigation by fostering individual behavioral change toward more sustainable energy practices (Middlemiss 2008, 2011; Heiskanen et al. 2010; Seyfang 2010). CBE initiatives are said to influence their members' energy-related behavior, notably by activating social norms.[10] From an institutional perspective, a "community" is a social institution characterized by high entry and exit costs and non-anonymous interactions among members (Bowles and Gintis 1998, 2002). In addition, interactions among community members are more frequent and extensive than interactions with 'outsiders'. These structural characteristics of interactions contrast with those of other institutions, such as markets, at least in their idealized forms. Market interactions are characterized by ephemerality of contact, anonymity among interacting actors and ease of entry and exit. In contrast to markets, by facilitating direct personal interactions, communities effectively encourage the formation of norms, such as interpersonal trust, group identification, solidarity, reciprocity, reputation, personal pride, vengeance, etc.

Norms have proven to be powerful and cost-efficient mechanisms to encourage energy conservation (Allcott 2011; Nolan et al. 2008).[11] Gadenne et al. (2011) showed that environmental concern, combined with social norms and community influence, can positively contribute to environmental behaviors. Ek and Söderholm (2008) also found that social or moral norms can affect the purchase of green electricity. In addition, different qualitative studies suggest that some communities encourage low-carbon lifestyles by stressing the associated social rewards for climate-beneficial actions (Middlemiss 2008) or by turning the social dilemma they

---

[10]Social norms are "customary rules of behavior that coordinate our interactions with others. Once a particular way of doing things becomes established as a rule, it continues in force because we prefer to conform to the rule given the expectation that others are going to conform" (Young 2008: 647).

[11]Social psychologists distinguish between two types of norms: injunctive norms and descriptive norms. The former involve perceptions of which behaviors are typically approved or disapproved of by other people and provide points of comparison, e.g., concerning others' energy consumption, while the latter involve perceptions of other people's typical behaviors (Schultz et al. 2007).

represent into assurance games where members can be assured that others will participate (Heiskanen et al. 2010). Furthermore, CBE initiatives may lower information costs related to energy efficiency technologies and conservation behaviors and therefore contribute in overcoming some of the informational and behavioral barriers to energy efficiency constituting the so-called "energy efficiency gap" (Gillingham and Palmer 2014).[12] Indeed, CBE projects raise their members' awareness about sustainable energy practices through communication channels and information provision. Again, norms are likely to play a role in this respect as the trustworthiness of the sources of information can positively affect the effectiveness of a message (Stern et al. 1986; Laskey and Syler 2013). Finally, several studies emphasized the effects of trust networks and peer behaviors on the adoption of sustainable microgeneration technologies, such as photovoltaic panels (e.g., Bollinger and Gillingham 2012). CBE initiatives are able to inform and influence consumer decision-making because of the trust networks they hold through long-standing linkages with key individuals in local communities (Noll et al. 2014).

Clearly, economic incentives play a role as well in overcoming the collective action problem. Indeed, by allowing citizens to become the residual claimants on the financial surplus generated by RE assets and on the decision-making power, CBE initiatives contribute to trigger investments in and public support of sustainable technologies at the community level. Thus, CBE initiatives combine both social norms and standard economic incentives to foster contributions to the global public good of averting climate change. The respective weight given to norm-driven behaviors and economic incentives is not necessarily the same in all CBE initiatives and depends on several factors, including how "market" and "community" institutional dimensions are prioritized within the business model of CBE initiatives, spatial characteristics of membership and the stage of development that CBE initiatives have reached (Bauwens 2016).

The case of Connexus energy illustrates particularly well the roles of CBE initiatives in enhancing social norms and, ultimately, encouraging sustainable energy-related behaviors. Connexus Energy is the largest cooperatively-owned supplier in Minnesota (it supplies electricity to about 125,000 households). In partnership with the energy efficiency company Opower, it has launched one of the longest behavioral intervention programs for energy efficiency in the United States. Home Energy Reports were sent to the 80,000 participating households and contained two components: an Action Steps Module providing household-specific energy conservation tips and a normative comparison of the household's energy use to that of similar neighbors. During the three years since the start of the program, households of Connexus have collectively reduced their consumption by about 30,000 MWh and avoided CO_2 emissions equivalent to 350 air flights in the US (Laskey and Syler 2013). Another empirical example is provided by the web

[12]The energy efficiency gap describes the existence of unexploited 'profitable' investment options in energy saving technologies and practices.

platform EnergieID,[13] based in Flanders (northern Belgium). It enables energy users to follow their energy consumption and compare it with that of other similar households, thereby activating social norms. The website was put online in 2011 and offered free monitoring tools for individuals. In 2014, EnergieID was formalized as a cooperative, with the view of creating a common platform in which users can share their data with different service providers in a secured and anonymous way. EnergieID collaborates with partners such as municipalities, other cooperatives and service providers. For instance, the Flemish cooperatives BeauVent and Ecopower encouraged their members to use the platform and a more formal partnership was established in 2012 with the creation of groups called “BeauVent” and “Ecopower” on the website. The members who register in these groups are invited to report their electricity consumption each month. After one year, BeauVent and Ecopower analyze these consumption figures and provide members with a personalized report about their consumption and how they can reduce it further. They incentivize people to report their consumption by offering a prize (generally a device related to energy efficiency, such as a consumption monitor, a LED light, etc.) at the end of the period.

Regarding trust in institutions involved in energy, which can also be described as a specific kind of social norm, the literature on CBE initiatives shows that these initiatives are typically characterized by a high degree of trust (Walker et al. 2010). Similarly, it has been shown that cooperatives are generally perceived as more trustworthy than investor-owned firms, given their constraint on the profits distribution and their democratic governance (Hansmann 1996; Ole Borgen 2001). In addition, citizen ownership contributes to the trust capital of CBE initiatives and cooperatives as it provides the guarantee to noncontrolling stakeholders that the firm is managed by people who share their interest (Spear 2000). This is consistent with the findings that horizontal networks, where people have equivalent status and power, engender trust because they facilitate exchanges of information and face-to-face communication, whereas hierarchies tend to inhibit information flows due to asymmetric power relationships (Kasperson et al. 1999). Finally, the local anchorage of CBE initiatives and cooperatives reduce the social distance between stakeholders, further consolidating trust. As a result of this high trust capital, there is evidence that community-based or cooperative ownership enhances social acceptability of controversial RE facilities, such as onshore wind power (Bauwens 2015; Maruyama et al. 2007). Comparative research has shown that a high degree of citizen involvement in wind energy projects is positively correlated with high deployment rates (Bauwens et al. 2016; Toke et al. 2008). If citizens are the residual claimants on the organization’s surplus and decision-making power, they are likely to feel more fairly treated and would be more willing to accept or support the outcome.

Finally, the ways CBE initiatives and cooperatives could contribute to overcoming the challenge of vested interests are less obvious, because this challenge is generally of a systemic nature that cannot be solved at the operational level,

[13]http://www.energieid.be.

whereas most of the time the main mission of CBE initiatives and cooperatives is to implement sustainable energy projects on the ground. The notion of polycentric systems is crucial here. The Governance Systems variables affecting energy systems are the outcome of interactions between political, industrial and civil society actors located at higher levels of decision-making and thus local CBE initiatives taken individually are not likely to influence these decisions. However, as Ostrom (2005) notes, local communities often spontaneously form larger associations in order to deal with larger issues. The creation of federated structures is a way of enhancing the bargaining power of small players such as CBE initiatives in the face of incumbent energy actors. Indeed, the latter are smaller in number, have relatively homogeneous interests and are able to coordinate their resources to resist any threatening change. In contrast, CBE initiatives are dispersed, generally focus on very local issues and have limited resources and political power. Several studies have acknowledged the difficulties experienced by grassroots initiatives in surviving in increasingly hostile environments, not to mention the obstacles to scaling up their impact and challenging mainstream actors (Bauwens et al. 2016; Seyfang et al. 2013). Coordinated actions may thus be seen as an attempt to reach a more balanced distribution of political power in energy markets, which is still very biased in favor of large-scale players. While decentralization of governance in energy systems is sometimes conceived as a panacea, the emergence of coordinated actions among cooperative initiatives calls for a more polycentric approach, according to which "various scales need to be taken into account when designing regulatory answers and setting up governance arrangements" (Goldthau 2014: 136). In this perspective, although decentralized energy systems obviously exhibit a strong local component, federated structures highlight the importance of the ability of local initiatives to transcend their local experience in order to form networks at higher levels and articulate their interests to national and international strategies.

An Example of Federation of RE Cooperatives: REScoop.eu

The creation of a federation of RE cooperatives at the European level, like REScoop.eu,[14] can typically be interpreted as a way to integrate the local level with the national and international ones. This federation was established in 2011 by a consortium of 12 cooperatives and 2 national federations, with the objective of supporting the development of RE cooperatives. By 2016 active membership had risen to more than 1200 cooperatives. The mission of REScoop.eu encompasses three main activities. First of all, REScoop.eu seeks to gather and centralize information and knowledge from individual initiatives. It has, for example, identified and contacted more than 2400 existing RE cooperatives across Europe and created a database containing basic information about 693 of such organizations. On the basis

[14]See http://www.rescoop.eu.

of this inventory, REScoop.eu has been able to produce a number of documents including best practice case studies, guides and handbooks targeting new initiatives. The second key activity conducted by REScoop.eu is the exchange of information within the network. This exchange takes place both through a web-based platform and through personal interactions. Twenty-five "mentors", i.e. representatives of well-established RE cooperatives, were also identified in the network in order to actively support emerging initiatives across Europe. Finally, REScoop.eu also conducts communication and advocacy activities toward external audiences, such as policymakers, citizen groups, corporations, NGOs and the media. Accordingly, the missions of REScoop.eu encompass most of the missions of intermediary actors identified by Geels and Deuten (2006) and extended by Hargreaves et al. (2013). In addition, this network has formally set membership standards, which include the ICA principles mentioned above and additional ecological, social and ethical common principles in the charter of REScoop.eu. Therefore, the network has defined the set of basic rules shared by all initiatives belonging to the network. These rules are not blueprints, however, and do not preclude local initiatives from developing their own additional rules. Although the European federation is still in infancy, it does share some of the essential characteristics of a polycentric system. Firstly, local cooperatives form a multiplicity of local autonomous decision centers that are able to put their different methods into practice and to make operational decisions independently from each other. Secondly, despite this autonomy in the implementation of local actions on the ground, the European federation has also defined some guiding principles under the form of a charter, which provides a framework of overarching basic rules, including the cooperative principles, which are supposed to be shared by all members. Currently, however, it does not have clear monitoring mechanisms against free riding.

Conclusion and Recommendations for Policymakers

Acknowledging the complex interdependencies between the various components of energy systems is essential for designing effective responses to the urgent challenges posed by the ongoing energy transition. Elinor Ostrom's contribution to the literature on social–ecological systems fits well with complexity theories and, in particular, with hierarchical perspectives on complex systems. This chapter argued that present and future energy demand and supply systems can be adequately conceptualized as SESs and that Ostrom's polycentric approach holds great promise for analyses of their institutional resilience.

Institutional resilience of energy systems and their capacity to adapt to changing conditions are crucial factors in tackling present and future energy and climate change challenges. However, different obstacles undermine this resilience capacity. The present chapter focused on three of these hindrances: the collective action problem in the diffusion of more sustainable energy technologies and practices, the lack of trust in conventional energy actors and the existence of strong vested interests within the energy industry. It showed that community-based energy initiatives, as parts of larger polycentric systems, may greatly help overcome these barriers and, thereby, enhance the institutional resilience of energy systems. More specifically, it was argued that CBE initiatives and RE cooperatives present institutional features encouraging the activation of social norms and a high trust capital, therefore enabling them to offer effective solutions to avoid free riding and enhance trust in energy institutions and organizations. The creation of federated polycentric structures may also offer a partial response to the existence of vested interests in favor of the status quo.

While social and environmental contexts within which policy interventions take place become more and more complex and uncertain, a polycentric approach appears to offer a flexible and adaptive framework for the governance of the low-carbon transition. In this approach, a crucial role for policymakers is to create favorable conditions for the self-organization of local communities. They must also ensure the coordination of the whole system of initiatives and guarantee the enforcement of the overarching set of rules common to all by sanctioning defection from compliance. As Mansbridge (2014) noted, they should also provide a relatively neutral source of information, manage potential conflicts and facilitate negotiations between lower level governance units. All in all, the vision of the State's role in the polycentric governance perspective is that of a "supportive" State. It should cope with complex problems of modern life, such as energy-related challenges, by relying on the massive amount of social capital contained in local communities and let local self-organized initiatives flourish, thereby enforcing individuals' and communities' feeling of autonomy and self-determination.

References

Agrawal, A. 2001. Common property institutions and sustainable governance of resources. *World Development* 29 (10): 1649–1672.

Aligica, P.D., and V. Tarko. 2014. Institutional resilience and economic systems: Lessons from Elinor Ostrom's work. *Comparative Economic Studies* 56 (1): 52–76.

Allcott, H. 2011. Social norms and energy conservation. *Journal of Public Economics* 95 (9–10): 1082–1095.

Alston, L.J., G.D. Libecap, and B. Mueller. 1999. *Titles, Conflict, and Land Use: The Development of Property Rights and Land Reform on the Brazilian Amazon Frontier*. Ann Arbor: University of Michigan Press.

Anderies, J.M., M. Janssen, and E. Ostrom. 2004. A framework to analyze the robustness of social-ecological systems from an institutional perspective. *Ecology and Society* 9 (1): 18.

Andersson, K.P., and E. Ostrom. 2008. Analyzing decentralized resource regimes from a polycentric perspective. *Policy Sciences* 41 (1): 71–93.

Arthur, B. 1994. *Increasing Returns and Path Dependence in the Economy*. Ann Arbor: The University of Michigan Press.

Batley, S.L., D. Colbourne, P.D. Fleming, and P. Urwin. 2001. Citizen versus consumer: Challenges in the UK green power market. *Energy Policy* 29 (6): 479–487.

Bauwens, T. 2015. Propriété coopérative et acceptabilité sociale de l'éolien terrestre. *Reflets et Perspectives de la Vie Economique* LIV (1): 59–70.

Bauwens, T. 2016. Explaining the diversity of motivations behind community renewable energy. *Energy Policy* 93: 278–290.

Bauwens, T., and N. Eyre. 2017. Exploring the links between community-based governance and sustainable energy use: Quantitative evidence from Flanders. *Ecological Economics* 137: 163–172.

Bauwens, T., B. Gotchev, and L. Holstenkamp. 2016. What drives the development of community energy in Europe? The case of wind power cooperatives. *Energy Research & Social Science* 13: 136–147.

Birchall, J. 1997. *The International Cooperative Movement*. Manchester: Manchester University Press.

Black, J. 2008. Constructing and contesting legitimacy and accountability in polycentric regulatory regimes. *Regulation & Governance* 2 (2): 137–164.

Bollinger, B., and K. Gillingham. 2012. Peer effects in the diffusion of solar photovoltaic panels. *Marketing Science* 31 (6): 900–912.

Borzaga, C., and J. Defourny. 2001. *The Emergence of Social Enterprise*. London & New York: Routledge.

Bowles, S., and H. Gintis. 1998. The moral economy of communities: Structured populations and the evolution of pro-social norms. *Evolution and Human Behavior* 19 (1): 3–25.

Bowles, S., and H. Gintis. 2002. Social capital and community governance. *The Economic Journal* 112 (483): F419–F436.

Bulkeley, H., and M. Betsill. 2005. Rethinking sustainable cities: Multilevel governance and the 'urban' politics of climate change. *Environmental Politics* 14 (1): 42–63.

Bulkeley, H., and K. Kern. 2006. Local government and the governing of climate change in Germany and the UK. *Urban Studies* 43 (12): 2237–2259.

Byrnes, B., C. Jones, and S. Goodman. 1999. Contingent valuation and real economic commitments: Evidence from electric utility green pricing programmes. *Journal of Environmental Planning and Management* 42 (2): 149–166.

Carlson, J.M., and J. Doyle. 2002. Complexity and robustness. *Proceedings of the National Academy of Sciences* 9 (1): 2499–2545.

Carpenter, S., B. Walker, J.M. Anderies, and N. Abel. 2001. From metaphor to measurement: Resilience of what to what? *Ecosystems* 4 (8): 765–781.

Clement, F. 2010. Analysing decentralised natural resource governance: Proposition for a "politicised" institutional analysis and development framework. *Policy Sciences* 43 (2): 129–156.

Cvetkovich, G. 1999. The attribution of social trust. In *Social Trust and the Management of Risk*, ed. G. Cvetkovich, and R.E. Lofstedt, 53–61. London: Earthscan.

David, P.A. 1985. Clio and the economics of QWERTY. *American Economic Review* 75 (2): 332.

Devine-Wright, P. 2007. Energy citizenship: Psychological aspects of evolution in sustainable energy technologies. In *Governing Technology for Sustainability*, ed. J. Murphy, 63–86. London: Earthscan.

DTI/Ofgem. 2006. Distributed energy: A call for evidence for the review of barriers and incentives to distributed electricity generation, including combined heat and power. URN 06/2043. Department of Trade and Industry.

Ek, K., and P. Söderholm. 2008. Norms and economic motivation in the Swedish green electricity market. *Ecological Economics* 68 (1–2): 169–182.

Ekins, P., N. Strachan, I. Keppo, W. Usher, J. Skea, and G. Anandarajah. 2013. *The UK Energy System in 2050: Comparing Low-Carbon, Resilient Scenarios*. London: UKERC.

Eltham, D.C., G.P. Harrison, and S.J. Allen. 2008. Change in public attitudes towards a Cornish wind farm: Implications for planning. *Energy Policy* 36 (1): 23–33.

Eyre, N. 2013. Decentralization of governance in the low-carbon transition. In *Handbook on Energy and Climate Change*, ed. R. Fouquet, 581–597. Cheltenham, UK: Edward Elgar Publishing.

Fici, A. 2013. Cooperative identity and the law. *European Business Law Review* 24 (1): 37–64.
Fouquet, R. 2010. The slow search for solutions: Lessons from historical energy transitions by sector and service. *Energy Policy* 38 (11): 6586–6596.
Gadenne, D., B. Sharma, D. Kerr, and T. Smith. 2011. The influence of consumers' environmental beliefs and attitudes on energy saving behaviours. *Energy Policy* 39 (12): 7684–7694.
Geels, F., and J.J. Deuten. 2006. Local and global dynamics in technological development: A socio-cognitive perspective on knowledge flows and lessons from reinforced concrete. *Science and Public Policy* 33 (4): 265–275.
Gillingham, K., and K. Palmer. 2014. Bridging the energy efficiency gap: Policy insights from economic theory and empirical evidence. *Review of Environmental Economics and Policy* 8 (1): 18–38.
Goldthau, A. 2014. Rethinking the governance of energy infrastructure: Scale, decentralization and polycentrism. *Energy Research & Social Science* 1: 134–140.
Greenberg, M.R. 2014. Energy policy and research: The underappreciation of trust. *Energy Research & Social Science* 1: 152–160.
Greenberg, M.R., F.J. Popper, and H.B. Truelove. 2012. Are LULUs still enduringly objectionable? *Journal of Environmental Planning and Management* 55 (6): 713–731.
Groot, K. 2014. The impact of distributed generation on European power utilities. In *Distributed Generation and Its Implication for the Utility Industry*, ed. F. P. Sioshansi. Cambridge: Academic Press.
Gunderson, L.H. 2000. Ecological resilience—In theory and application. *Annual Review of Ecology and Systematics* 31 (1): 425–439.
Hansmann, H. 1996. *The Ownership of Enterprise.* Cambridge: The Belknap Press of Harvard University Press.
Hardin, G. 1968. The tragedy of the commons. *Science* 162 (3859): 1243–1248.
Hargreaves, T., S. Hielscher, G. Seyfang, and A. Smith. 2013. Grassroots innovations in community energy: The role of intermediaries in niche development. *Global Environmental Change* 23 (5): 868–880.
Heiskanen, E., M. Johnson, S. Robinson, E. Vadovics, and M. Saastamoinen. 2010. Low-carbon communities as a context for individual behavioural change. *Energy Policy* 38 (12): 7586–7595.
Hodbod, J., and W.N. Adger. 2014. Integrating social-ecological dynamics and resilience into energy systems research. *Energy Research & Social Science* 1: 226–231.
Holling, C., D. Schindler, B. Walker, and J. Roughgarden. 1995. Biodiversity in the functioning of ecosystems: An ecological synthesis. In *Biodiversity Loss: Economic and Ecological Issues*, ed. C. Perring, K.-G. Mäle, C. Folke, C. Holling, and B.-O. Jansson, 44–83. Cambridge, UK: Cambridge University Press.
Huijts, N.M.A., E.J.E. Molin, and L. Steg. 2012. Psychological factors influencing sustainable energy technology acceptance: A review-based comprehensive framework. *Renewable and Sustainable Energy Reviews* 16 (1): 525–531.
Huybrechts, B., and S. Mertens. 2014. The relevance of the cooperative model in the field of renewable energy. *Annals of Public and Cooperative Economics* 85 (2): 193–212.
International Cooperative Alliance (I.C.A.). 1995. Statement on the cooperative identity. *Review of International Cooperation* 88 (3): 3–4.
International Energy Agency (I.E.A.). 2015. *Key World Energy Statistics 2015*. Paris: International Energy Agency.
Irwin, A. 1995. *Citizen Science: A Study of People, Expertise, and Sustainable Development.* London: Routledge.
Janssen, M.A., J.M. Anderies, and E. Ostrom. 2007. Robustness of social-ecological systems to spatial and temporal variability. *Society & Natural Resources* 20 (4): 307–322.
Jobert, A., P. Laborgne, and S. Mimler. 2007. Local acceptance of wind energy: Factors of success identified in French and German case studies. *Energy Policy* 35 (5): 2751–2760.

Kasperson, R.E., D. Golding, and J.K. Kasperson. 1999. Risk trust and democratic theory. In *Social Trust and the Management of Risk*, ed. G. Cvetkovich, and R.E. Lofstedt. London: Earthscan.

Koestler, A. 1973. The tree and the candle. In *Unity Through Diversity, Part 1*, ed. W. Gray, and N.D. Rizzo, 287–314. New York: Gordon and Breach Science Publishers.

Koster, A., and J. Anderies. 2013. Institutional factors that determine energy transitions: A comparative case study approach. In *Renewable Energy Governance*, ed. E. Michalena, and J. M. Hills, 33–61. London: Springer.

Künneke, R.W. 2008. Institutional reform and technological practice: The case of electricity. *Industrial and Corporate Change* 17 (2): 233–265.

Kuzemko, C., M. Lockwood, C. Mitchell, and R. Hoggett. 2016. Governing for sustainable energy system change: Politics, contexts and contingency. *Energy Research & Social Science* 12: 96–105.

Laskey, A., and B. Syler. 2013. The ultimate challenge: Getting consumers engaged in energy efficiency. In *Energy Efficiency: Towards the End of Demand Growth*, ed. F.P. Sioshansi. Oxford: Academic Press.

Longo, A., A. Markandya, and M. Petrucci. 2008. The internalization of externalities in the production of electricity: Willingness to pay for the attributes of a policy for renewable energy. *Ecological Economics* 67 (1): 140–152.

Löschel, A., J. Johnston, M.A. Delucchi, T.N. Demayo, D.L. Gautier, D.L. Greene, J. Ogden, S. Rayner, and E. Worrell, 2009. 'Stocks, flows, and prospects of energy.' In *Linkages of sustainability*, eds. Graedel, T.E., and E. van der Voet. Cambridge, MA: MIT Press.

Mansbridge, J. 2014. The role of the state in governing the commons. *Environmental Science & Policy* 36: 8–10.

Marshall, G., D. Connell, and B.M. Taylor. 2013. Australia's Murray-Darling Basin: A century of polycentric experiments in cross-border integration of water resources management. *International Journal of Water Governance* 1: 231–251.

Maruyama, Y., M. Nishikido, and T. Iida. 2007. The rise of community wind power in Japan: Enhanced acceptance through social innovation. *Energy Policy* 35 (5): 2761–2769.

McGinnis, M.D. 2011. An introduction to IAD and the language of the Ostrom workshop: A simple guide to a complex framework. *Policy Studies Journal* 39 (1): 169–183.

McGinnis, M.D., and E. Ostrom. 2014. Social-ecological system framework: Initial changes and continuing challenges. *Ecology and Society* 19: 2.

McShea, D.W. 2001. The hierarchical structure of organisms: A scale and documentation of a trend in the maximum. *Paleobiology* 27 (2): 405–423.

Meinzen-Dick, R. 2007. Beyond panaceas in water institutions. *Proceedings of the National Academy of Sciences* 104 (39): 15200–15205.

Middlemiss, L. 2008. Influencing individual sustainability: A review of the evidence on the role of community-based organisations. *International Journal of Environment and Sustainable Development* 7 (1): 78–93.

Middlemiss, L. 2011. The effects of community-based action for sustainability on participants' lifestyles. *Local Environment* 16 (3): 265–280.

Mitchell, M. 2009. *Complexity: A Guided Tour*. Oxford: Oxford University Press.

Moe, E. 2010. Energy, industry and politics: Energy, vested interests, and long-term economic growth and development. *Energy* 35 (4): 1730–1740.

Moe, E. 2015. *Renewable Energy Transformation or Fossil Fuel Backlash: Vested Interests in the Political Economy*. London: Palgrave Macmillan.

Morçöl, G. 2014. Self-organization in collective action: Elinor Ostrom's contributions and complexity theory. *Complexity, Governance & Networks* 1 (2): 9–22.

Mumford, J., and D. Gray. 2010. Consumer engagement in alternative energy—Can the regulators and suppliers be trusted? *Energy Policy* 38 (6): 2664–2671.

Nagendra, H., and E. Ostrom. 2012. Polycentric governance of multifunctional forested landscapes. *International Journal of the Commons* 6 (2): 104–133.

Nolan, J.M., P.W. Schultz, R.B. Cialdini, N.J. Goldstein, and V. Griskevicius. 2008. Normative social influence is underdetected. *Personality and Social Psychology Bulletin* 34 (7): 913–923.

Nolden, C. 2013. Governing community energy—Feed-in tariffs and the development of community wind energy schemes in the United Kingdom and Germany. *Energy Policy* 63: 543–552.

Noll, D., C. Dawes, and V. Rai. 2014. Solar Community Organizations and active peer effects in the adoption of residential PV. *Energy Policy* 67: 330–343.

Nomura, N., and M. Akai. 2004. Willingness to pay for green electricity in Japan as estimated through contingent valuation method. *Applied Energy* 78 (4): 453–463.

Nordhaus, W.D. 1994. *Managing the Global Commons. The Economics of Climate Change*. Cambridge, MA: M.I.T. Press.

Ohler, A.M., and S.M. Billger. 2014. Does environmental concern change the tragedy of the commons? Factors affecting energy saving behaviors and electricity usage. *Ecological Economics* 107: 1–12.

Ole Borgen, S. 2001. Identification as a trust-generating mechanism in cooperatives. *Annals of Public and Cooperative Economics* 72 (2): 209–228.

Olson, M. 1965. *The Logic of Collective Action: Public Goods and the Theory of Groups*. Cambridge: Harvard University Press.

Olson, M. 1982. *The Rise and Decline of Nations*. New Haven and London: Yale University Press.

Ostrom, E. 1990. *Governing the Commons: The Evolution of Institutions for Collective Action*. Cambridge: Cambridge University Press.

Ostrom, E. 1999. Coping with tragedies of the commons. *Annual Review of Political Science* 2 (1): 493–535.

Ostrom, E. 2000. Crowding out citizenship. *Scandinavian Political Studies* 23 (1): 3–16.

Ostrom, E. 2005. *Understanding Institutional Diversity*. Princeton: Princeton University Press.

Ostrom, E. 2009. A general framework for analyzing sustainability of social-ecological systems. *Science* 325 (5939): 419–422.

Ostrom, E. 2010. Polycentric systems for coping with collective action and global environmental change. *Global Environmental Change* 20 (4): 550–557.

Ostrom, E. 2012. Nested externalities and polycentric institutions: Must we wait for global solutions to climate change before taking actions at other scales? *Economic Theory* 49 (2): 353–369.

Ostrom, V., C.M. Tiebout, and R. Warren. 1961. The organization of government in metropolitan areas: A theoretical inquiry. *The American Political Science Review* 55 (4): 831–842.

Pahl-Wostl, C., and C. Knieper. 2014. The capacity of water governance to deal with the climate change adaptation challenge: Using fuzzy set qualitative comparative analysis to distinguish between polycentric, fragmented and centralized regimes. *Global Environmental Change* 29: 139–154.

Platteau, J.-P. 2004. Monitoring elite capture in community-driven development. *Development and Change* 35 (2): 223–246.

Platteau, J.-P., and F. Gaspart. 2003. The risk of resource misappropriation in community-driven development. *World Development* 31 (10): 1687–1703.

Powers, K.J., and F. Thompson. 1994. Managing coprovision: Using expectancy theory to overcome the free-rider problem. *Journal of Public Administration Research and Theory* 4 (2): 179.

Samuelson, P.A. 1954. The pure theory of public expenditure. *The Review of Economics and Statistics* 36 (4): 387–389.

Sandler, T. 2004. *Global Collective Action*. Cambridge: Cambridge University Press.

Schultz, P.W., J.M. Nolan, R.B. Cialdini, N.J. Goldstein, and V. Griskevicius. 2007. The constructive, destructive, and reconstructive power of social norms. *Psychological Science* 18 (5): 429–434.

Seyfang, G. 2010. Community action for sustainable housing: Building a low-carbon future. *Energy Policy* 38 (12): 7624–7633.

Seyfang, G., J.J. Park, and A. Smith. 2013. A thousand flowers blooming? An examination of community energy in the UK. *Energy Policy* 61: 977–989.
Shepsle, K.A. 1989. Studying institutions: Lessons from the rational choice approach. *Journal of Theoretical Politics* 1 (2): 131–147.
Simon, H.A. 1962. The architecture of complexity. *Proceedings of the American Philosophical Society* 106 (6): 467–482.
Slovic, P. 1993. Perceived risk, trust, and democracy. *Risk Analysis* 13 (6): 675–682.
Smith, A. 2012. Civil society in sustainable energy transitions. In *Governing the Energy Transition: Reality, Illusion, or Necessity*, ed. G. Verbong, and D. Loorbach. New York: Routledge.
Sovacool, B.K. 2011. An international comparison of four polycentric approaches to climate and energy governance. *Energy Policy* 39 (6): 3832–3844.
Sovacool, B.K. 2012. Energy security: Challenges and needs. *Wiley Interdisciplinary Reviews: Energy and Environment* 1 (1): 51–59.
Sovacool, B.K., and M.H. Dworkin. 2014. *Global Energy Justice: Problems, Principles, and Practices*. Cambridge: Cambridge University Press.
Sovacool, B.K., R.V. Sidortsov, and B.R. Jones. 2014. *Energy Security, Equality, and Justice*. Abingdon & New York: Routledge.
Spear, R. 2000. The co-operative advantage. *Annals of Public and Cooperative Economics* 71 (4): 507–523.
Spear, R. 2004. From co-operative to social enterprise: Trends in European experience. In *Trends and Challenges for Co-operatives and Social Enterprises in Developed and Transition Countries*, ed. C. Borzaga, and R. Spear, 31. Trento: Edizione.
Stern, N. 2007. *The Economics of Climate Change: The Stern Review*. Cambridge and New York: Cambridge University Press.
Stern, P.C. 2014. Individual and household interactions with energy systems: Toward integrated understanding. *Energy Research & Social Science* 1: 41–48.
Stern, P.C., E. Aronson, J.M. Darley, D.H. Hill, E. Hirst, W. Kempton, and T.J. Wilbanks. 1986. The effectiveness of incentives for residential energy conservation. *Evaluation Review* 10 (2): 147–176.
Tarko, V. 2015. Polycentric structure and informal norms: Competition and coordination within the scientific community. *Innovation: The European Journal of Social Science Research* 28 (1): 63–80.
Toke, D., S. Breukers, and M. Wolsink. 2008. Wind power deployment outcomes: How can we account for the differences? *Renewable and Sustainable Energy Reviews* 12 (4): 1129–1147.
Unruh, G.C. 2000. Understanding carbon lock-in. *Energy Policy* 28 (12): 817–830.
von Bertalanffy, L. 1969. *General System Theory: Foundations, Development, Applications (Revised Edition)*. New York: George Braziller Inc.
Walker, G., and P. Devine-Wright. 2008. Community renewable energy: What should it mean? *Energy Policy* 36 (2): 497–500.
Walker, B., C.S. Holling, S.R. Carpenter, and A. Kinzig. 2004. Resilience, adaptability and transformability in social–ecological systems. *Ecology and Society* 9 (2): 5–14.
Walker, G., P. Devine-Wright, S. Hunter, H. High, and B. Evans. 2010. Trust and community: Exploring the meanings, contexts and dynamics of community renewable energy. *Energy Policy* 38 (6): 2655–2663.
Wiener, J.B. 2007. Think globally, act globally: The limits of local climate policies. *University of Pennsylvania Law Review* 155 (6): 1961–1979.
Wilson, D.S., E. Ostrom, and M.E. Cox. 2013. Generalizing the core design principles for the efficacy of groups. *Journal of Economic Behavior & Organization* 90 (Supplement): S21–S32.
Wiser, R.H. 1998. Green power marketing: Increasing customer demand for renewable energy. *Utilities Policy* 7 (2): 107–119.
Wynne, B. 1996. May the sheep safely graze? A reflexive view of the expert-lay knowledge divide. In *Risk, Environment and Modernity: Towards a New Ecology*, ed. S. Lash, B. Szerszynski, and B. Wynne, 44–84. London: SAGE Publications Ltd.

Yandle, T., and Dewees, C. 2003. Privatizing the commons…twelve years later: Fishers' experiences with New Zealand's market-based fisheries management. In *The Commons in the New Millennium: Challenges and Adaptations*, ed. N. Dolsak, and E. Ostrom, 101–27. Cambridge: MIT Press.

Young, H.P. 2008. 'Social norms'. In *The New Palgrave Dictionary of Economics*, eds. Durlauf, S.N., and L.E. Blume. Basingstoke: Palgrave Macmillan.

Chapter 7
Energy Conservation Policies in the Light of the Energetics of Evolution

Franco Ruzzenenti and Paolo Bertoldi

Abstract With more energy efficiency it is possible to do the same—or even more—with less energy. This is why energy efficiency is prompted by many as an absolute remedy for the evils of energy use, such as the environmental pressure or the security of supply. Nevertheless, historically energy consumptions at the world level have always been growing in spite of—or perhaps because of—an increasing level of energy efficiency. Some scholars have called this paradox the rebound effect. The rebound effect (REE) is an unintended consequence of the introduction of more energy-efficient technology. It occurs when the reduction in energy consumption is less than that expected from the magnitude of the increase in energy efficiency. REE and backfire are caused by behavioural and/or other systemic responses to efficiency gains in production or consumption (Maxwell et al. in Addressing the rebound effect, a report for the European Commission DG Environment, 2011). However, this paradoxical nexus between energy efficiency and energy consumption is not only confined to human-made systems: nature exhibits a same type of linkage among energy efficiency, energy growth and complexity. To what extent can the energetics of evolution help us in understanding this conundrum and forge a doable energy policy aimed at reducing energy use by fostering energy efficiency? In this chapter we will analyse current areas of improvement in energy policy targeting energy efficiency in the light of the rebound effect and we will try to advance a different policy framework, based on a deeper understanding of this phenomenon.

F. Ruzzenenti (✉)
Department of Management and Quantitative Sciences,
Parthenope University of Naples, Naples, Italy
e-mail: ruzzenenti@gmail.com

F. Ruzzenenti
Institute of Sociology, Jagiellonian University, Krakow, Poland

P. Bertoldi
European Commission, Directorate-General Joint Research Centre,
Unit C.02 Energy Efficiency and Renewables,
Via E. Fermi 2749, 21027 Ispra, VA, Italy

N. Labanca (ed.), *Complex Systems and Social Practices in Energy Transitions*,
Green Energy and Technology, DOI 10.1007/978-3-319-33753-1_7

Introduction

There is clear trend in Nature that goes from less complex to more complex structures, less efficient to more efficient and, at the same time, from less energy dense to more energy dense.[1] If we apply this lesson from Nature to society and economy, the first implication is that efficiency is certainly useful to reduce the energy input of single activities but may not be sufficient to achieve the goal of decoupling economic growth and energy consumption. If we make an analogy between ecology and economy and consider that a more complex artefact or system acquires a competitive advantage, we unavoidably come to the conclusion that the GDP growth is a trace of complexity growth. Or, at least, inverting the terms of the equation: if complexity grows, GDP grows. If this analogy holds, evolution tells us that it is impossible, in the long run, to grow in complexity (and GDP) without growing in energy consumption. The implications for energy conservation policies are manifold and tremendous. If efficiency alone does not contribute to decarbonize the economy, it is recommendable to change our strategy. More than ever, the challenge of mitigating climate change demands that we revisit the use of energy and the role of energy efficiency in our economies. Drastic changes in consumption patterns will be necessary to achieve long-term CO_2 emission reductions and stabilize atmospheric concentrations. Together with an energy efficiency improvement, the objective now is an absolute reduction in energy demand. The strategy to adopt is to privilege innovation, new technology, new services and new ways of doing business and, make full use of the price signal through energy or carbon taxation. We need an "energy conservation revolution" to respond to the important challenges faced by our societies. However, only modest steps have been done in this respect. We need to understand why, if we are to do better in the future. If developed nations do not do it, how can we even think that developing nations will not duplicate the mistakes that we made in our past and that still constitute a burden for our economies? The first part of this chapter will briefly survey the European environmental and energy policy aimed at reducing carbon emissions based on energy efficiency (the "third pillar") and will then quickly introduce the rebound effect and its implications on such a policy. We will then review some of missed opportunities and wrong policies approaches to energy efficiency and then we will address current myths about energy efficiency and the rebound effect. In the last sections the conundrum posed by the rebound effect in the light of the energetics of natural evolution will be analysed and it will be showed that energy efficiency and energy growth are wedded together even in the domain of nature. Finally, some conclusions will be drawn on the nexus between energy efficiency and complexity with the aim of providing policy indications suitable for integrating energy efficiency and energy consumption reduction strategies.

[1]Energy density is generally defined as the amount of energy flowing per unit of time and unit of volume. For further information, see for example (Chaisson 2002).

The Third Pillar: Energy Efficiency in the European Environmental and Energy Policy

The International Protocol to mitigate greenhouse gas emission agreed in Kyoto in 1997 has ended its first commitment period and yet China has never signed the agreement, USA never ratified it and Canada withdrew from it.[2] The fate of Kyoto mainly relies on EU and Japan, the last big emitters taking part to it. On 22 April 2016, 174 countries signed the Paris Agreement, a global agreement on climate change, and began adopting it within their own legal systems. This Agreement sets the goal of limiting global warming to less than 2 °C compared to pre-industrial levels with the aim of reaching 1.5 °C, but it does not impose emission targets for countries. The EU has been at the forefront of international efforts towards a global climate deal. Following limited participation in the Kyoto Protocol and the lack of agreement in Copenhagen in 2009, the EU has been building a broad coalition of developed and developing countries in favour of high ambition that shaped the successful outcome of the Paris conference. The EU was the first major economy to submit its intended contribution to the new agreement in March 2015. It is already taking steps to implement its target to reduce emissions by at least 40% by 2030. One of the pillars of the EU strategy to reduce carbon emissions, together with renewable energy sources (RES), is energy efficiency. The European policy-makers introduced specific targets for the year 2020 in 2007. In the energy sector the 2020 targets were based on the three pillars leading European energy policy: Security of supply, competitive markets and sustainability. The 2020 targets call for a binding 20% reduction in CO_2 emissions compared to 1990 levels (now raised to 40% by 2030), for a binding 20% target concerning the amount of consumed energy coming from renewables and for a not binding 20% increase in energy efficiency, formulated as a maximum primary and final energy consumption to not be exceeded in 2020.

Energy efficiency describes how much useful work, activity or service can be generated for each unit of energy consumed. From this simple definition, two important observations can be made about the nature of energy efficiency. First, what is 'useful' output is sometimes inherently subjective. What is judged useful by one person may be judged wasteful by another. If personal utility is subjective, then it is not possible (on a neo-classical understanding of market-based consumer behaviour) to sanction high, wasteful or 'conspicuous' energy consumption. If the consumer is willing to pay, then the consumption is assumed to be justified. Second, improving energy efficiency does not necessarily mean using less energy. Energy efficiency creates a range of direct benefits, or impacts, which range from less energy use to deliver the same service (energy savings), to same energy use to deliver more output (energy productivity). Indeed, with rebound effects (see below) it is possible that

[2]On September 3rd 2016, at the G20 summit in Hangzhou, China and USA agree to ratify Paris climate deal and delivered a joint declaration to put the pact of Paris into force before the end of the year. This is a major step to change the secular position of these countries on climate change and a promising breakthrough.

energy efficiency may trigger more energy use over time, due to a combination of direct and indirect effects, as the energy productivity effect of energy efficiency stimulates additional growth and energy consumption. This leads to a clear economic benefit, but also to a clear increase in greenhouse gas emissions policy.

By using less energy to create the useful services that people demand, less greenhouse gas emissions and other pollutants are emitted, less primary fuels are demanded, less new energy supply infrastructure is required, and energy costs are reduced. Compared with other solutions, energy efficiency and energy conservation is generally less expensive (often profitable after a short payback period), and more readily available. While energy efficiency is not the only solution to the challenges of an unsustainable energy system—and must be complemented by other policies such as those that encourage the use of renewable energy sources, fuel-switching toward low carbon fuel, technology development including carbon capture and sequestration, low energy/carbon spatial planning—efficiency and energy conservation should be the first priority in moving towards a sustainable energy system. Despite the benefits of energy efficiency itself, which are generally understood and not challenged by policy-makers, energy efficiency policy is generally also sometimes weakly supported by all stakeholders, from policy-makers, to the end-users. The main reason for this appears to be the pervasive but incorrect view that the market will deliver whatever level of energy efficiency is justified.[3] To this must be added the wider reluctance of governments to intervene in market processes, misconceptions about the nature of energy efficiency policy, the diversity of end-use products and markets, and the political economy. Despite this view, it is clear that the market systematically undersupplies energy efficiency relative to that which is economically optimal, and undersupplies it to an even greater degree relatively to that which is required for a sustainable energy system. Persuading policy-makers to exercise their powers to make stronger and more effective efficiency and conservation policy is a crucial necessity.

Short Discussion on the Rebound Effect

Economic analysis suggests four categories of possible rebound effects in response to the implementation of an improvement in energy efficiency.

(1) Direct rebound effect: For the buyer of a more energy-efficient technology, the effective price of the energy service produced with it is now lower and this encourages increased consumption of the service. The likelihood that this effect occurs and is substantial varies with the type of energy service involved. For household purchases of various energy technologies, large direct rebound

[3]The difference between the actual and the optimal efficiency level is often referred to as the efficiency gap. The energy efficiency gap can depend from many factors, from market barriers, to social behaviors and justify the policy intervention due to the lost opportunity (IEA 2007).

effects are quite unlikely due to the satiation of demand. Once basic needs and comfort levels are satisfied in relation to such services as refrigeration, carpet cleaning and space heating, a reduction in their prices is unlikely to lead to more consumption of them. In other cases there is greater scope; for example, improvements in fuel technologies may play a role in decisions to buy larger and more powerful automobiles. In industry, substantial direct effects depend on the extent to which technologies allow fuel to be substituted for other inputs in production processes and on the effect of improved energy productivity on a nation's international competitive position—on the potential for reduced energy costs to allow firms to expand their markets without taking business away from other firms in the same country. For a given firm, the size of the productivity effect will depend on the proportion of its total production cost accounted for by energy and on the market price elasticities of the goods being produced.

(2) Income effects on other goods: A household undertaking an efficiency improvement will use less energy and this will free a portion of the income that was being spent on energy; some or all of this freed income will be used to buy other goods and services, the production of which will require energy. Similarly, firms will have a source of cash to use to expand their activities or distribute to employees and owners, who will spend some or all of it. However, the original reductions in household and business spending on energy also show up as a reduction in income received by the sellers of energy, meaning that some or all of shareholders, employees and input suppliers of energy companies will now have less income to spend. Thus, for the economy as a whole, one effect can offset the other. While this offset is not likely to be exact, the net effect of the redirection of income and spending flows can be either positive or negative and will in general be very small. Rebound effects of this sort are therefore likely to be negligible.

(3) Energy price feedbacks: The effects of improvements in energy efficiency can be spread throughout the economy through price effects. The most interesting question in this regard is what happens to the physical quantities of fuels saved as a result of the widespread use of a given improvement in energy efficiency. Fuel and electricity companies will find themselves with excess supplies, which they may try to market by lowering their prices. In the economist's idealized model of a competitive economy, prices would adjust until excess supplies are totally used up—the rebound effect would in that case be total.

(4) Long-run effects on productivity, consumer tastes and economic structure: in this category are the effects suggested by efficiency skeptics when they argue that a focus on changing technology in order to solve environmental problems affects how people live and what they buy. Lower energy consumption can also affect decisions made by entrepreneurs to introduce new products. Thus the long-term effect might be to increase purchases of energy-using goods and services and to be more dependent on them than before energy efficiency was improved. For instance, more fuel-efficient cars presumably make people more willing to live far from their place of work, which could mean that higher

energy efficiency would lead to more fuel use in the long term than would occur if people had less fuel-efficient cars and lived closer to their work.

Efforts have been made to estimate direct rebound effects for particular categories of energy services, though the kinds of data needed for thorough empirical studies are not readily available and estimates are therefore rough and vary within wide ranges. The important result of such studies is that estimated direct rebound effects tend to be small, though at levels significant enough to be taken seriously. For instance, a survey of studies of data from the United States (Greening et al. 2000 reports estimates for household rebound effects in space heating in the range of 10–30%, space cooling 0–50%, lighting 5–12%, household appliances zero and automotive transport 10–30%. In sum, direct rebound effects appear to be relatively small—a direct rebound effect of, say, 10–20% signals a direct reduction in energy consumption of 80–90%. However, the possibility that the total rebound effect is much larger depends on the feedbacks that occur through the policy-induced energy price reductions and changing consumer tastes referred to in categories 3 and 4 above, but we are not aware of any estimates of the magnitude of these effects. Nevertheless, as we note below, it is an historical fact that energy demand in IEA member countries has continued to grow since the 1970s despite oil-shock induced price rises and decades of energy efficiency policies and programmes.

Some Examples of Missed Opportunities, Wrong Policies and Rebound Effects

- We know how to build houses that consume much less energy than houses built 30 years ago. However, newer houses are generally larger that the houses they replace, again leading to higher levels of energy consumption overall.
- Modern cars are often more energy efficient than the older cars they replace. However, there are more cars on the road; we drive faster on the highways and experience more congestion in cities, travel more, and our cars are equipped with more and more energy consuming devices like air conditioners, on board computers, etc. How can we expect to see energy demand in road transport go down?
- Efficient lighting will increase the number of lamps in a household and the burning hours, for examples additional lamps used to light the garden or the building facade at night.
- Since 1995, Europe has a mandatory energy label on refrigeration appliances. The label displays a scale of 7 of energy efficiency categories, from A (most energy efficient) to G (less energy efficient). The energy efficiency rating takes into account the size of the different compartments, as well as their indoor temperatures, and benchmarks against the energy consumption of the appliances. Despite these efforts to calibrate and compare refrigeration appliances in a unique format, there is a bias. It is easier for a larger unit to obtain a better energy efficiency category. Larger units therefore appear to consumers to be more energy efficient, even when they consume more energy.

- It has been reported that some rebates provided by an electric utility company were given to purchasers of plasma screen TVs, because the standby power level of the appliance was supposed to be efficient. By doing so, the utility is encouraging the replacement of a regular 80 W TV set by a 300 W plasma screen TV set.

Facts and Myths on Energy Efficiency

In spite of the historical evidence of growing energy demand amid energy efficiency increase, the rebound effect (REE) is not widely and thoroughly accepted. Indeed, it has always provoked a heated dispute over its magnitude and relevance for the goals of energy conservation policy. The attitude towards the REE by supporters of energy conservation approaches exclusively based on efficiency swings from a mild play-down to a harsh dismissal. In our view, five major critiques to the REE can be identified. Let us now try to have a closer look at the foundations of these critiques, which can be grouped in the following five types:

1. The epistemological flogger
2. The methodological reductionist: *ceteris paribus*
3. The empirical reductionist: the good country or the good period
4. The counterfeiter: the right metrics
5. The Aristotelian: growth it is the *final cause*.

A remarkable example of the epistemological critique has been recently provided by Cullenward and Koomey, when, in their article, they expose a long and detailed critique to the model on direct REE in the US industrial sector presented by Saunders to conclude the consensus view on energy efficiency as an effective means for energy conservation policy should not be *altered* (Cullenward and Koomey 2016). It is interesting that Cullenward and Koomey deploy the same kind of argument to diminish the role of REE used by many *global warming negationists* when they claim that the greenhouse effect does not exists because the *model of IPCC is flawed by methodological imperfections*.[4] Indeed, it is impossible to prove formally the nexus between anthropogenic emission and climate change, as there is not a full-fledged, universally accepted model of the phenomenon—if any is possible at all. However, even assuming that the nexus has not been modelled, this does not prove that nexus does not exists. Notably, very few *epistemological floggers* apply they critical verve to the link between cancer and tobacco, for which a model does not exist either. Conversely, the *epistemological reductionist* does not deny the validity of the model and uses the results to claim that *the effect is negligible.* Although it is widely recognized that there are several kinds of REE and relative models which cannot be *simply added together to give a combined effect*, many still

[4]http://www.nature.com/news/2010/100202/full/463596a.html.

wave the results of single models, whose application is limited to one kind of effect or sector, to downplay the significance of REE (Gillingham et al. 2013). Interestingly, in this case, the methodological limitations are beneficial for the sake of the survival of energy efficiency. In order to assess the RE, it is customary to apply extensively and copiously the concept of the *ceteris paribus* and thereby, if the contingent measurement of REE is smaller than 100% (backfire) and often it is, the deduction is that energy efficiency succeed in reducing energy consumptions. Nevertheless, the REE is a complex phenomenon—or an emergent one, with the jargon of the Breakthrough institute, like cancer or climate change, and in emergent phenomena the whole is greater than the sum of its parts. A different form of *ceteris paribus* is that of projecting the past energy efficiency level into the present economy to state that hadn't been for energy efficiency improvements, we would be using much more energy. *Prime facie* this is a very convincing argument but at a closer look this is fallacious: the size of our cities, the gigantism of our firms or the miniaturization of our devices, the pace and the localization of our activity and production (for example, globalization and outsourcing) were all shaped by ever-increasing energy efficiency (see Chap. 5). It would be just inappropriate to mechanically apply the former efficiency level to the current structure of economy and society as if these were unrelated to it. How much energy would consume a smart phone running with the transistors of the 1980s? The answer is simple: it would not exist (or it would not fit in one hand). Energy efficiency had had a structural impact on the system, not just a functional one (Ruzzenenti et al. 2015a). The empirical reductionist is not interested on the methodological methods concerning the REE modelling and only focuses his attention on the *empirical evidence* suggesting that energy efficiency has led to declining energy consumptions. This evidence is generally obtained by restraining the scope on a particular area or period. In Fig. 7.1 trends in per capita energy use of major aggregates are shown.

During short periods, generally concurrently with recession, the growth tendency is reversed. However, the long trend, as depicted by the dotted lines, is unmistakeably positive. Yet someone may claim that the longer trend is too much biased toward the past and that after the 1980 in many developed countries economic growth and energy use decoupled.

This result is normally achieved by looking at the so called *energy intensity*, that is, the GDP output for unit of energy input. This latter metrics is often taken as a measure of energy efficiency, but it is not. Economists are not used to dimensional analysis and this is why many view the ratio between GDP and primary energy use as an *efficiency*.[5] However, the problem is not just methodological, it is also

[5]Despite many scholars still employ energy intensity (energy consumption over gross domestic product) as a measure of efficiency, in the European Union most of experts agree that E/GDP is a very rough indicator of efficiency and it is often used ODEX as a much more refined indicator. In addition decoupling is an indication that GDP and energy grow are not growing at the same rate, usually the grow rate of energy is smaller than GDP. But recently (since 2008) in the EU energy consumption is declining notwithstanding a GDP increase, this is a new trend never experienced before.

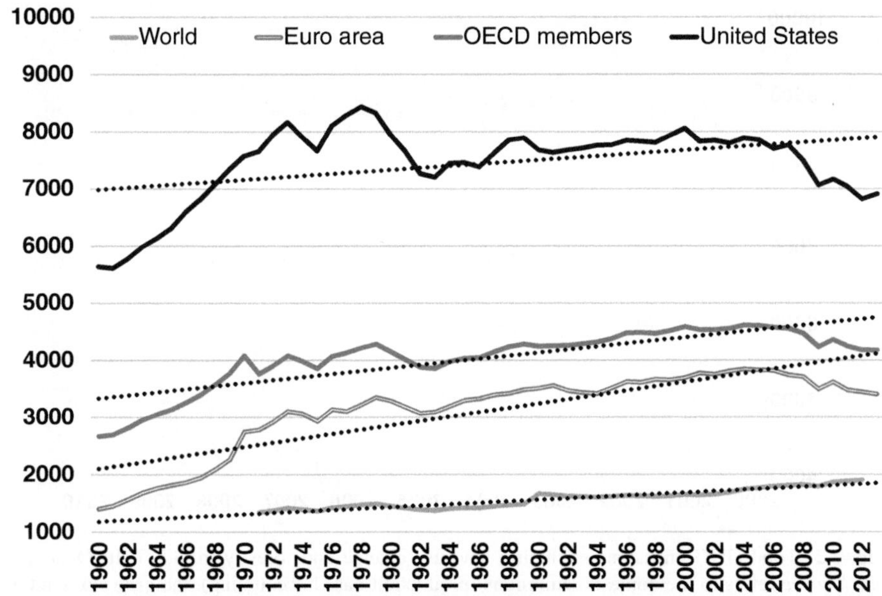

Fig. 7.1 Energy use (kg of oil equivalent per capita). *Source* World Bank

practical. For the sake of climate change mitigation what actually matters is the *global level of emissions,* rather than the per capita or the carbon content (and energy content) of every unit of GDP. It might be a solace the awareness that we are producing more value per unit of energy consumed but that does not mean that we are reversing the trend and decreasing the energy balance of our economy. It means indeed that in the advanced societies and developed economies the *leverage of energy* is growing. Indeed, a hair dryer absorbs 1 kW and a laptop just 100 W, but the amount of information transmitted by the latter device is incommensurably higher. What is the energy consumption of a giant company of the dot economy compared to a dinosaur of the metallurgy? In Chap. 4 we presented the energy budget of some of the largest corporations around the world to show that Arcero-Mittal, the leader in the World in steel production with a gross revenue comparable to Google (79 vs. 74 billions of dollars) consumes 100 times more energy, but can we live without steel?

Some reductionists, to conceal the fact that the energy demand of our societies is still growing despite the fact that we can produce more GDP for energy input, often look at the promising trend of carbon emissions in the last decades. One remarkable example is that of U.S., where the growth trend seems to have slowed down and even reverted (Fig. 7.2). Is this a signal that we are *decoupling* energy and economy as many believe? Feng et al. showed that this reduction of the carbon footprint of U.S. primary energy production was mainly obtained by switching from coal to natural gas and an unintended effect of recession (Feng et al. 2015). However, there

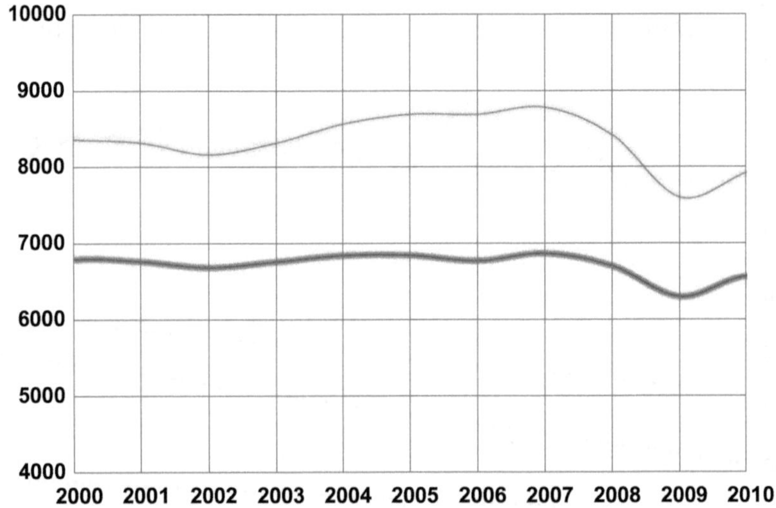

Fig. 7.2 Total GHG emissions excluding land-use change and forestry in U.S. ($MtCO_2e$). The *black curve* refers to the domestic emissions (production-based accounting), the *grey curve* refers to the domestic emissions minus the exported emissions, plus the imported emissions (consumption-based accounting). *Source* Elaboration of CAIT (WRI 2015); EORA (Lenzen et al. 2012, 2013)

is a more stringent and important issue we tend to overlook when analysing the energy use and carbon footprint of a single economy: globalization and international outsourcing. Most of OECD countries are net importer of both material and energy embodied in goods (Ruzzenenti et al. 2015b). The second curve of Fig. 7.2 shows the net carbon emissions (consumption based) of U.S economy compared to the domestic emissions. Figure 7.2 shows that: (1) almost a third of emissions are due to trade; (2) when the imports and the exports of CO_2 embodied in goods are taken into account the steady trend of the last decades vanishes and is replaced by a growing trend that stops only because of recession.

The problem is that if we widen the scope and look at the very long trend in carbon emissions, any hint of reverting trend occurred along time looks just as small fluctuations (Fig. 7.3).

Even for the European Union (Europe 28), when disentangling the former soviet economies, the trend of the core, developed countries (Europe 15) is but declining. And even if we would be able to significantly revert this trend, the road back to a pre-industrial level of emissions would be very long. However, many economists maintain that this is the effect of *growth* rather than efficiency. Albeit this might appear a circular explanation, if not tautological, the debate on the causes of growth is longstanding and problematic. Is growth an endogenous process of the economy, only determined by the availability of resources and the prolific fantasy of human beings (innovation), or exogenous, caused by the rate of savings and the rate of

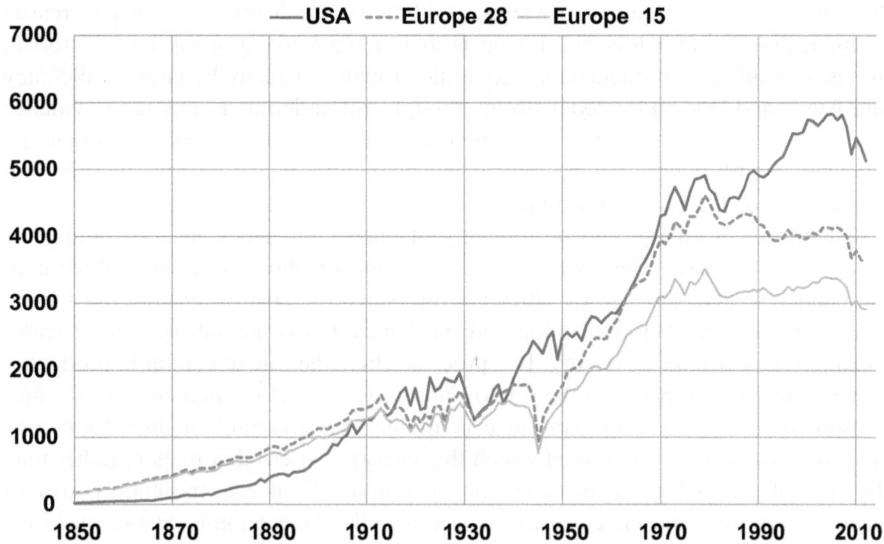

Fig. 7.3 Total CO_2 emissions excluding land-use change and forestry ($MtCO_2$). *Source* CAIT (WRI 2015)

technical change. The famous exogenous growth model of Robert Solow explains economic growth of U.S. economy by means of technical change, encoded in the so called Solow-residuals[6] (Solow 1956). Nevertheless, in the model, the residuals remain unexplained and technological change is thus *exogenously* introduced in the model in order to fit the curve. The most famous attempt to *endogenize* growth was made by Lucas, who actually translated the technical change into *human capital* and, as clearly shown by Solow, did not change the *exogenous* nature of the model[7] (Lucas 1988). It was Ayres and Warr who were able recently, at least until the 1970s, to explain economic growth of the U.S. economy back to the 1900, *endogenously* relying only on capital, labour, energy and efficiency (measured by exergy) and thereby eliminating the Solow-residuals (Ayres and Warr 2005). The notion that the model explains very well growth until the 1970s perhaps depends on the fact that both globalization entangled the production process internationally, thus making any analysis based on a single country flawed, and also that, as

[6]The "Solow-residuals" are the observed divergence from the predictions of the model, based only on capital and labour force, and the real GDP. These residuals, according to Solow, encapsulate the role of technological progress, which is an endogenous factor in his view (thus, "unexplained" by the endogenous variables of the model).

[7]Lucas and other scholars advocating endogenous growth tried to establish a relationship between technological change—the exogenous, augmenting factor of economic growth, which is expressed by a fitting parameter in the equation, by some kind of variables measuring the "human capital" of economy, like the number of new patents, educated people or skilled workers over unskilled workers, etc.

previously envisaged, the energy leverage in the post-industrial economy increased dramatically. Nevertheless, the lesson is that, when looking at the long historical trends, it is difficult to understand economic growth separately by energy efficiency and Ayres and Warr provided a strong, though still underrated, important evidence of this link. The reason—or at least one reason, why the role of energy and energy efficiency is still overlooked by most of the economics is that energy is a little part in the costs of the production function of firms. But this is a misconception of the actual impact of energy and prominently of energy efficiency in determining the *structure* of the costs. The price of energy only mirrors the cost of the production of the energy source, but energy efficiency measures the cost of *substituting labour force with capital.* Automatization and mechanization is the substitution of *endo-somatic* with *exosomatic* energy. In a process where the role of capital in producing value increases compared to the role of labour, it is the *embodied energy* (the past exosomatic energy used up to produce an artefact or a structure) and thus *the energy efficiency* (the conversion rate at which this energy is used) that matters rather than the cost of producing the *primary energy source.* What was the energy cost of slaves? Very little and the economic cost was null. The economic cost of unskilled workers was positive, at least, yet the energy cost was the same. The economic costs of trained and specialized labour were much higher than that of unskilled workers, although the energy costs were the same. This increasing cost was proportional to the costs of the facilities and artefacts enabling the productivity of their work that is the *embodied energy* in the factory. Landes noted that what made internal combustion more attractive than external combustion engines were the high costs of skilled and rare stokers rather than the cost of oil compared to coal (Landes 1969). Energy efficiency is the *invisible hand* that makes labour and capital more productive. Dahmus developed an interesting long term, historical analysis of 10 economic sectors in world economy to show that energy consumption has always increased together with energy efficiency and that this latter seldom, and only for limited, small intervals, has actually led to a reduction in energy use (Dahmus 2014). In Fig. 7.4 we show four major sectors and the positive correlation between energy efficiency and energy consumption is striking. In our opinion, this is just another, compelling evidence of the nexus between energy efficiency, energy use and economic growth. The rebound effect is intrinsic to economic growth and energy efficiency has led to a reduction of energy consumption only during limited periods or for specific sectors or economies. This is an *energy imperative,* though difficult to accept and complex to understand, that we must acknowledge when developing our energy policy.

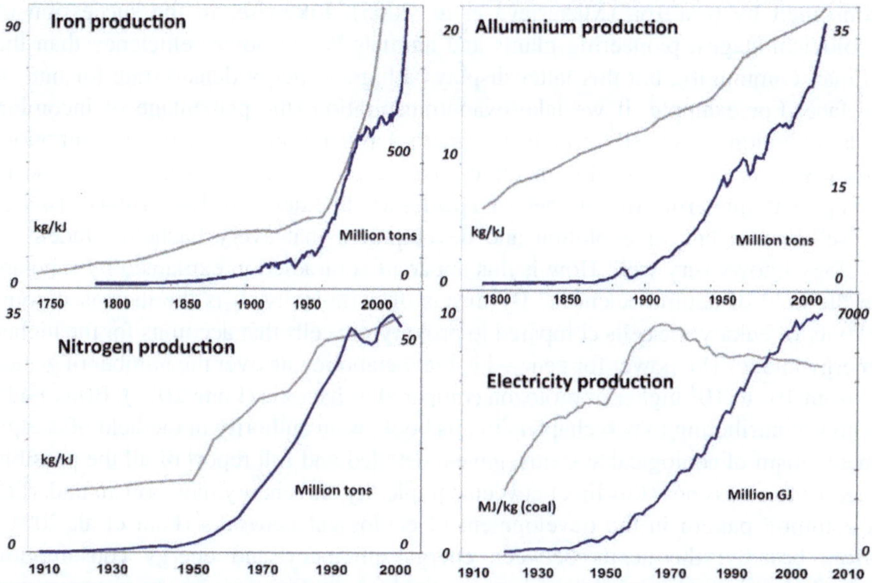

Fig. 7.4 Plots of energy efficiency (*grey*) versus energy consumptions (*blue*) in four sectors of the world economy. Worldwide pig iron production, measured as the mass of pig iron produced, and efficiency, measured as the mass of pig iron produced per unit of coke consumed in smelting. Worldwide aluminium production, measured as the mass of aluminium produced, and efficiency, measured as the mass of aluminium produced per unit of electricity consumed in the smelting process. Worldwide nitrogen fertilizer production, measured as the mass of nitrogen produced, and efficiency, measured as the mass of nitrogen produced per unit of energy consumed in the Haber-Bosch process. Amounts of electricity generated from coal and efficiency of the related production process. Modified from: Dahmus J 2014

The Energetics of Evolution: Energy Efficiency, Energy Growth and Complexity

If we were only able to look beyond our garden, the small world made of technology, economy and society as developed since the industrial revolution of the nineteenth century, and we would look at the longer term perspective of some billions of years offered by nature and evolution, we could probably learn a lot, even about the relationship between energy efficiency and energy consumption. Indeed, the biological record of the Earth presents us an unequivocal evidence of rebound effect lasting four billions of years. The aerobic respiration of eukaryotic cells is the most efficient form of energy conversion. It produces 686 kcal per mole of oxidant, to be compared with 643 kcal produced in case of nitrate reduction, 190 kcal produced in case of sulphate reduction and 8.3 kcal produced in case of methanogenesis during anaerobic respiration processes. Yet, eukaryote cells consume more energy than the less evolved prokaryotic cells. The mean endogenous (basal) mass-specific metabolic rate of prokaryotes is indeed 8 W/kg to be compared with 10–12 W/Kg consumed by protozoa (Makarieva et al. 2005). Likewise, in the succession of ecological stages, pioneering plants and animals have a lower efficiency than the climax community, but this latter displays a higher energy density rate for unit of surface. For example, if we take evapotranspiration (the percentage of incoming solar radiation dissipated through evaporation) as a measure of energy consumption, from Sahara to Amazon this increases from 2 to 70% along with an increase in complexity in terms of number of species (Schneider and Kay 1994). This is a well-known law of evolution and development that every bachelor student of biology knows very well. How is this apparent contradiction explained by scholars in the field of natural sciences? By means of complexity. It is the increased complexity of eukaryotic cells compared to prokaryotic cells that accounts for the higher energy intake. The power for gene—i.e. the metabolic rate over the number of genes, is from 10^3 to 10^4 higher in protozoa compared to bacteria (Lane 2011). Brian Fath, who is contributing to two chapters in this book, is an authority in the field of energy metabolism of ecological systems, gave a detailed and full report of all the possible metrics to assess how this link between complexity and energy metabolism underlies a common pattern in the development of ecological networks (Fath et al. 2001). More broadly, the nexus between energy efficiency and energy consumption (metabolism) pertains evolution, as suggested for the first time by the visionary and enlightening thought of Lotka (1956):

> **The Law of Evolution Adumbrated as a Law of Maximum Energy Flux**. This at least seems probable, that so long as there is abundant surplus of available energy running "to waste" over the sides of the mill wheel, so to speak, so long will a marked advantage be gained by any species that may develop talents to utilize this "lost portion of the stream". Such a species will therefore, other things equal, tend to grow in extent (numbers) and its growth will further increase the flux of energy through the system. It is to be observed that in this argument the principle of the survival of the fittest yields us information beyond that attainable by the reasoning of thermodynamics.

In this fundamental passage Lotka *adumbrates* what now for many scholars is his famous *Lotka's power principle of thermodynamics,* which can be summarized as follows: when resources are abundant, natural selection will award those species that are *faster* in utilizing the energy. In other words, evolution, under some conditions, will forward power maximization of living systems and this will divert more energy through the system. Nevertheless, Lotka envisages a second mechanism, which in his view is not in contradiction with the maximum power tendency of living systems, and it is that of efficiency maximization (economy in husbanding resources). This second mechanism will occur in natural selection when resources become *scarce*.

As to the other aspect of the matter, the problem of economy in husbanding resources will not rise to its full importance until the available resources are more completely tapped than they are today. Every indication is that man will learn to utilize some of the sunlight that now goes to waste. The general effect will be to increase the rate of energy flux through the system of organic nature, with a parallel increase in the total mass of the great world transformer, of its rate of circulation, or both.

It is worth noting that in the conceptual framework of Lotka, energy efficiency and power are two sides of the same coin, working synergistically to divert an ever-increasing flux of free energy into the living system. Natural selection is the process that assures that the energy flows increase: the most efficient and fastest in tapping the available free energy. The best organisms (i.e. the most efficient and powerful) are also the most complex, hence, the arrow of evolution is toward an increasing level of complexity. The attempt to pave the way for a thermodynamics of life by Lotka was seminal and forerunner. After him, several contributions followed, along his line of thought or seeking different approaches—like those of Prigogine, Odum, Schneider and Kay, Chaisson, all aiming at explaining evolution in the light of the nexus between energy growth and complexity (Odum and Pinkerton 1955; Nicolis and Prigogine 1977; Schenider and Kay 1994; Chaisson 2002). In Chap. 4 we have provided a description of the view of *finite-time thermodynamics* on the link between energy efficiency and power to show that real thermodynamic machines work at sub-optimal efficiency in order to maximize power, adumbrating a *law of power maximization* that bridges inanimate and animate power, *exosomatic* and *endosomantic* energy, living systems and economy. Is this the shade of a fundamental law of evolution still far from being understood or a law of thermodynamics altogether? Lotka himself was very cautious in viewing his theory as a *fifth law of thermodynamics,* suggesting the history of science is full of *dead-fifth-laws*. Science, indeed, from *vitalism* to ether, developed fundamental laws in the lack of clear explanations. The question whether the so called *maximum power law* (or the *minimum entropy production,* for Prigogine) is to be considered a law of thermodynamics or could be explained within a new and thoroughly different conceptual framework goes beyond the scope of the present analysis. Indeed, this is a salient property of biological evolution and economic development that hints to one of the most intriguing and still unexplained question for science: why is there a tendency in nature toward more energy density rate (more efficiency and more power) and higher complexity? Furthermore, by acknowledging this immanent trend shaping the flow of energy and information across nature and society, we cannot but think that energy growth is the inescapable fate of *energy efficiency*, in the context of growing complexity and available energy resources.

Concluding Remarks and Policy Indications

So far we have tried to explain why the rebound effect, besides being one of the fundamental open questions for science, represents an impending threat to any energy conservation or climate change mitigating policy aimed at reducing energy consumptions by promoting and pursuing energy efficiency. Of course, this is not to say that energy efficiency should not be pursued or, even worse, hindered, but that it should not be considered alone as a means or, even worse, a solution to achieve energy conservation. Being inefficient to pursue less energy consumption is just as absurd as impractical. Our energy policy must achieve both: energy efficiency and energy conservation, with the knowledge that the two are in conflicts, in the long run. We are hereby proposing two possible strategies to reconcile energy efficiency and energy conservation: (1) decoupling energy efficiency and pure squandering; (2) when the rebound effect acts, conjugate energy efficiency with counterbalancing measures. In other words, take serious measures to enhance energy efficiency and curb energy demand growth. These measures can be: fiscal, regulatory or social, that is, based on communication and awareness. The awareness on the rebound effect is still restrained to the scientific milieu (and even there, to a limited number of scholars) and developing a public conscience on this subject would probably help in implementing effectively energy policies based on efficiency.

Avenues for an Enhanced Energy-Efficient Future

The time has come to design energy policies as a contributor to absolute reduction in energy demand. For this, energy efficiency will have to become more than a minor element of a wider energy policy package. Very likely, energy efficiency and energy conservation should come in a global policy package that comprises all dimensions, such as the technology, the price signal, the behaviour, etc. Furthermore it must be fully integrated not only within energy policies in general, but more importantly, into policies at the international, national and sectoral levels, including in city planning, transport, housing, building, industry and wider fiscal policies. It is when a house is being designed and built or when a decision to link two cities with a road or with a railway is being taken, or when an appliance is being manufactured, that energy efficiency and energy conservation can best be delivered. There certainly exist many different ways to revisit energy efficiency. The following five points are proposed to structure the efforts to be made.[8] Each point corresponds to a given dimension of the renewed ambition for a more energy-efficient economy; that is, aiming for an absolute reduction in energy demand. They each represent a component of the policy package.

They are of course complementary and do overlap at some level. They are:

1. Enhanced knowledge
2. Information, education and motivation
3. Stimulate research and development
4. Set energy efficiency norms
5. Use price signals.

1. *Enhanced Knowledge*

Analysing where and why we use energy (what form, which quantity, etc.…) is a prerequisite to any sound programme. The two oil shocks in the 70s taught us how to collect information on oil production, and we do so in real time. Statistical

[8] On these points see also Lebot et al. (2004).

analyses on the supply side have become a routine everywhere. They are used to understand where the market is, where the prices for supplying electricity, oil, gas or coal go. Energy efficiency, by contrast, suffers from a lack of data that would enable both a global picture, as well as a detailed view at the level where policy-makers or market actors could make informed decision in order to maintain or choose an energy-efficient path. This dimension comprises efforts to be made on data collection on the end-use sector, develop energy efficiency indicators and understand the respective impact of human behaviour and technology in a given energy service. Although some important steps have been maide toward a development of an energy efficiency database, like in the case of the Odyssee-Enerdata project,[9] there is still much to do on a national level. One major hurdle in the research in the file of energy efficiency is that national statistics are inconsistent, if not lacking. Governments should therefore take responsibility for maintaining and enhancing research on that side of the economy of energy and provide statistics standardized to European criteria. Europe, on the other hand, should develop further a common platform for such a system of data collection.

2. *Information, Education, Motivation*

Information, education and motivation are often quoted as pillars of any energy efficiency and energy savings programmes. However the time has come to revisit them in the market environment that we described earlier, acknowledging for instance the excess of advertising of all sorts in our daily life, in order to identify how to build a proper communication campaign. As an illustration, a concrete and simple idea would be to oblige advertisers to display the level of energy efficiency performance of an appliance, a car, a building, when the product is being advertised. In Europe, appliances, cars and buildings are progressively being labelled under the same format (7 categories from A—more energy efficient to G—less energy efficient). The category could be displayed as mandatory information on the advertisement support. Some retailers already do so in their commercial brochures. The 20 years of anti-smoking campaigns in OECD countries can teach a lot to energy efficiency advocates as to how to transform bad habits and adopt more responsible ones. First, direct promotion of cigarettes and cigars have been banned from any advertising campaign, then messages such as "smoking kills" have been place on the packages. Many countries have adopted some format for labelling appliances and cars. An extension of that could be to oblige the manufacturers and the retailers to display similar information. To push the idea even further, we could envisage that the energy efficiency category, identified in Europe with a coloured arrow could be tattooed on the appliances or the cars for the second-hand market. A further improvement in the energy labelling system could be achieved by signalling embodied emissions, by means of LCA (Life Cycle Assessment). This could enhance the awareness of the consumer on the global energy budget linked to the

[9]http://www.enerdata.net/enerdatauk/solutions/data-management/odyssee.php.

production of a certain service or good rather than just its use, with a further positive feedback on the appraisal of its origin and value chain of concern.

3. *Stimulate Research and Development*

Many supply side options for producing energy have been heavily supported by public research funding and activities. More should be done to promote research and development activities aimed at improving the energy efficiency of end-use technologies. For instance, top of the Solid State Lighting (SSL) 1 achieves an energy efficiency of 100 lumens/W. It is recognized that in theory, SSL efficiency could reach twice that figure. Encouraging R&D activities to explore further how energy efficiency could be improved and to design a new generation of SSL above 150 lumens/W may have an overall important impact on our economies. In addition, and with much less public funding than the nuclear fusion research, multiple R&D programmes could encourage the design of new generation of energy-efficient end-use technologies in the field of combustion, enhanced heat exchange, enhanced electricity transformation (DC/DC, AC/AC, and AC/DC), reduce motor losses, enhanced motor drives, cooling compressors, lighting, computing, telecommunication as a complement to R&D efforts in renewable energy. As said earlier, there is a need to reinforce research activities on the socio-economic impact of past and present energy efficiency programmes including the consumer behaviour (social norms, individual and companies investment decisions, etc.) and the rebound effect. This is to better understand relations and elasticity between energy efficiency, energy price and energy consumption in order to introduce or redesign, for instance, sound financial incentive instruments such as a tax on energy to assure that energy conservation and related greenhouse gas reduction are achieved.

4. *Set Energy Efficiency Norms, Develop Energy Savings Standards and Codes*

Let us take the case of a house or an appliance. When being designed and built, the home builder or the appliance manufacturer has to respect safety norms. They do so by default. Safety norms have been designed sometimes long ago, often times through international standards. They have been set at levels that protect human life from accident, from casualty. The whole society accepts the costs of meeting the safety norms. In effect, they are insurances that we collectively pay to protect ourselves and future generations. Safety norms do save human life. Energy saving norms can be designed and implemented to alleviate planet earth's risk vis-à-vis climate change. Hence energy conservation norms should be generalized in all sectors of the economy. New buildings should be by default energy efficient, same as new cars or new end-use equipment. As discussed earlier, energy efficiency is not enough and energy savings must become the policy goal. This can be translated when setting regulation, codes, norms and standards. For instance, for a new refrigerator, a house, or a car—and on top of a mandatory energy label and a minimum energy efficiency requirement—policy-makers should also think about setting a maximum energy consumption target, regardless of the size of the product or the service that is provided. A new house could not consume, for instance, more than 10,000 kWh in primary energy per year (any excess to be met with on-site renewable energies), comprising all end-use; a car no more than 1,500,000 gCO_2 per year (if one drives more should buy a much more efficient car); a refrigerator no more than 100 kWh/year; etc. This would counteract the tendency of current energy efficiency regulations that make larger energy systems (appliances, houses) appear more energy efficient than smaller ones. For each end-use and each energy system, maximum consumption limits should be introduced. There is no reason for not implementing specific energy savings regulation for some existing energy consuming systems such as buildings. In Europe, the Energy Performance Building Directive (2010/31/EU) introduced the obligation for Member States to establish minimum energy performance requirements for new and existing buildings. Recent building codes in Denmark and France are below 50 kWh of primary energy/m^2. It is argued that it is the only path for France to bring the building sector close to the 2050 greenhouse gas official target. However, even these targets could still allow buildings to continue to consume more energy over time. In the long run, CO_2 maximum budget for each person/household/buildings could be introduced, leaving choice on how to meet it. It could be that people/building going beyond their allocated limit would have to pay to a fund that could be used to help the fuel-poor households to achieve low energy bills through energy efficiency measures or trading of allowances could be set up. There are numerous synergies between a renewed policy for setting energy efficiency and energy savings regulations and an enhanced scheme for energy labelling as described in previous sections. In Europe, the Directive 2010/31/EU also introduces the obligation for Member States to lay down the necessary measures to establish a system of certification of the energy performance of buildings. As most of the energy and climate change challenges that we are facing are global, energy saving norms (or standards or codes or regulations, whatever their nature) should be designed through international collaboration. To the least, international benchmarking of energy efficiency or energy savings norms can stimulate and influence the decision of analysts and policy-makers. Also, standards, codes, norms and energy savings regulations could first be implemented in government procurement—this would allow the market, in a second step, to prepare for the energy efficiency requirement on a wider scale.

5. *Use Price Signals*

There exists an extensive literature on the impact of price signals on energy consumption. Of course, much more should be done to reinforce the role and the impact of the consumer's reaction to the price signal. The price of energy should at least reflect the known environmental externalities. As the cost to access conventional energy is likely to growth in the decades to come, countries could introduce a progressive tax on non-renewable energy resources. For instance a 2% tax per year for the next 20 years could help our economy progressively accommodate for the foreseen increase of fossil fuel. The amount collected could easily be recycled by government back to the economy in investment in energy efficiency policies and clean energy technologies. Hence the introduction of such tax can be neutral to the global economy. The tax collected on fuel transport could be recycled for building and maintaining clean public transport system, tax collected on electricity could fuel demand-side management programmes and energy-efficient measures and technologies. Tax collected on stationary fossil fuel system could be invested in building renovations. Of course, since taxation affects the overall economy and can disturb market competition, it should best be applied in a coordinated way across all nations. International taxation of energy products could start with taxing kerosene for air travel. The more energy efficiency labelling is enforced on energy consuming systems and equipment, the easier it is to invent variable Value Added Taxes (VAT) according to the energy performance or to organize some rebates schemes: the less energy-efficient system are taxed heavier than the average ones and the money collected could alleviate the cost of the most energy-efficient system. Similarly, labels and norms facilitate the obligations that governments can impose on energy utility companies to deliver energy savings at their clients' level, as it is currently being implemented in Europe in the frame of an energy efficiency directive. There exist many other possibilities to reinforce the role of price signal in order to support overall energy savings strategies.

References

Ayres, R.U., and B. Warr. 2005. Accounting for growth: The role of physical work. *Structural Change.*

Chaisson, Eric J. 2002. Cosmic evolution: The rise of complexity in nature. Harvard University Press.

Cullenward, D., and J.G. Koomey. 2016. A critique of Saunders' 'historical evidence for energy efficiency rebound in 30 US sectors'. *Technological Forecasting and Social Change* 103: 203–213. doi:10.1016/j.techfore.2015.08.007 (ISSN 0040-1625).

Dahmus, J.B. 2014. Can efficiency improvements reduce resource consumption? *Journal of Industrial Ecology* 18 (6): 883–897.

Fath, B.D., B.C. Patten, and J.S. Choi. 2001. Complementarity of ecological goal functions. *Journal of Theoretical Biology* 208 (4): 493–506.

Feng, K., S.J. Davis, L. Sun, and K. Hubacek. 2015. Drivers of the US CO_2 emissions 1997–2013. *Nature Communications.*

Gillingham, K., M.J. Kotchen, D. Rapson, and G. Wagner. 2013. The rebound effect is overplayed. *Nature* 493: 475–476.

Greening, L., D. Green, et al. 2000. Energy efficiency and consumption—The rebound effect—A survey. *Energy Policy* 28: 389–401.

IEA. 2007. *Mind the gap. Quantifying principal-agent problems in energy efficiency.*

Lane, N. 2011. Energetics and genetics across the prokaryote-eukaryote divide. *Biology direct* 6: 35.

Landes, D.S. 1969. *The unbound prometheus: Technological change and industrial development in Western Europe from 1750 to the present.* Cambridge University Press.

Lebot, B., P. Bertoldi, and P. Harrington. 2004. Consumption versus efficiency: Have we designed the right policies and programmes? In *Proceedings of the ACEEE 2004 Summer Study on Energy Efficiency in Buildings*. Washington: American Council for Energy Efficient Economy.

Lenzen, M., K. Kanemoto, D. Moran, and A. Geschke. 2012. Mapping the structure of the world economy. *Environmental Science and Technology* 46 (15): 8374–8381. doi:10.1021/es300171x.

Lenzen, M., D. Moran, K. Kanemoto, and A. Geschke. 2013. Building Eora: A global multi-regional input-output database at high country and sector resolution. *Economic Systems Research* 25 (1): 20–49. doi:10.1080/09535314.2013.769938.

Lotka, A.J. 1956. *Elements of mathematical biology*. New York: Dover Publications, Inc. (first publication: *Elements of physical biology*, The Williams and Wilkins Co., Inc, 1924).

Lucas, R.E. 1988. On the mechanics of economic development. *Journal of Monetary Economics* 22 (1): 3–42.

Makarieva, A.M., V.G. Gorshkov, and Li B. Bai-Lian. 2005. Energetics of the smallest: do bacteria breathe at the same rate as whales? *Proceedings of Royal Society B* 272: 2219–2224.

Maxwell, D., P. Owen, L. McAndrew, K. Muehmel, and A. Neubauer. 2011. *Addressing the rebound effect, a report for the European Commission DG Environment*, 26 April 2011 (http://rebound.eu-smr.eu/).

Nicolis, G., and I. Prigogine, 1977. *Self-organization in nonequilibrium systems*, vol. 191977. New York: Wiley.

Odum, H.T., and R.C. Pinkerton. 1955. Time's speed regulator: The optimum efficiency for maximum power output in physical and biological systems. *American Scientist* 43 (2): 331–343.

Ruzzenenti, F., F. Picciolo, and R. Basosi. 2015a. Rebound effect and structural change (Chapter 17). In: *Energy security and development—The global context and Indian perspectives*. ed. Reddy, B. Sudhakara, and Ulgiati, Sergio. Berlin: Springer (ISBN 978-81-322-2064-0).

Ruzzenenti, F., A. Joseph, E. Ticci, P. Vozzella, and G. Gabbi.2015b. Interactions between financial and environmental networks in OECD countries. *PLoS ONE* 10 (9):e0136767. doi:10.1371/journal.pone.0136767

Schneider, E.D., and J.J. Kay. 1994. Life as a manifestation of the second law of thermodynamics. *Mathematical and Computer Modelling* 19 (6): 25–48.

Solow, R.M. 1956. A contribution to the theory of economic growth. *The Quarterly Journal of Economics*, 65–94.

WRI. 2015. *CAIT country greenhouse gas emissions: Sources & methods*. World Resources Institute.

Part III
Complex Systems, Social Practices and Issues Generated by Reification

Part III
Complex Systems, Social Practices and Issues Generated by Reification

Chapter 8
Ontological Fallacies Linked to Energy, Information and Related Technologies

Nicola Labanca

Abstract People and the socio-technical systems they constitute are being literally identified with motors and information processors in the present age of complex systems. This means, among others, that (a) energy and information are seen as actual ontological entities whereon the survival and the evolution of social aggregates depend and that (b) these entities entirely frame mainstream research and policy approaches aiming to increase the sustainability of human activities. This chapter (1) discusses how and why these extremely useful conceptual entities should be considered as metaphors when applied to study societies; (2) identifies some main social dynamics causing that these metaphors are instead constantly taken literally; (3) describes the implications of this literal interpretation for the ongoing energy transition. In particular, it shows how the literal interpretation of these metaphors reinforces dependency on abstract resource units supplied by energy and information technologies as well as a continuous growth in their consumption. In addition, this chapter illustrates how the unwanted effects of this literal interpretation can be effectively escaped by researchers, policy makers and all people involved in the current energy transition by focusing on the design and implementation of policy actions where the installation of technical solutions with reduced energy input and/or emissions is made complementary or subordinated to a reorganization of the energy outputs. This chapter also shows how this can allow exploiting an otherwise neglected huge variety of context dependent policy solutions relying on people capacities and on their more active involvement in policy making.

N. Labanca (✉)
European Commission, Directorate-General Joint Research Centre,
Unit C.02 Energy Efficiency and Renewables,
Via E. Fermi 2749, 21027 Ispra, VA, Italy
e-mail: Nicola.labanca@ec.europa.eu

N. Labanca (ed.), *Complex Systems and Social Practices in Energy Transitions*,
Green Energy and Technology, DOI 10.1007/978-3-319-33753-1_8

Introduction

Complex systems are being constructed on a massive scale principally thanks to the large-scale employment of information computer technologies allowing people to perform an increasing number of tasks. These systems are made of huge amounts of standardized information, energy and material units circulating within highly interconnected and intricate networks. They result from a series of scientific and technological advancements that are leading to a complete reorganization of societies based on the properties of very abstract entities which cannot be sensed by people and have been named energy and information by physicists, engineers, biologists, ecologists and cyberneticians. This reorganization reframes the way in which the problems caused by the increasing burden on the environment by human activities have to be faced.

The present chapter examines this transformation under a very particular point of view. Rather than endorsing the theoretical perspective informed by the concepts and operational definitions whereby these systems are being socially constructed, the reasoning presented hereunder aims to put these concepts under the microscope and to show how their deconstruction can serve to identify important biases and alternative policy approaches to energy sustainability. This reasoning has been inspired by a series of paradoxical situations generated by technologies and the scientific discourse supporting their large scale utilization. These paradoxical situations are determined by the fact that the energy and information concepts hold rigorously only within laboratories, whilst scientists and people in general tend to mistake these concepts for entities actually populating everyday life when they literally identify human aggregates and societies with motors and information processors. The thesis supported in this chapter is that the literal interpretation of what should instead be considered as metaphors is responsible for a huge process of homogenization and reification. Rather than being considered as metaphors with potentially different referents in different contexts of everyday life, input–output systems represented by motors and information processors are literally identified with the aggregates made by people and their technologies while energy and information are changed into standard and abstract resource units to be consumed everywhere to accomplish any kind of activity. The literal interpretations of these metaphors reinforce each other and take place on a large scale within present competitive market settings basically because it is constantly validated by the huge technological apparatus whereby complex systems are being socially constructed. These interpretations shape our imaginary and the way we organize our societies and, in so doing, increase our dependence on the supply of abstract resource units and artificially reinforce a process of continuous growth in natural resources consumption.

As explained in this chapter, the misplaced concreteness attributed to given amounts of energy and information units, whenever they are considered as the input that people necessarily need to perform any kind of activity, is the result of an interplay between linguistic phenomena and sequences of technologically mediated activities by which it is constantly validated. This particular interplay results from attempts to bridge what actually is an unbridgeable gap separating experimental setups and conceptual artefacts developed by scientists from what people do and say during their everyday life. Due to this type of separation between spaces, materials, and roles of involved actors, the dynamics generated by these attempts can be described through the language and the mechanisms of myth.[1]

This being said, it is probably necessary to stress that the author of this chapter does not certainly want to suggest a technophobic view. His aim is rather to show how the study of the phenomena described can help identify important complementary approaches to mainstream policies for energy sustainability and relevant weaknesses thereof.

The first two sections of the chapter are hence dedicated to discuss how energy and information play the role of central metaphors that are being constantly taken literally in the present age of complex systems and which are the consequences of this aberration for social organizations. The third section discusses how this constant literal interpretation and its validation by existing energy and information technologies[2] can inhibit the expression of a human faculty named practical knowledge and is associated with the exertion of a radical monopoly by these technologies on the practices that people can elaborate to provide for their necessities. The fourth section puts these phenomena under a more general framework and shows how they result from a particular type of counterproductivity generated by technologies largely employed by people that has been studied and identified for the first time by a series of scholars during the 1970s. This section also discusses how the main characteristics of this counterproductivity have been changed by the social transition from instrumental tools to complex systems that has been taking place since the mid-twentieth century, as described in the first chapter of this book. The fifth section discusses instead how and why the constant literal interpretation of energy and information metaphors and myths should be escaped and which are the implications for the solutions that people can identify to increase the sustainability of their lifestyles. Finally, the sixth section of the chapter draws some main conclusions from the questions previously addressed and discusses some relevant policy implications.

Energy Metaphors

Whenever we turn a switch on to get light, fill our automobiles with gasoline, install a PV panel on the roof of our house or attend a debate where scientists and politicians discuss about the need to construct new power plants, we implicitly

---

[1]The reasons for this can be understood by reading, for example Durkeim (1915).

[2]By energy technology it is generally meant here any energy end-use technology or technology used to produce, transmit or distribute some type of energy carrier (e.g. electricity, gas, etc.).

receive a confirmation concerning the indispensable presence in our daily life of a protean and scarce resource that has been named energy. This confirmation is also given whenever we are identified with human motors and rewarded by counting the number of the specific outputs we produce through our job or by measuring the number of hours we spend within an office or while supervising the proper operation of a machinery. Even our leisure, when organized according to more or less tight time schedules for the activities we plan, is informed by a mindset and social imaginary that has consolidated through the intensive production and large scale application of steam engines and motors starting from the mid-nineteenth century. Despite the implicit validation provided by most of our habitual practices and by contemporary public discourses concerning energy, the actual nature of this entity has been highly debated and has been redefined several times by scientists. The physicist and Nobel Prize winner Richard Feynman has for example stressed the importance of acknowledging that we "have no knowledge of what energy is" and that energy "is an abstract thing in that it does not tell us the mechanism or the *reasons* for the various formulas".[3] Percy Williams Bridgman, another physicist and Nobel Prize winner, has pointed out that "the energy concept has no meaning apart from a corresponding process. One cannot speak of the equivalence of the energy of mass and radiation unless there is some process (not necessarily reversible) by which one can get from mass to radiation".[4] The objective of this section, however, is not certainly that of contesting the validity of the outcomes of the scientific activities related to energy and, for example the validity of the law of energy conservation to which no exception seems to have been so far identified within laboratories.[5] The above quotations are just reported to point to the difficulties arising when it is pretended to identify the operational definitions produced for energy with something whose existence can be derived from experience and the

[3]See Feynman (1964).

[4]See Bridgman (1961).

[5]It remains, however, extremely interesting to follow the evolution and the transformations undergone by the formulations of the energy conservation principle, for example in the account provided by Mirowski (1989). This account illustrates how. during the first decades of the nineteenth century, the verification of the constancy of the ratios between the amounts of heat, mechanical work, electricity, radiation, etc., that can be generated from given amounts of these same physical quantities within specific physical transformation processes conducted under very controlled conditions provided the experimental basis to interpret these transformations as *conversion* processes whereby amounts of a same and conserved ontological entity were converted from a form to another. This account also shows how the energy conservation principle has undergone after that time a series of reformulations that have transformed it into a consequence of specific fields symmetries and of time homogeneity up to the quite recent and serious cosmological theories that have proposed that the universe has been generated from a vacuum fluctuation and that there is no globally conserved energy (Mirowski 1989, p. 129). What would then remain of the ontological entity named energy and of its conservation principle would just be the previously mentioned conversion factors widely employed by engineers to calculate how much mechanical work, heat, electricity, etc., can be generated within specific transformation processes of given natural resources together with the possibility of measuring these different physical quantities by using a same unit of measure.

objective of this section is rather that of hopefully contributing to better understand the social impacts and the problems generated by the interpretation of given social dynamics in terms of the physical laws verified by scientists within their laboratories. Analogies and similitudes populating discourses whereby scientists interpret and explain their intellectual advancements often obscure the fundamental difference existing between laboratories and everyday life and the perverse effects that can be generated when this difference is neglected by experts and policy makers asked to intervene over societies. It is for this reason that a good part of this chapter is dedicated to discuss how this difference can be kept hidden and to describe the specific social dynamics, whereby the myth of the actual presence of an ontological entity named energy is kept alive when persons and societies are seen as kind of engines and the perverted outcomes of this mindset are not recognized.

When analysed in relation to its social impacts, energy comes to represent a very particular type of abstraction together with a handful of other conceptual artefacts produced by science. Strange as it may seem, the main peculiarity of this concept and some of its present and most relevant social impacts can be highlighted by studying a roundtrip that this concept has started during the nineteenth century. This roundtrip has brought energy from the vernacular to the laboratories of physicists and engineers and has taken this concept back to everyday life under the guise of a *metaphor* that is *constantly* being taken *literally*. Put it bluntly, the result of this roundtrip is an aberration that consists in assuming that when we, e.g. say "Mr. Smith is a Lion" we constantly and completely overlap and identify the person of Mr. Smith with a Lion, as if we would constantly be under the influence of a kind of tribal ritual where Mr. Smith plays the Lion or as we would constantly feel as we felt when we were children and acted the part of a Lion in the game of the Savanna wild animals with other children transforming themselves into as many wild animals.[6] In these ritual transformations the Lion becomes so magnified to completely blur and overcome Mr. Smith's identity in the interactions we have with him, as if we would forget about the context where Mr. Smith is not a Lion, or as if we would lose the capability to distinguish the context where Mr. Smith lives his everyday life from the context and the intercourses where the statement "Mr. Smith is a Lion" actually holds. In this situation, the Lion (and not Mr. Smith) is the authority defining and controlling how we have to interact with it.

Because of the peculiar relationship existing between things of everyday life and concepts developed within laboratories and because of its roundtrip between these two worlds, it can be assumed that *energy* is transmogrifying ourselves and our environment as the Lion of the example has overlaid and reduced Mr. Smith's unique identity and the potentially infinite types of different intercourses and

[6]It may be interesting to observe that the disability consisting in not being able to distinguish among different contexts is associated with schizophrenia (see Bateson et al. 1956). It has then to be pointed out that the example and the considerations presented here result from a specific way of intending metaphors, myths and rituals. Put it shortly, metaphors are seen as small myths telling a story that is embodied by people through a ritual. This interpretation of metaphors, myths and rituals can be found, e.g. in Vico (1744), *La Scienza Nuova.*

interaction contexts that it is potentially possible to establish with him. One of the main enacting "ritual ceremonies" whereby this is happening worldwide has very likely been set-up during the nineteenth century when a specific ritual has started developing around the metaphor "Labour is Energy". This is most probably the central metaphor whereby the energy concept has irradiated over Western societies and has transformed and reorganized *labour* and most of human activities organized around this concept according to conservation and degradation principles. Through this central metaphor most of the social settings, skills and artefacts that people had previously developed to provide for their necessities through labour started (and still are) being transmogrified to generate *work*[7] according to laws and principles defined in laboratories for the energy concept. The tracks of this transformation are for example reflected in the highly intensified commodification of labour and in the wide diffusion of labour theories of value that could take place with the progressive assimilation of labour to the abstract notion of energy as defined and employed by physicists and engineers starting from the mid-nineteenth century. On the material side, these tracks can be found in the impressive diffusion of technologies generating work from natural resources according to precise conversion factors and in all the sequences of human activities organized around these technologies since that time. These sequences of activities have, among others, played the role of ritual actions and ceremonies whereby the literal interpretation of the above-mentioned metaphor has been constantly validated and the distinction between humans and motors has been progressively obscured.[8] The validation of the literal interpretation of the energy metaphor does not, however, exclusively come from activities organized around motors and steam engines. As it can be inferred, for example from Mirowski (1989) this validation is actually the result of a mutual reinforcement and validation process involving metaphors of motion produced by physicists, metaphors of value produced by economists and metaphors of body produced by biologists. Put in other words, the constantly literal interpretation of the energy metaphor has to be seen as the result of the joint and mutually reinforcing social constructions of invariants and conservation principles taking place in the fields of physics, biology and economics. As pointed out by Mirowski (1989), the structures of explanation produced in these three different fields have probably always been homeomorphic and would legitimize each other even in the face of possible disconfirming evidences produced in each field. Labour theories of value reflected

[7]The word "work" is being used here to refer to a motor/machine like conception of the organization of productive activities that prevailed after the invention of the energy concept. The word "labour" is instead used to refer to a pre-existing conception of the human activities whereby goods and services were provided within economies. The decision to use these two words to denote these two radically different ways of intending human activities is not accidental. Compared to work, labour denotes indeed a type of activity where the body of persons and their physical effort is more directly involved and needed. A similar distinction is also present in French (travail vs. oeuvre), Italian (travaglio vs. lavoro), German (Geburtswehen vs. Arbeit).

[8]For an historical account concerning how this identification between human and motors has developed, see Rabinbach (1992).

perfectly this interconnection and mutual reinforcement at least one century before the advent of complex systems could suggest that the information flows developing within biological organisms and ecosystems could be identified with those occurring within monetary systems and human made energy distribution networks.[9] An homeomorphism and mutual reinforcement between metaphors related to conservation and variation principles applied in economics, biology and physics would have been at work well before the formal analogy among the dynamics of money, information and energy flows within complex systems could be established. According to Mirowski (1989), this mutual confirmation of metaphors was indeed already operating when the institution of money was disconnected from any reference to a particular commodity and became the abstract representation of pure value, when the dual concepts of the organism and of natural selection were established within the evolution theory of Darwin and when the energy conservation and degradation principles were established by physicists and engineers around the mid-nineteenth century.

Besides this mutual legitimization occurring among metaphors produced in different areas, further elements to understand the social construction occurred around of the literal interpretation of the energy metaphor have to be found in the studies conducted by Ivan Illich and Uwe Poerksen in relation to the evolution of some terms used by science.[10] These studies allow inferring that the previously described transformations could not have been possible without a roundtrip undertaken by the word "energy" itself. This round trip started in the vernacular where energy was used to refer to "the vigor of an utterance, the force of an expression, the quality of a personal presence",[11] at least until the sixteenth century. It continued then through the laboratories of physicists and engineers where energy was associated during consecutive phases with a magnitude remaining constant during collisions of rolling balls and springs oscillations, with a primordial entity obeying conservation and degradation principles, with states of electromagnetic fields, with field symmetries, with time homogeneity. It is through these metamorphoses that energy has become more and more esoteric and distant from what can be experienced. At the same time, however, physicists have made energy come back to common speech by popularizing the supposed *real nature* of the various conceptual artefacts they have progressively associated with this word and by supporting their theses through the material artefacts that were being massively constructed by referring to the energy concept. The account provided by Illich and Poerksen explains how this roundtrip has caused a colonization of the vernacular by an abstract concept whose precise meaning cannot be discerned anymore and that, contrary to other abstractions generated within languages and practices, makes impossible that people can use it with precision by adapting it to the different

[9]For an explanation concerning how this identification is realized, see for example Goerner et al. (2015).

[10]See Illich (1983) and Poerksen (1995).

[11]Illich (1983).

contexts of their everyday life.[12] The following sections of this chapter will discuss some of the consequences of this roundtrip and how the intensive technological and economic development that has accompanied it have led to the creation of energy metaphors which are constantly being interpreted literally and which are transforming people into consumers of a phantom entity.

As several concepts and ideas have been already introduced in the brief reasoning so far presented, it is however probably better to pause a bit and provide some more element to hopefully make it clearer.

First, it may be worth highlighting that the metaphors being considered are not just a mere figure of speech and the elements involved in the considered social dynamics are not merely linguistic. These metaphors represent indeed small myths telling a story that is validated through specific ritual actions. In the example of the metaphor "Mr. Smith is a Lion" it is assumed, for example that this small story is being actually lived by Mr. Smith and/or is being told by a speaker who has actual had interactions with Mr. Smith. Similarly, the metaphor "labour is energy" is not a simple figure of speech. It is deeply interwoven with energy technologies and is validated and kept alive by other central metaphors produced by biology and economics and by all the sequences of actions that since the mid-nineteenth century are being widely generated and undertaken around engines and that may lead to organize people activities as if they were motors consuming the conserved and degradable entity whose properties have been established by physicists and engineers in their laboratories. Human actions and practices are, therefore, assumed to have primary role for the generation, validation and maintenance of these types of metaphor.

Second, despite the metaphor of Mr. Smith and the Lion initially used as an example is of a different type, the metaphors and the ritual actions of interest in this chapter are always made by one term (e.g. energy) that refers to a concept and the related operational definition produced by science by rigorously applying the scientific method, whilst the other term (e.g. labour) refers to a series of embodied interactions with the physical world experienced by people during their everyday life and is assumed to be progressively identified with the term representing the abstract concept defined by science through the previously described process of literalization. Behind these specific types of metaphor there is hence, on the one hand, an abstract scientific conceptual artefact which one term of the metaphor refers to and, on the other hand, the object of a variegated and personal experience

[12] For further information concerning how energy and other abstractions undertaking a similar roundtrip can be characterized and distinguished by the abstractions generated within common languages see (Poerksen 1995). Before describing 30 criteria allowing characterizing a series of science abstractions like energy, Uwe Poersken explains how they differ from other abstractions, like e.g. the concept of "love" as used within common parlance. He explains how the meaning of the word "love" can be expanded to embrace a wide range of meanings (from affection within families, to physical love, to pleasure at a piece of music, to the love of humanity, etc.) allowing the speaker to employ it in a series of different ways depending on the context where it is used. The word energy does not instead allow the speaker to define it. It disempowers the speaker, it cannot be replaced by pantomime or gesture, it is like a lego block that can be put everywhere within speech, its meaning is not affected by the context where it is used, etc.

made by people which has socially coalesced into the word constituting the other term. The two words constituting the metaphors studied here have, therefore, a completely different nature and history. One word (e.g. the word "labour" in the example above) still keeps track of a variety of meanings that allowed the speaker to use it very precisely to denote very different situations and practices, the other (e.g. the word "energy") results instead extremely impoverished in content because of the roundtrip and reshuffling operated by science. As first highlighted by Uwe Poersken, this latter type of words generally disempowers the speaker, creates the need for expert help, leads to silence, cannot be represented by gesture or pantomime, is not affected by the context where it is used, etc.[13]

Third, the above considerations and the way in which human labour has been progressively aligned and organized according to energy conservation and degradation principles, while all the different meanings and practices associated with the word labour have been progressively reduced to and identified with an activity produced by an engine, are assumed to represent a specific case of the following general perverse effect of applications of scientific artefacts: the extensive use of technologies developed around an abstract concept of science can serve and be used to constantly and extensively validate the literal interpretation of an associated metaphor. In this way, this concept can function as a kind of black hole attracting into its orbit and causing the progressive disappearing of myriads of alternative practices while the different meanings coalesced into one term of the metaphor and denoting these practices progressively disappear and are drained into the void generated by the term used to connote this scientific concept. This void is due to the fact that this scientific concept has typically no analogous in everyday life and the term used to connote it has, therefore, a mere symbolic value.

Fourth, together with the extensive technical applications they can contribute to develop, scientists and experts producing the above mentioned abstractions can contribute to validate the literal interpretation of the above mentioned metaphors by popularizing the real nature of these abstractions. Curiously, however, these abstractions may be object of continuous redefinitions by scientists and experts themselves; as happened, for example in case of energy. Despite the abstract conceptual and material artefacts developed by scientists belong to a world which is definitely separated from everyday life, these scientists and laboratories have however not to be considered as living within kind of case glasses keeping them separated from the real world. They are clearly and constantly influenced by the cultural milieu, where they live and their artefacts have to be considered as the result of a particular type of social construction.[14] The separation existing between

[13]See explanations provided in the previous footnote.

[14]It is a particular social construction because, contrary to other concepts and categories developed by societies, the possibility of an involvement of common people during the construction and validation of these conceptual artefacts remains very limited. Despite this limited involvement, these latter conceptual artefacts can remain in use within common language due to the indirect validation they can receive from scientists and from technological applications employed on a large scale.

the world of their conceptual artefacts and the world where people usually live is mainly determined by the fact that the scientific method whereby these artefacts are produced imposes that their properties must hold everywhere, for everybody and at any time. The specificity of all the particular cases that can be identified during everyday life escapes by definition the reductions operated to produce the abstractions of science. This is generally why the properties of these abstractions can be observed only under the very controlled and specific conditions created within laboratories.[15]

Fifth, the reasoning being presented implies that energy is engaging all the members of societies into a constantly literal interpretation of a central metaphor. It might be objected that many persons cannot actually be engaged in this literal interpretation because they do not have a proper cognition of what energy is in so far as, for example, they do not know the difference between energy (kWh) and power (kW) or they do not know-how the associated conservation principle is defined. The point, however, is that the knowledge of the physical properties of energy and of the related units of measure is not a necessary condition for engagement. The simple physical involvement in the ritual actions represented by the sequence of actions they undertake around existing technologies together with the fact that societies and experts validate the assumption that these technologies and the social aggregates constituted around them function like engines consuming energy inputs is alone sufficient for engagement. On the other hand, it might be objected that energy does not engage in any metaphor at all, because it is a real entity of which people can have direct and physical experience; for example when they receive an electric shock or when their body is heated by solar radiation. The answer to this objection is that the direct experience of these transformation processes does not certainly allow per se to infer the presence of a universally conserved and continuously degraded entity that has been named energy by scientists and whose existence is nowadays questioned by the scientists themselves.

Sixth, the aberration being described in this chapter does not relate to the presence of the metaphors being discussed. It rather concerns their *constant* literal interpretation and the social constraints impeding that a metaphorical character can be associated with some abstractions generated by science. Rather than as the result of a social construction, these abstractions are in fact presented and interpreted as always existed natural entities that have been detected by science. The energy concept is, for example extremely useful in so far as it used to study specific

[15]Science is axiomatically rooted on abstractions supposed to hold anywhere, at any time and (in principle) for anybody because of the irrevocable application of the principle of repeatability and reproducibility of observed events that must be observed by the scientific method. Due to the strict observance of this principle, science cannot typically tell or suggest anything in relation to how explain particular and unique events. Relationships with single and unique entities can be explained by science only by referring to qualities that are shared with other entities, i.e. by neglecting what makes these entities unique. On the temporal side, not repeatable events occurring in a specific instant are considered by science either as never happened or (in case they produce durable and detectable changes) are considered as the result of pure "chance".

transformation processes of matter and radiation and to design, manage and construct technologies relying on these processes. This concept should, however, be considered a "reality" only within a specific technical and scientific environment. When energy is instead constantly interpreted as a real entity populating everyday life, the survival of any person and biological being become subordinated to it, energy considerations can become one of the main driver of political decisions concerning how people should organize their lives, the extensive usage of technologies supplying energy becomes automatically legitimized and our dependence on them is reinforced. When, on the contrary, they can be seen as a metaphor, the installation of energy technologies can be subordinated to the political decisions concerning plenty of alternative practices that people can design and implement to provide for their necessities. It should not be very difficult, for example to realize how the number and the diversity of solutions that can be identified and implemented to solve a number of issues change dramatically depending on whether people are seen as energy consumers or as active persons endowed with the practical knowledge needed to provide for their necessities. How labour activities can be organized in a city in such a way that the highest possible number of person in a city can receive profit from these activities to enjoy their lives? How houses and buildings in a city should be rebuilt or retrofitted, which of them can instead be left how they are in order to allow a comfortable way of life to everybody without increasing the burden on the available natural and economic resource budget? Which transformation processes of available resources is better to implement to provide for the necessities of people living in a city? How might people's transit be organized to make it more sustainable and less stressful? Answers to these question provided by mindsets identifying cities and persons with a kind of energy motor will be very likely informed by principles of measurability, efficiency that can be applied everywhere and that would necessarily increase homogenization[16] and dependence on specific technologies, whilst mindsets recognizing the metaphorical nature of energy will more likely consider all the specificities of the situation at stake and will potentially be able to generate a much wider variety of context specific answers which have nothing to do with energy consumption. The solutions generated in this way will nevertheless generally have important energy impacts and will be identified by starting from considerations related to what people do and say in relation to their lives. Needless to point out that possible solutions elaborated under the former mindset can be implemented mostly by technicians and experts, whilst the knowledge concerning the specific practices that can be implemented in a specific context can reside within all people living in this context and this people can therefore play a fundamental and much more active role in their design and implementation.

[16]Homogenization is seen here as the somehow inevitable outcome of the application of solutions based on an identification with energy motors and informed by measurability and efficiency principles.

Most of the issues above mentioned would certainly require much more detailed discussion and actually concern a variety of scientific abstractions developed in conjunction with widely employed technologies. They concern, e.g. abstractions and operative definitions that have been produced by scientists for *time*, *space*, *speed*,[17] etc., and how these abstractions have been applied in relation to a large series of human activities and practices related, for example to time management, urban planning and mobility, buildings design, etc.

All of these concepts and applications would deserve a specific "metaphorical" study because they are having deep impacts on how sustainability issues are currently framed by researchers and policy makers. The following section will nevertheless refer the above considerations to a further key scientific concept that is closely connected to the energy concept and that has determined a radical shift of paradigm in how social practices are conceived and organized since the mid of the past century: *information*.

Information Metaphors

Computers represent the main central metaphor whereby we have been taken in the present age of complex systems. The current social imaginary concerning the nature of the world around us and the way we interact with it is deeply shaped by computer technologies and is progressively changing us into *multitaskers* while the universe is being interpreted as a bits processor.[18] The main abstraction whereby we are brought into the world of this metaphor is *information*. As happened with energy, information has undertaken a round trip started from everyday life during the second decade of the twentieth century when it still had several meanings and could take additional ones depending on how it was used in a context. At that time, its three basic meanings were generally related to *instruction* (in the domain of education), *inquiry*, *investigation* (in jurisprudence) and *message, report, evaluation* (probably in the area of institutional assignments).[19] It then happened that it reached the laboratories of cyberneticians and biologists around the mid-twentieth century and was subsequently taken back to everyday life extremely impoverished in content as an abstract entity that can be measured through probabilities.[20] This

[17]Scientific developments concerning the notion of *time* and related social implications are closely interwoven with those of the energy concept. For information on this point see, for example Perulli (1996). Some considerations on how the social construction of energy and the social transformations that led to conceive time as a resource implicate each other are provided also in the first chapter of this book. Concerning the notion of *speed*, Illich et al. (1996) provide important elements to perform the proposed study. Concerning the notion of space, Bauman (1998) can instead represent a good starting point.

[18]On this point see, for example Chiribella et al. (2011).

[19]See Poerksen (1995), p. 38.

[20]See the first chapter of this book for further explanations.

roundtrip has brought societies within the rituals animated by a relatively new phantom-named information.

In the light of what was mentioned so far, it is not so bizarre to assume that the nature and the impacts of the transformations induced by these rituals can be grasped by studying the elements of the central metaphor that can be identified by looking at how information has been defined by cybernetics.

Cyberneticians tell us that "information is a difference which makes a difference".[21] It can probably be stated that by redefining information in this way, cybernetics has led to the technical implementation of the latest vision of "change"[22] provided by science. This is the metaphor that deserves to be studied to understand the central ritual in which people are collectively engaged by complex systems.[23] To do so, it is necessary to try to grasp the nature of this particular type of information and how it has been modifying the way in which we interpret change and the role of persons within it.

The anthropologist Gregory Bateson has contributed to this end in important ways and has, among others, provided a series of fascinating examples taken from the phylogenesis of biological organisms, from the phenomenology of perception, from linguistics, from sociology, etc., whereby this transformation can be understood. Gregory Bateson has explained how meaning and information content would be purely *relational* and constitute an infinite series of hierarchically organized contexts.[24] According to Bateson, this type of information regulates the functioning and the evolution of all natural aggregates involving living entities. Through this type of information, any entity of the natural world comes ultimately to be defined and regulated by infinite chains of relationships (i.e. differences) with other entities. In the world made of the information created by cyberneticians, there are no objects with proper and intrinsic characteristics. There are only relationships. Starting from the smallest elementary bricks constituting any natural object up to the largest aggregates available in nature, we only find dualities, i.e. relationships between two irreducible entities, which are non-entities when taken alone.[25] Natural objects would appear just when artificial delimitations are created within complex systems by defining an "inside" and an "outside". The descriptions of complex systems

[21]See for example Bateson (1972) and what discussed in the first chapter of this book for further details.

[22]Change can indeed be considered as a "difference which makes a difference".

[23]People can somehow be considered as collectively engaged in a ritual when their habitual actions are informed by given central ideas. The rituals constituted by complex systems are generated through the actions accomplished by using the technologies whereby these systems are being socially constructed (e.g. computer technologies) and are informed by the central ideas whereby these technologies are conceived, used and imagined.

[24]See the first chapter of this book for further information.

[25]This specific property of information relates also to how sequences of 0s and 1s come to constitute computer programs. 0 and 1 are indeed the two non-entities defined by their mutual relationship that can be used to represent any kind of elementary difference starting from which complex and hierarchical organized pieces of computer codes can be produced.

dynamics that can be provided in this way (i.e. a description in terms of delimited entities and related functions within an external environment) are assumed to result just from a decision/intervention by an *observer*.

It is for this reason that complex systems actually engage people into a ritual aiming at associating objects populating everyday life and their related functions with the entities populating an underworld made of the information bits that constitute the complex systems which these objects are supposed to belong to and that regulate the evolution of these systems. It is through this ritual that change is being reinterpreted. With complex systems, the "classical" explanation of change formulated by evolutionary theory and relying on a combination of variation (due to stochastic processes) and selection is indeed somehow assumed to belong to an outside world created by the observer. These two explanatory principles are being revised by complex systems theorists by interpreting *observed* stochastic variations as the result of thermodynamic processes generated by energy and matter flows occurring within a kind of underworld and obeying phenomenological principles that can be equivalently described and interpreted through information theory and thermodynamics.[26]

Nevertheless, besides this underworld made of matter and energy flows where information and energy metaphors are mutually validated, an upper world made of functions supposed to evolve and adapt continues to remain. Our attention has hence to focus at the microscopic interface existing between these two worlds in order to understand the ritual in which we are engaged by taking literally the information metaphor associated with complex systems. It is indeed at this interface that functions performed and observed during everyday life are being identified with the abstract information that can be managed by computers and that can be possibly associated with underlying energy and matter flows.

It has indeed to be stressed that functions observed in natural entities are being artificially jointed to bits of information and, thanks to information, to energy and matter flows. As already mentioned in the first chapter of this book, the establishment of these joints in any department of knowledge and human practice is what essentially characterize the present age of complex systems. These joints are being

[26]Variations appearing in the world of the observer are supposed to be generated by dynamics studied by science addressing far from equilibrium open systems. Put is shortly, these dynamics are described in terms of structures (i.e. structured/not random patterns of energy and matter flows) emerging through the dissipation of energy gradients. It is as if steep gradients applied to open systems would give open systems "a certain tension that creates a condition of an accident waiting to happen" (see Allen et al. (2003), p. 331. Thinking of a fluid within a box and of the convection currents generated through it because of a temperature difference applied at the two opposite extremities of the box may help to visualize what being described here). This accident generates then a kind of cascade through positive feedback loops whereby structures of energy and matter flows are created. These energy and matter flows would then tend to dissipate the previously mentioned applied gradient. The creation of these energy and matter flows can be equivalently described and studied by information theory in terms of probabilities and creation of information. It is through the establishment of this equivalence between probabilities and energy and matter flows that information theory can incorporate and confirm thermodynamics (for a detailed account of how this incorporation takes place see, for example, Ulanowicz (1997), pp. 63–71).

established, for example in ecology and energy studies when each observed function reproduced by biological entities is associated with the consumption of given amounts of energy and matter, or within biological studies when it is pretended that molecular biology (studying life with a thermodynamic/informational posture) and organismal biology (studying life in terms of evolution and adaptation of functions) can be merged, or when computerized prostheses are created to enable people to perform given functions through the elaboration of information, or when personal time is managed by prioritizing and allocating given amounts of measured time units to each daily activity. Present information technologies allow performing these allocation tasks at such level of detail that the constructed joints often appear as the reproduction or the detection of something established by nature itself. Unfortunately, however, they always generate or are generated through a *discretization* and a reduction to *standardized* functions of the continuous spectrum of unique functions that human beings and nature can generate[27].

This discretization and standardization is the sign of the artificial character of an underworld made of energy and matter flows supposed to generate functions reproduced within complex systems. This underworld comes indeed to represent an artificial layer that is interposed between people and what people see and do in the material world. Its energy and matter flows come to constitute a kind of artificial membrane impeding a direct and fleshy coupling with this world and impeding to generate an infinite variety of functions while interacting with it. The constant literal interpretation associated with the information metaphor consists in accepting the constant presence of this artificial membrane constituted by a cybernetic version of information (and by the associated energy and matter flows). It consists in considering this membrane as a natural entity and in accepting the limitations[28] determined by its interposition in the interactions that we have with the world.

[27]As discussed in the first chapter of this book, this *discretization* and *standardization* can be seen as the result of an (at least partly) arbitrary resolution of an otherwise unsolvable allocation problem. This problem inevitably arises whenever the amount of resources (e.g. energy, time, matter, etc.) consumed by a given organism or person to perform given functions has to be established. The solutions that can be found to this problem are inevitably associated with a discretization and standardization of these functions. Complex systems somehow always invite to take decisions in relation to these types of unsolvable allocation problems and make people blind to the distortions they generate in this way.

[28]Limitations may relate to the types of explanations that can be provided by referring to dynamics occurring within this membrane when studying natural and social phenomena. As far as the reproduction of human functions is concerned, it has to be stressed that the creation of the artificial prosthesis represented by this membrane through suitable interfaces and information technologies can clearly highly potentiate a number of single and specific human functions. We daily experience this by using computer technologies to purchase goods, write letters, communicate with friends and colleagues, etc. The limitations being discussed relate mostly to the isolation from the external world generated by this membrane and to the variety and the character of the functions that is possible to reproduce thereby. The kind of limitation effect on human functions being described resembles in some respects to the effects produced by a magnifying lens. While magnifying single and particular details, this lens inhibits indeed the vision of all the details allowing constructing the whole picture of the object being observed.

What has been highlighted so far in relation to the influence exerted by energy and information metaphors seems to be the inevitable consequence of the following general social phenomenon concerning science and technologies: when technologies commonly employed by people become the object of intense scientific reflection, they become able to *speak* and to convey messages on the large scale that societies end-up accepting as certainties. They tell very convincingly what specific human activities are and what are some of the specific processes that constitute the world around us. This has happened, for example, when mechanical sciences generated around human tools during the twelfth century have led to conceive the world and the beings populating it as constituents of a gigantic clock or, as previously discussed, when energy physics generated around engines during the nineteenth century led to conceive people and the whole universe as motors, or with the developments occurred within cybernetics and biology with information theory and information technologies during the past century.[29] Due to the intensive and large scale employment of the technologies conveying them, the power of these messages can become so strong as to generate a social blindness in relation to how they actually cause distortions within societies while legitimizing extensive processes of homogenisation determining the disappearance of a large series of discourses and practices which are typically specific of the cultural context where they are generated and which better fit it. These dynamics are extremely reinforced by the fact that the metaphorical nature of the abstract concepts whereby the functioning of technologies is explained remains hidden and people do not realize, for example, that abstractions like energy and information do not actually belong to their everyday life. These abstractions can certainly serve to engineers and technicians to build extremely useful technologies. Given the way in which they are developed and used, they should however be seen as belonging to just one specific sphere of everyday life and become subordinated to the myriad of concrete processes and solutions people can actively develop to comfortably provide for their necessities. The social factors impeding the recognition of the metaphorical nature of the above mentioned abstractions are ultimately the same factors impeding to recognize what makes people fundamentally different from motors and information processors. They pave the way to a colonization by technologies and impede to identify some perverted implications of this colonization that will be further analysed in a subsequent section of this chapter. Before doing that, it is necessary to further discuss how metaphors come to play a fundamental role in social organizations and how they and the material artefacts whereby they are kept alive can inhibit the exertion of a specific human capability contributing to constitute the vital force of any social aggregate.

[29]These transformations have been discussed under Chap. 1 of this book.

The Role of Practical Knowledge and the Effects of Its Inhibition Within Social Organizations

Metaphors are the fundamental and necessary element of any scientific advancement[30] and play a central role in the social construction of concepts and principles constituting languages, rules, material artefacts, institutional settings and know-how whereby societies can be organized.[31] Their relevance for social organization is the consequence of the funding role played by a particular type of *complementarity* within human affairs.

Emotional and rational, analogical and logical, unconscious and conscious, profane and sacred, particular and general, feminine and masculine are useful couples of adjectives that can serve to connote and characterize the two polarities of this particular dual structure. Out of the two polarities, the former is usually more primary (i.e., it pre-exists to the latter) and generates the latter whilst constantly embedding it. The former polarity is more closely associated with human bodies, senses and feelings of persons, whereas the latter may be seen as a result of a social process of abstraction, objectification, hypostatization. The way in which this generation takes places can be described archetypically by referring to the myth of *Eros* and *Psyche* (Neumann 1971).

The generative power and the primacy to be acknowledged to the first polarity are a consequence of the fact that the human body and perception are the necessary precondition for the social production of any kind of abstraction (either this abstraction is represented by the terms constituting a given language, or by a concept, a rule, a principle, or an institution) whose meaning can be communicated among people and can represent a point around which individual persons coalesce in a social aggregate without generating violence on some of its members. In order to perform this aggregation function, this abstraction has indeed to be generated by the body and to remain embodied in all the individuals constituting a society. The meaning of words, rules, concepts or ideas used to organize a social aggregate has to be felt by the flesh of all of its members. These abstractions could not be maintained and used without violence within societies if they were not embodied, and the evidence of this embodiment is that the concepts generated in this way can always be referred and adapted by people to the context and the particular circumstances where they are used.

Based on the above, it can be concluded that the type of *complementarity* to be maintained between the previously mentioned two poles relates to the process whereby one pole is generated from the other and can be returned to the other. This kind of process is always bent over and submitted to the primacy of the particular. It results from possibility of exerting the human habit that Aristotle (Nicomachean Ethics 6.8) named *Prudentia or Phronesis* (i.e., the wisdom of prudence and practical thought) and that allows being guided from and adapting general rules to

[30]On this point see, for example MacCormac (1976).

[31]On this point see, for example Lakoff and Johnson (1980).

the particular case when producing good actions. This habit involves primarily human senses, not reason. It requires a kind of opening to the possibility of being constantly surprised by the disclosing of unexpected developments and the capacity to cope with them by taking own perception and feelings as ultimate guide.[32] The exertion of this habit corresponds to the exertion of a practical knowledge described also in (Heidegger 1994) and it can be probably assumed that practical knowledge acts within a kind of infinitely recursive cycle: it is what initially allows people properly generating general rules and abstractions from the particular and then allows taking the general rules back to the particular case and the particular case to the general rules during the circumstances of everyday life.

Nevertheless, it has also to be observed that, despite their original union, the two polarities at stake in this process constitute two separate and completely different worlds governed by radically different rules and principles. The two worlds where the two aforementioned polarities live are completely separated and disjointed. Given this radical separation, the inhabitants of one out of the two worlds can indeed be described in the language of the inhabitants of the other world only by using metaphors. Given this otherworldliness, the only reasonable statements that can be produced about the former inhabitants by the latter inhabitants are statements like "Mr. Smith is a Lion" with Mr. Smith and the Lion respectively belonging to each of these two radically different worlds. The fundamental role played by practical knowledge derives from this unescapable separation. Practical knowledge can indeed be seen as a kind of boat connecting the shores of these two worlds.[33]

[32]It may be interesting to observe that the capabilities required for the exertion of this habit are the same that Aristotle attributed to artisans. While advancing with their works, artisans are indeed supposed to be able to adapt the ideas they have in their minds to the specificities and particularities unexpectedly emerging within the matter and the materials being used. The ultimate guide for the making of their activity is not reason, but perception (see Mitcham (1994), p. 122; Carl Mitcham produces this description of artisans' activity based on what reported in *Nicomachean Ethics* 2.9.1109b23; cf. 2.2.1104a1-9). Same capabilities were considered also essential for politicians. Politics was indeed assumed to be concerned with action and deliberation about particulars. Grounding in law was assumed to be necessary, but law alone could not serve to do justice. Judges, for example had certainly to be educated by the law, but they were also supposed to perfect and complete it while applying it. Judges and politicians were in this respect the functional equivalent of artisans (see Mitcham (1994), p. 125; Carl Mitcham produces this description of politicians and judges based on what reported in *Nicomachean Ethics* 3.3, 10.9 and in *Politics*, 2.8.1269a10).

[33]This particular role of practical knowledge can probably be verified also within human languages. It can indeed be probably assumed that practical knowledge allows converting own feelings and sensations into utterances that can be understood by others and allows interpreting utterances produced by others by converting them into own feelings and sensations. A noun or a sentence can after all be considered as one part of a metaphor, the other part being constituted by the feelings and the sensations of the speaker pronouncing it. The exertion of practical knowledge for understanding languages could then be identified with the act of interpretation as performed by the listener during the process whereby he understands the words pronounced by a speaker and connects in this way to the speaker's internal world. Due to the way in which science can attach particular operative meanings to some words, the above mentioned process can be inhibited (see, what mentioned in subsequent parts of this section).

As long as generated metaphors are not taken literally, it can be assumed that the realization of this connection entails the same capabilities at stake with hermeneutical *interpretation* (i.e., the exertion of practical knowledge entails the same capabilities and involvement at stake when performing texts translations, interpretation of historic events, etc.). Interpretation brings foreign worlds under the interpreter's spatial and temporal perspective. Interpretation implies that foreign worlds are brought within the context of the interpreter (i.e., it automatically brings foreign worlds within the geographical and historical space where the interpreter lives). Interpreters are (or should be) always within an "as if" condition, i.e. they are (or should be) aware that their interpretations consist in the construction of metaphors and that these metaphors should not be taken *literally*. The two worlds bridged by their interpretations should never be considered as strictly equal: they are analogous. This characterization of the role of practical knowledge indicates the kind of awareness to be cultivated in relation to abstractions and concepts used within societies and provides important insights concerning a particularly relevant type of perversion which often passes unnoticed.

The previously mentioned generation process may indeed be perverted and the primacy of the first polarity may be disregarded. Societies can be organized based on the properties of abstract concepts like the ones described in the previous sections and made completely dependent on the dynamics associated with the supply of energy units, information bits or monetary units by allowing that the technologies validating the existence of these entities take the radical monopoly of most human activities. This phenomenon is completely similar to what happens on a smaller scale to some social practices in those places where, for example you "cannot move any longer without wheels, you cannot eat without a refrigerator, or you choke unless you do not turn on the air conditioner"[34] and people's lives become completely subordinated to the abstract dynamics that can be generated by car mobility, food refrigeration and artificial cooling of living environments. Although extremely useful, technologies can in some cases completely redefine, reduce and homogenize the meaning of human practices linked, e.g., to transit, to eating or to human comfort by taking the radical monopoly of these practices and by obliging people to live according to the dynamics of the abstract concepts whereby they are conceived. In case of energy and information, the exertion of this type of radical monopoly is much more than a mere possibility due to the literal interpretation and mutual reinforcement of associated metaphors occurring within competitive market settings and to the diffuse presence of technologies validating this interpretation. People's lives are nowadays completely dependent on the dynamics generated by the joint evolution of aggregated energy demand and supply within existing energy markets. In a similar way, they are made more and more dependent on the various large scale information flows and decision systems

[34]These examples have been taken from Illich (1983).

wherein people, machines and the environment are increasingly integrated.[35] There is indeed a very close formal similitude and mutual reinforcement among the dynamics exhibited by energy, information and monetary flows due to the way in which energy, information and money have come to constitute complex systems by becoming the ultimate resource units that have to be consumed and exchanged to perform any kind of human activity.[36] The type of dependence that has been nowadays established on existing financial markets offers indeed another fundamental evidence of the abstractness of the rules regulating these artificial complex systems, of how these rules can be highly disembodying and generally correspond to a blind and passive submission to myths and assumptions of very different nature which are not anchored in what persons can feel and practically make and verify.

The consequences of this submission are, however, anything but immaterial. Their degree of materiality is the same of the immense technological apparatus and the huge amount of natural resources supplying the standardized energy and information units whereby most of us live today. Their origins have to be ultimately found in the impossibility of putting the dynamics related to the consumption of these artificial units under some type of social control, this impossibility being due to the fact that every action we accomplish is nowadays mediated by this consumption.

Despite the possible best intentions of people submitting to their dynamics, the complex systems so created become actual and autonomous delegates of the administration and regulation of human violence and desires. This is basically what happens when the metaphors whereby these systems are created are constantly taken literally and, due to a social blindness and misplaced concreteness, societies are made increasingly dependent on technologies validating this interpretation. As idols, they may tend to dis-embed[37] from any type of social control. The occurrence

[35]These information flows may concern, e.g. the elaboration of risk profiles based on the continuous monitoring of socio-technical systems and related feedback loops needed to timely respond to emerging threats (wars, nuclear accidents, environmental accidents, etc.), or management systems for transport networks used within large cities and regions, or the elaboration of consumers profiles whereby investment decisions are taken by companies, or the monitoring of the GDP of national economies, etc.

[36]See the first chapter of this book for further information. Important insights concerning the mutual reinforcement taking place among the energy, information and money metaphors can be found in Mirowski (1989).

[37]This idea of dis-embeddedness has been taken from (Polanyi 1944). Polanyi argues that the large scale application of the international gold standard and the transformation of land, labour and money into fictitious commodities that can be sold within a market regulated by Adam Smith's "invisible hand" has been at the root of the upheavals and violent disorders that took place in the North Atlantic Community and its periphery at the beginning of the twentieth century and has led to the World War I and the subsequent Great Depression. According to Polanyi these disorders would be the consequence of a "double movement" of long duration made of the expanding application of the above mentioned abstractions on the one hand and of the spontaneous resistance to the pressure they generate within civil societies on the other hand. As suggested by other scholars, in this chapter it is being assumed that this double movement of long duration can be generated also in other social spheres where scientific abstractions are largely applied (e.g. within social arrangements established to regulate energy and natural resource consumption).

of these situations of dis-embedding can most probably be considered as the result of the inhibition of people's practical knowledge and the impossibility to generate the associated virtuous hermeneutical cycle whereby persons can improve their present and future conditions.

Counterproductivity of Technological Myths

The social phenomena so far described can be analysed in more detail by taking the studies performed by Ivan Illich,[38] Jean-Pierre Dupuy, and Jean Robert[39] as a starting point. To the knowledge of the author of this chapter, these phenomena have been so far investigated only by these scholars and mostly in relation to *instrumental* artefacts, i.e. in relation to artefacts industrially conceived as means to allow any person to achieve a specific end (e.g., cars, medicines, but also institutions like schools and hospitals in so far as these entities can be seen as instruments conceived to allow people mobility, allow solving specific health problems or achieving conditions of equality through education). Given their seminal influence on the ideas presented in the previous sections and the light they can help shed on the previously described dynamics associated with energy and information technologies, the outcomes of these studies will be briefly summarized hereunder.

The above mentioned scholars performed their researches during the 1970s, when the social attention was intensely focused on the impacts of the industrial production on the environment and on the availability of natural resources. During that time, they decided to direct their attention to how an intensive production of given technological instruments could turn these instruments into the *main obstacle* to the achievement of the ends for which they had been conceived by causing an overall decrease in the number of persons able to reach these ends. They were interested in a type of instrumental counterproductivity which had nothing to do with negative environmental impacts or reduced marginal returns of investments on given technologies. This type of counterproductivity rather concerned the dynamics whereby, e.g. cars can become the main obstacle to people's mobility, medicines and hospitals can become a cause of iatrogenic diseases or schools can become a social threat for learning capabilities of people. Thanks to their studies, these scholars managed to identify three different and interwoven social dynamics contributing to this type of counterproductivity that were respectively named *technical*, *social* and *cultural* counterproductivity.

The *technical* counterproductivity is basically generated by overlooked and unwanted impacts of technological instruments on the *systems* where they are massively put in circulation. These impacts generally consist in the creation of situation of overcrowding and obstruction or are generated by the presence of

[38]See Illich (1976).

[39]See Dupuy and Robert (1976), p. 55.

interconnections and correlations existing among parts of these systems which are overlooked for various reasons when instruments and the possible infrastructures needed for their employment are fabricated and used. These situations may occur, for example, whenever too many vehicles are put in circulation within a road network, or when too many graduated students are produced by a scholastic institution in specific disciplines compared to existing demand in related job sectors, or when the extensive and prolonged use of specific medicines produces side effects which are worse than the diseases to be cured for the organisms wherein they are injected, or when the way in which an hospitalization system is established or managed generates an unexpected circulation and exchange of diseases among patients, etc.

The *social* counterproductivity associated with the intensive production of technological instruments is instead due to a confusion similar to that usually arising between stock and flux variables that leads to mistake these instruments for human capabilities. This confusion would lead to erroneously identify existing possibilities to express personal capabilities (concerning for example how we manage to move, how our bodies manage to keep themselves in good health conditions, how we learn, or how we establish relationships with other persons and the environment) with the presence of given instruments (e.g., cars, hospitals, medicines, schools, telephones, etc.) and would cause the intensive technological multiplication of the latter. Instruments would indeed allow developing personal capabilities in the targeted areas only up to a given threshold value achieved by the intensity of their industrial reproduction. Whenever this threshold is exceeded, instruments would markedly constrain and limit the development of these capabilities by causing people to prefer receiving or buying things rather than doing things by themselves. This situation increases, among other, the chances that a condition of technical counterproductivity is achieved.[40]

Finally, the *cultural* counterproductivity concerns directly myths and metaphors accompanying the extensive use of instruments. As already explained in the previous sections, due to the way in which they are conceived and produced, the extensive use of instruments would automatically put them at the centre of rituals, mythopoetic ceremonies and social liturgies generating existing certainties concerning human action and extinguishing the added value that persons can give to their design, fabrication and employment. It is in this way that for example cars would become able to convey a specific and standardized message concerning what has to be intended for mobility, refrigerators would tell with the authority of science what food conservation is, air conditioners would define what has to be meant by comfort, etc.

[40]The presence of these critical thresholds has been subsequently questioned by the same scholars who hypothesized their presence. This probably happened when they acknowledged that the achievement of given ends can be socially delegated to different instruments. Besides schools, the task of educating people has been, for example socially delegated also to TV programmes, the Internet, etc.. This possibility probably renders the concept of counterproductivity threshold practically inapplicable.

These three factors comprised the main conclusion of the above mentioned scholars in a time when instruments were probably still the main human artefact informing the current ideas about human action and related social and environmental impacts.

The situation was however already radically changed in the age of complex systems we were already entering when these scholars were conducting the above mentioned studies.[41]

Complex systems are indeed made of single types of often highly interdependent artefacts whereby an increasing number of different functions can be performed by people thanks to flows of abstract and standardized units of information, energy, matter, money, time, etc. This characteristic radically changes the social dynamics whereby situations of counterproductivity can be generated.

Nowadays, the extensive use of monitoring and information processing technologies associated with these systems can for example contribute to solve major problems linked to the technical counterproductivity that was generated by instruments. This can be done by redirecting, rearranging, and rescheduling the production of their outputs or by avoiding unwanted instruments outputs interactions in a way that looks very similar to what can be done by modern GPS systems to solve potential situations of traffic congestion or by internet based monitoring systems to prevent the outbreak of dangerous epidemics, natural disasters, or accidents.

Situations of *technical* counterproductivity do not nevertheless disappear with complex systems. Rather than to single ends achieved by given instruments, these situations relate nowadays to how complex systems and persons integrated therein may become the main obstacle to the achievement of the only comprehensive end these systems can allow achieving, this end being represented by systems survival through expansion, i.e. through increased complexity. This counterproductive situation is generated specifically by the way in which existing complex systems rely on the flow of highly standardized resource units and tend to become a single global system made of tightly interconnected parts to increase their chances of survival. Small perturbations originating into one part of the system may indeed become constantly able to generate kinds of avalanches whose effects quickly propagate throughout the whole system causing a deep restructuring and putting system survival at risk.

The *social* counterproductivity of instruments is also deeply reshaped by complex systems. Contrary to instruments, the artificial prostheses integrating people into complex systems mediate any kind of action and cannot therefore be abandoned. The inhibition of personal capabilities caused by an existing confusion between the possibility of expressing these capabilities and the presence of given artefacts that characterized instrumental social counterproductivity cannot therefore be a problem anymore. This problem is completely overcome due to the fact that persons are *de facto* constantly integrated into specific types of artefact for any type

[41]For a description of this transition see the first chapter of this book.

of action they accomplish. With complex systems, the expression of personal capabilities may at most relate to *informed decision making* concerning the possible consequences of a predetermined finite number of decision options that can be taken within these systems. Either people have to decide whether eating pasta is better than eating vegetables for their metabolism, or they have to decide which of the products available in their supermarkets is less harmful for the environment, or they are asked to take decisions concerning their life and death by medical geneticists,[42] or they have to manage private companies, states budgets or ecosystems, the various involved complex systems always frame the so-called informed decision making process in terms of a choice among a series of pre-established evolution patterns to be made based on the knowledge of given risk probabilities. The possibility of expressing personal capabilities and autonomy is in this way reconfigured in terms of decisions to be taken within a kind of poker game where players can know in advance the probabilities of the outcome of their choices. The social counterproductivity of decision systems framed in this way has to be found in how they inhibit any kind of personal initiative by limiting it to a selection among different choices that could be performed also by a computer. At the same time, these decision systems can make persons blind in relation to the hypotheses whereby probabilistic scenarios are elaborated and to how these hypotheses could be put into question by how people can reconfigure their habits to face the problems at stake. Probabilities create the illusion that the statistical categories defined for their calculation are something real that can deterministically influence the evolution of the specific case at stake. On the other hand, complex systems constantly provide evidences concerning the impossibility of predicting the evolution of associated dynamics neither by using probabilistic nor deterministic methods.[43]

The confusion existing between stocks and flux variables that had characterized the instrumental *social* counterproductivity does not result any longer from a confusion between personal capabilities and the presence of given material artefacts. With complex systems this confusion moves and relates to each of the single units of information, energy, money, time, etc. assumed to flow within these systems. Their definition creates indeed the illusion that these units represent actual and stable entities, whilst information and the probabilities whereby it is calculated remain an abstract difference between non-entities, energy is a variation that only materializes during transformations, the value represented by money is an abstract estimate produced around exchange processes, time units are assumed to measure a pure flow, etc.

Complex systems reshape finally also the *cultural* counterproductivity of instruments, in so far as the metaphors animated by these systems change. Rather than being related to what single instrument types tell to people concerning their activities, these metaphors relate altogether to the abstract resource units flowing within the complex systems that people and their material artefacts have started constituting.

---

[42]On this point see for example Samerski (2002).

[43]On this point see, for example, Taleb (2001).

It is amazing to think of how their literal interpretation can serve to legitimize a complete reorganization of societies and how the constant integration into the artificial membrane created by complex systems implies the inhibition of any form of control. The association of any activity performed by people with the combined consumption of abstract units of information, energy, money, etc. is alone sufficient to render the complex of these units scarce in so far as their flow obey a conservation principle implying that they cannot be generated at will from nothing and the abstractness of these units put people at the dependence of the technologies and experts that can supply them. This scarcity is at the same time responsible for a continuous intensification of production activities and for continuous efforts to make this production more efficient.[44] Rather than to reduce the overall

[44]The interplay of conservation principles artificially established for different abstract resource types is crucial for the generation of this situation of scarcity and the associated social dynamics combining increased resource efficiency and increased consumption. The following imaginary story may perhaps help clarify this point. Let us assume that on our planet there is still a land whose inhabitants are not aware of the fact that whatever action they accomplish requires the consumption of some units of energy, time, information or money. Let's then assume that one person of this land decides to move, e.g. to New York City, London, or Rome. Once arrived in one of these cities, the first thing he learns is probably that he cannot do anything without some amount of money and that he is not allowed to get this money *for free*. The second thing he learns is that he has to work to get this money and that this means that during each day he will have to spend given amounts of time while producing something that can be valued and rewarded with the money he needs. Unfortunately, however, the amounts of time he can spend each day are *limited* and this will oblige him to divide his time into a part that can be used to get money and a part that can be used to spend this money and do what he wants to do. In addition, most of the things he could do while working, or during his leisure or while at home, necessarily require the utilization of complex devices consuming units of energy (that, overall, is available in *limited* amounts and cannot be got for *free* either) because these devices can do in less time what he or other persons could do and people cannot spend too much of their time while producing what they want, otherwise the time remaining to use it will be too few. The third thing he might probably have to learn is that he has to be trained and receive the necessary information to properly use the previously mentioned machines, to be able to work, to get money and to be finally able to do what he wants to do. Unfortunately, these information cannot be get for *free* either. If he is lucky, it might be allowed to download this information from the internet. He might even manage to work and get money and the things he wants through the internet. The information and the things he gets through the internet however also require money, energy and other types of resources units. Overall, he would then learn that he must find a way to maximize the amounts of money, energy, time, information and other indispensable resource units he receives or employs because his survival and well-being depend on them. This, however, will require that he has to be very efficient while using these resource units, because these resources, when taken individually or all together, are necessarily scarce. The anxiogenic condition described in this very simplified and imaginary account could not certainly be referable to all persons living in the cities taken as an example. Some of them might have important amounts of the previously mentioned resources and manage to escape this condition. There might even be entire cities and nations depending on these resources where all the inhabitants could manage to escape this condition. The above mentioned conservation principles imply, however, that situations of particular abundance in one part of the system where these principles are enforced determine situations of exacerbated scarcity in another. The previously mentioned dynamics have indeed to be referred to the whole system, the system at stake being probably already represented by our whole planet.

consumption, these efforts, however, generally serve to destine what saved to other end-uses within a process of continuous consumption growth.[45]

As also showed by the examples produced in the first chapter of this book, the nature of these dynamics typically transcend the possibility that people can intervene to change them. With complex systems we are always within a bigger context that determines how things will ultimately go. Somehow, complex systems make any responsible action impossible. Besides the previously mentioned causation mechanisms associated with variation and selection, complex systems entail a type of causation that reminds a kind of Aristotelian *causa finalis* by which we are guided and to which we cannot nevertheless give sense. Its effects are indeed intrinsically unpredictable. Within complex systems we are invited to live within an oxymoron. We have to prepare for and learn to manage the unpredictable. The adjectives used to describe what complex systems call for are: resilient, flexible, adaptable, etc.

Escaping Energy and Information Myths

As repeatedly pointed out, metaphors play a central role within languages, rules, material artefacts, institutional settings, and know-how that organize societies. They allow connecting the subjective experience and feelings of persons with the general and abstract concepts and rules whereby people can form social aggregates. They can be considered as small stories and myths that people can embody through repeated actions necessarily mediated by given material artefacts. This embodiment could not however take place if these metaphors were not generated by referring to entities that people can experience during everyday life and if people could not report and adapt these stories to their personal experience through their practical knowledge. Practical knowledge is indeed another fundamental element whereby societies are established. This faculty can be seen as a human characteristics allowing people to enter (or create) and exit (or adapt) the metaphors, myths and rituals they create to administer societies. Practical knowledge somehow represents the possibility that, as persons, we are *socially* given either: (1) to produce and rigidly stick to the cultural constraints represented by concepts and rules established within societies, or (2) to modify and adapt these constraints to our specific situation by taking our body and our feelings as ultimate guide. The proposed view sees practical knowledge as the manifestation of a kind of vital and grounded force that animates and complements while remaining irreducible to any rational account of its functioning. If the type of flexibility associated with the exertion of this human faculty could not be expressed within a society, this society would soon or later become violent or generate violent reactions by its members. Persons and their

[45] On this point see, for example Jarvis et al. (2015).

specific situations come indeed before any abstract rule that can be established to administer a society.

As previously mentioned, the exertion of the human faculty associated with practical knowledge can, nevertheless, be inhibited whenever people constantly live within a myth, i.e., whenever the metaphors underlying this myth are constantly taken *literally*. Considering that this is what probably happens in case of the energy and information metaphors validated by current complex systems, the question then is whether it is possible to stop taking these metaphors literally and understand in this way the social implications of their literal interpretation. It is indeed through this change of posture that the influence of these metaphors can be taken to the foreground and alternatives to the technological practices validating them can be found. By taking the first metaphor used in this chapter as an example, it can be stated that the main route to stop taking a metaphor literally consists in managing to enter an observation perspective wherefrom this metaphor can become a *similitude* and an "as if" can hence be added in front of the statement "Mr. Smith is a Lion". By doing so, we automatically put ourselves in a different context, we put ourselves outside and take distance from the world of the metaphor. In this way we become able to observe Mr. Smith under a perspective that allows identifying in which respect Mr. Smith is different from a Lion and in which sense he can be assimilated with a Lion. Put in other words, the possible close similitude between Mr. Smith and the Lion notwithstanding, we must become able to *speak* about Mr. Smith whilst *not* speaking about the Lion. When we manage to speak of a metaphor in terms of an "as if", this is the sign that we are performing an act of *interpretation and translation*; we are exiting the world of the metaphor and we are taking Mr. Smith into our world, into the place and the time where we are staying. This means that, in order to exit a metaphor that is being taken literally, it is necessary to take either a spatial or a temporal distance from the place where the metaphor is being lived. In case of energy, this means that we have either to attempt to move back to the past in a time when this metaphor did not hold and societies were not organized according to energy principles, or to move to a possible still existing social context where people are not organized according to energy principles. There is, however, also a third and very interesting possibility. Given the categories used within the proposed account, this third approach can be called *profanation*[46] *of energy*. To profane energy means to bring the materials, the technical apparatus, the institutional settings and the technical skills out of the "sacred" world where liturgies and rituals around energy are administered by experts and give them back to the profane life of ordinary people. This is probably the most practical approach to experience the *exiting* and understand the implications of living *within* the energy metaphor. The more the administration of so-called energy resources, of related technologies and technical skills are left to the management of ordinary people, the more it can

[46]The description of the proposed approach in terms of a *profanation* has been derived from a series of considerations on how persons should relate themselves with technological artefacts as formulated in (Agamben 2009).

become possible to observe the flourishing of a diversity of practices whereby people can provide for their necessities by using natural resources without becoming energy addicted and causing an unnecessary depletion of these resources. The decision to call it "practical" approach is not accidental, as this approach corresponds to the possibility of developing and exerting the type of practical knowledge previously described.

The same type of considerations applies to the information metaphor and the value metaphor validated by current monetary systems. As previously mentioned, the literal interpretation of energy, information, and value metaphors reinforce each other within existing complex systems. These metaphors keep people within a separate sphere from where it becomes more and more difficult to exit. As pointed out by Jeremy Rifkin,[47] in an age that has been consecrated to information, people receive some comfort by believing that their efforts to generate and exchange larger amounts of information allow them to increase their autonomy and organization and that, as for natural evolution, they can strengthen in this way all their social relationships thanks to the increase of their interactions and of systems complexity. Nevertheless, the general loss of the sense of agency caused by the integration in these systems is nowadays common experience and the progressive integration into dynamics that escape any form of personal and social control takes with it the impossibility of experiencing and appealing to any sense of citizenship and responsibility. The case of renewable energy systems can represent one extremely relevant case of mutual reinforcement among the energy, information and value metaphors generating these complex dynamics. Modern information technologies allow nowadays to manage huge numbers of abstract energy units generated from extremely diffused natural sources and to distribute them within suitable energy distribution networks. The association of a monetary value with these units and their commercialisation within competitive market settings is rightly seen as the most promising approach to promote their extensive employment and reduce dependence on non-renewable energy sources and relative negative environmental impacts. The adoption on a larger and larger scale of this type of approach requires, however, an increased delegation to information technologies of the activities linked to production, transmission, distribution and end-use of energy sources together with a process of technologically managed adaptation to the discontinuities and variabilities in the supply of energy associated with the complex dynamics generated by these systems. Beside reducing harmful emissions, policy strategies reproducing this type of approach are destined to inevitably increase social dependence on complex systems dynamics and to contribute to constantly validate the literal interpretation of associated metaphors while causing a large scale homogenization in energy end-uses mostly due to the need of synchronizing energy demand to an energy supply that can unpredictably fluctuate during time. On the other hand, these strategies are completely blind to how people can develop context specific practices to temporarily and comfortably live disconnected from the energy

[47]See Rifkin (1998).

network and outside the energy metaphors. The consequences of this constant embedment in the literal interpretation of given central metaphors are relevant not only for our environment. They are relevant in particular for how they can negate to people the possibility of making direct experience of the world around them and to elaborate own solutions for the problems affecting their lives. The effects of this imprisonment within metaphors validating our constant dependence on the supply of energy and information units are completely similar to those generated by the marketisation of every aspect of our lives. The association of a monetary value with every dimension and activity of our daily life is transforming any place and social aggregate into a kind of museum and showcase where people cannot any longer make a direct and lively experience of the things around them. It is amazing to observe how this type of marketisation is changing cities (e.g. Venice, Florence and, in general, other cities declared Intangible Cultural Heritage of Humanity by UNESCO), regions (e.g. natural parks or oases) all over the world into museums where things and spaces are exhibited for consumption without allowing any actual experience of use and habitation.[48] The creation of these "sacred" spaces that people cannot profane by making any real experience of use or habitation is not an exclusive characteristic of capitalism and associated monetary systems. It corresponds to the previously described inhibition of practical knowledge that can be generally ascribed to specific types of technologies and, more in general, also to energy and information technologies. The curious thing is, however, that the main entities responsible for this inhibition are not the specific material instantiations whereby these metaphors are produced. Cars can, for example be substituted by bicycles without changing the metaphor according to which "transportation is the act of moving people from point A to point B in a given amount of time". The literal application of this metaphor within cities is alone sufficient to provoke radical modifications in the landscapes and to drastically limit the infinite ways of transit that can be adopted by people within these landscapes by expressing their practical knowledge. This metaphor does so by projecting all the possible conceivable ways of transit along the common metrics of travelled kilometers/hour either cars, or bicycles, or trains, or airplanes are used to move people. The same may happen, for example, when gas boilers and gasoline cars are substituted by heat pumps and vehicles consuming PVs' electricity. The production and the employment of these two different material arrangements can be animated by a same energy metaphor that can become the ultimate and main constraint shaping the way in which people address the issue of heating and transit. The different practices that people can develop to provide for their heating and transit are in this way generally subordinated to standard technical solutions that can be provided by experts of various kind.

This being said, it perhaps has to be stressed that technologies and associated metaphors are extremely necessary and useful. It is however also extremely necessary that they can be put among parentheses and subordinated to people

---

[48]These examples have been taken from Agamben (2007).

decisions. To do so, they have to be somehow read in transparency. They have to be kept at sufficient distance in order to become hopefully able to see how they are framing and constraining our ways of life. At the same time, however, they have to be read from the inside. We have to listen to the stories they tell us and we have to take them very seriously. These metaphors are extremely powerful. They speak to us and confirm themselves through an immense technical apparatus.

Conclusions and Policy Implications

Energy and information technologies can nowadays allow translating any human, mechanic and biological activity into extremely abstract flows of information, energy and matter units.

The evolution of any aggregate at any scale (from the smallest organisms populating an aquatic environment to the households living in a city, from the traffic congestions occurring in a road network to the winds circulating in the stratosphere) can be translated into flows of this kind and be monitored in the smallest detail.

While changing beings populating societies and ecosystems into intricate streams of abstract entities that can be managed and monitored by computers, these translations are also at the core of continuous and large scale reorganizations of the processes whereby people arrange their lives. The benefits that these processes and transformations can generate to the environment and to societies are huge.

The complex networks generated by energy and information technologies can in principle integrate on the large scale and up to the microscopic level any aspect of human and biological life. The possibility of associating a monetary value with the artificial units circulating within these networks creates incredible business opportunities which are increased by the fact that the links of these networks can be bidirectional and allow any of its nodes to manage and inject additional units into the networks. The classical distinction existing between producers and consumers of economic goods can be cancelled in this way. Persons integrated into these complex networks become goods prosumers whose activity is subordinated to the possibility of having access to the information, energy and matter flow circulating therein. With a computer and a 3D printer, persons seem to have become able to fabricate, use or sell whatever they want. Through suitable application service providers, people can transform their cars into a taxi, their houses into a hotel and their kitchens into restaurants where human and material flows can be redirected. At the same time, energy renewable technologies and computer technologies are assumed to enable everybody to provide for their own energy needs while selling extra energy units within complex distribution networks regulated by suitable market rules. The transformative power associated with energy and information technologies is enormous, and the fascination these technologies have on people, policy makers and corporate organizations is impressive.

It should however be taken into due account that the complex systems these technologies constitute can very easily disengage from any form of social control while generating artificial dynamics of consumption growth which can also have extremely negative and large scale impacts on persons and natural resources.

The thesis supported in this chapter is that a fundamental social factor contributing to the creation of these perverse dynamics has to be found in the huge process of homogenization and reification resulting from the fact that societies are identified with motors and information processors that, together with the energy and information resources whereby they function, play nowadays the role of central metaphors that are constantly taken literally The literal interpretations of these metaphors reinforce each other within present competitive market settings while being constantly validated by the huge technological apparatus whereby complex systems are being socially constructed. These interpretations shape our imaginary and inform the way communities organize becoming responsible for an increased dependency on abstract resources units and reinforcing growth in natural resources consumption.

This being said, it has however to be stressed that metaphors are necessary constituents of any social aggregate. Languages, rules and institutional settings established within societies rely typically on metaphors which contribute to structure personal feelings while connecting these feelings to abstract concepts around which social aggregates can coalesce. Nevertheless, what differentiates the energy and information metaphors validated by present complex systems is that they engage people into a literal interpretation that is practically impossible to escape and report to a personal experience.

In a time when we are constantly assimilated to computers and are invited to manage our lives and the environment as multitaskers working under a regime of resources scarcity, it becomes extremely difficult to explain how these metaphors can cause the extinction of a variety of sustainable practices while inhibiting the possibility of making actual experiences of use and habitation. The situation is then worsened by the fact that the problems caused by their literal interpretation cannot be identified even by scientists and experts asked to solve these problems. Their methods and the way they operate is indeed typically the result of this literal interpretation and represent therefore part of the problem. In a time when existing complex financial systems are escaping the control of national economies while increasingly generating situations of extreme poverty, professional managers mostly formulate solutions supposed to allow increasing monetary flows through increased monetary investments. In a time when working systems are expelling an increasing number of people inhibiting any possibility that they can provide for their necessities and actively contribute to the wellbeing of their societies, people and opinion leaders mostly just require that these systems provide more work, without discussing whether the output of this work is actually needed. In a time when fossil fuels have proved harmful for people and the environment, experts and scientists mostly point to a massive substitution with renewable energy sources and to the implementation of energy and material recycling systems that can hopefully guarantee a continuous growth in the power output of economies, without considering that this continuous growth is in itself a major sustainability problem.

These approaches are certainly important, but can serve to solve the problems at stake only to a very limited extent.

Possible way outs to this social entrapment so far described can only be found by acknowledging the axiomatic incompatibility of social organizations to the reductions operated by scientists (including economists) and the fundamental value of people's practical knowledge in the definition of solutions to increase their well-being and the sustainability of their ways of life. As mentioned at the beginning of this chapter, it is not a question of denying the extreme usefulness of the technical solutions provided by science. It is rather a question of subordinating their application to people's necessities and to the solutions they can develop through what they know and what they can experience about their context. This change of perspective requires that policy making re-gains a particular sensibility allowing that the implementation of abstract solutions is always subordinated to the specific needs of the specific context. Whilst this change of perspective is in principle easy to achieve under a theoretical point of view, it remains extremely difficult to implement in practical terms. It needs that, rather than just inputs, the outputs of human activities are also discussed and negotiated. In so far as policy solutions are informed by energy and information metaphors, they are mostly limited to technical solutions that can be implemented to reduce energy and material inputs or harmful impacts of these inputs. These solutions are, however, ultimately functional to continuous growth in resources consumption and, above all, somehow represent a way to bypass true political and democratic discussions concerning outputs. Technicalities concerning their design and implementation typically leads to the exclusion of people's contribution to their enforcement in very devious ways. People are typically just involved as passive consumers supposed to change their individual behaviour. The decision-making processes are in these cases mostly informed by individual and atomised considerations concerning investments needed by individuals and collectives to implement given technological solutions.

When sustainability issues are instead faced by taking an economy's outputs as a starting point, the problems at stake can in principle more easily become genuinely political. Energy and information metaphors can be put between parentheses and people can in principle stop taking them constantly literally. The nature of the policies that can be implemented to increase sustainability can also change radically in this way. When adopting the proposed perspective, it can be easily acknowledged that the variety of policy options that can be considered to increase sustainability increases dramatically. Decisions concerning, e.g. where and how to build a school or how to allow children reaching it that are taken by having the outputs of these activities as main focus, enable the implementation of plenty of context dependent solutions reducing the amount of inputs needed. Decisions taken within school institutions concerning, e.g. how to improve learning capabilities of people enable to envisage solutions not exclusively relying on the consumption of natural resources (like those related to the usage of computers, books, etc.). When, instead these decisions are taken by starting from the inputs, the related outputs result somehow implicitly frozen. They cannot either be changed or re-discussed, either because the presence of these outputs is erroneously considered as the expression of a not constrainable people's free will, or because a possible re-discussion of outputs is seen as potentially detrimental for the existing economic system. In addition, the implementation of most technological solutions aiming at reducing inputs is typically promoted to be replicated in the highest possible number of sites because of the economic interests of involved actors, this situation typically implying that this implementation can become counterproductive in several cases (e.g. is it really necessary that all houses, cars and other technical equipment are replaced by more efficient substitutes when some of them are very rarely used by their owners? Rather than pretending substituting gasoline cars with electric vehicles in every circumstance, could not it be better complementing this approach with an approach focused on the reduction of the needs for vehicles? In which circumstances specific human practices can be preferred to automated solutions for the provision of comfort within buildings?). For the way in which the energy and information metaphors have originated, there is always a close connection between energy, information and monetary considerations impeding that the need for given outputs and the possibility to increase well-being by reducing these outputs can be actually discussed. Objects and activities around us seem to be destined to be immediately transmogrified and converted into an associated energy, information and monetary content regulating their reproduction and impeding that they can be manipulated and adapted by people according to their will. This process of mutual reinforcement among energy, information and monetary metaphors is exemplary in the case of the ongoing transition to renewable energies occurring in several countries. Without computer technologies this transition could not have been even hypothesized. Through these technologies, it becomes possible to hypothesize that energy produced from highly distributed renewable energy sources can be managed and redirected to provide the energy inputs needed by whole cities and countries. Monetary values associated with the units of energy generated from renewables and managed through computer technologies are then supposed to provide the main leverage for this transition. The artificial dynamics of growth that can be generated in this way and the way in which they can dis-embed from any form of political control are probably considered by policy makers as a minor thing compared to the environmental and economic benefits expected from them. These general economic and environmental benefits often lead to neglect the risks generated by the progressive complexification and homogenization of the energy supply system (mostly oriented to supply electricity) associated with this transition. In addition, the need to promote the development of alternative practices that people may adopt to survive to the anything but infrequent system crashes that can be expected in this transition do not seem to be object of major concern. It is instead not so unrealistic to assume that the actual accomplishment of this transition depends on the possibilities that societies will be given to develop ways of life allowing them to temporarily and comfortably survive while disconnected from current energy, information and monetary systems. This condition of temporary disconnection would be enabled by the huge variety of context dependent solutions that people may elaborate to increase their well-being while reducing their dependence on these complex systems by relying on their practical knowledge.

References

Agamben, G. 2007. *Profanations*. Translated by Jeff Fort. The MIT Press.
Agamben, G. 2009. *What is an Apparatus?* Translated by David Kishik and Stefan Pedatella, Stanford University Press
Allen, T.F.H., J.A. Tainter, and T.W. Hockstra. 2003. *Supply-Side Sustainability*. Columbia University Press (ISBN 0-231-10586-X).
Bateson, G. 1972. *Steps to an Ecology of Mind*. University of Chicago Press (ISBN 0-22603905-6).
Bateson, G., et al. 1956. Toward a Theory of Schizophrenia. *Behavioral Science* 1: 251–264.
Bauman, Z. 1998. *Globalization: The Human Consequences*. Polity Press in association with Blackwell Publishers Ltd.
Bridgman, P.W. 1961. *The Nature of Thermodynamics*. Oxford University Press.
Chiribella, G., M.G. D'Adriano, and P. Perinotti. 2011. Informational Derivation of Quantum Theory. *Physical Review A*, 012311 (2011).
Dupuy, J.P., and J. Robert. 1976. *La Trahison de l'Opulance*. Presses Universitarie de France.
Durkeim, E. 1915. *The Elementary Forms of the Religious Life*.
Feynman, R. 1964. *The Feynman Lectures on Physics*. vol. I, 4–1.
Goerner, S.J., B. Fath, and D. Fiscus. 2015. Using energy network science (ENS) to connect resilience with the larger story of systemic health and development. Article available at https://journal.emergentpublications.com/article/using-energy-network-science-ens-to-connect-resilience-with-the-larger-story-of-systemic-health-and-development/.
Heidegger, M. 1994. *Being and Time* (trans: Macquarrie, J. and E. Robinson). New York: HarperSanFranscisco. (Original work published 1927).
Illich, I. 1976. *Medical Nemesis. The Expropriation of Health*. Pantheon Books
Illich, I. 1983. *The Social Construction of Energy*. Published in New Geographies 02—Landscape of Energy. Harvard University Press, 2009.
Illich, I., M. Rieger, and S. Trapp. 1996. *Speed? What speed?* Speeches given at the "Speed" Conference of the Netherlands-Design Institute available at http://www.pudel.uni-bremen.de/pdf/illich_rieger_trapp96speed.pdf.
Jarvis, A.J., S.J. Jarvis, and C.N. Hewitt. 2015. Resource Acquisition, Distribution and End-use Efficiencies and the Growth of Industrial Society. *Earth System Dynamics* 6: 1–14.
Lakoff, George, and Mark Johnson. 1980. *Metaphors We Live By*. Chicago: University of Chicago Press.
MacCormac, E.R. 1976. *Metaphor and Myth in Science and Religion*. Durham, North Carolina: Duke University Press.
Mirowski, P. 1989. More *Heat than Light. Economics as Social Physics, Physics as Nature's Economics*. Cambridge University Press.
Mitcham C. 1994. *Thinking Through Technology*. The University of Chicago Press.
Neumann, E. 1971. *Amor and Psyche: The Psychic Development of the Feminine*.
Perulli, A. 1996. *Il Tempo da Oggetto a Risorsa*. Franco Angeli Editore.
Poerksen, U. 1995. *Plastic Words—The Tyranny of a Modular Language*. The Pennsylvania State University Press.
Polanyi, K. 1944. *The Great Transformation*.
Rabinbach, A. 1992. *The Human Motor: Energy, Fatigue, and the Origins of Modernity*. New York: Basic Books.
Rifkin, J. 1998. *The Biotech Century. Harnessing the Gene and Remaking the World*. J P Tarcher (ISBN 0-87477-909-X).

Samerski, S. 2002. *Risk, Anxiety and the Myth of Informed Decision Making*. Lecture at the Institute for Criminology (Norway), April 17th, 2002. Text available at http://www.pudel.uni-bremen.de/pdf/SamerskiOslokorr.pdf.

Taleb, N.N. 2001. *Fooled by Randomness: The Hidden Role of Chance in Life and in the Markets*. Texere (ISBN 1-58799-071-7).

Ulanowicz, R.E. 1997. *Ecology, the Ascendent Perspective*. Columbia University Press, pp. 63–73.

Vico, G. 1744. *La Scienza Nuova*.

Chapter 9
Energy and Social Practice: From Abstractions to Dynamic Processes

Elizabeth Shove

Abstract Energy policies are typically organised around the supply, management and reduction of energy conceptualised as a singular resource and measured in standardised units like KWh or Mtoe. This kind of abstraction enables national and international institutions to collect and compare data on per capita consumption, the effect of efficiency measures, progress towards emissions targets and the like. The problem is that such approaches treat energy, and energy consumers, as topics of analysis in their own right, stripped from the historically and culturally specific situations in which demand arises. In this chapter I make the case for seeing energy demand as something that is intimately related to the conduct of social practices, and thus inseparable from the spatial and temporal ordering of society, and from the infrastructures and institutions involved. I argue that better understanding of the dynamic and recursive relation between supply, demand and social practice is both necessary and important, particularly given the increasing significance of renewable energy and related challenges of matching peaks in provision with those of consumption. This way of thinking has policy implications. Rather than seeking to maintain present ways of life (but with lower carbon energy supplies), I suggest that the longer term goal could and should be that of imagining and promoting technologies, practices and socio-temporal orders that are compatible with greater reliance on renewables and reduced demand, accepting that this is likely to entail the emergence of ways of living that are really very different from those with which we are familiar today.

Introduction

How energy is known, measured and understood is hugely important for the development of carbon reduction policies and for strategies adopted in pursuit of these goals. In collecting and analysing data on the production and consumption of

E. Shove (✉)
DEMAND Centre/Department of Sociology, Lancaster University,
Lancaster LA1 4YD, UK
e-mail: e.shove@lancaster.ac.uk

N. Labanca (ed.), *Complex Systems and Social Practices in Energy Transitions*,
Green Energy and Technology, DOI 10.1007/978-3-319-33753-1_9

energy, national and international organisations work with a limited palette of standardised metrics, the most common of which are Million Tonnes of oil equivalent (Mtoe), or million tonnes of carbon emissions. Units like these make it possible to aggregate and to thereby 'see' the extent of the problem and to quantify and evaluate the impact of steps taken in response. The paradox is that these arguably necessary methods and approaches prevent researchers and policy makers from engaging effectively with the multiple dynamics of energy demand or with the fundamentally different characteristics of renewable rather than fossil fuels.

There is nothing new in the suggestion that theories, methods and paradigms are inherently selective: in highlighting certain features they inevitably obscure others. It is also not surprising to discover that concepts and measures are products of their time: in the energy field, it is no accident that 'oil equivalent' is the common point of reference. As described below, historical reliance on fossil fuel has a bearing on the terms and parameters of energy-related analysis, and on conceptualisations of energy as a singular resource that can be allocated, managed and depleted.

It is now so thoroughly normal to represent and account for energy in these terms that it is easy to overlook the work that lies behind the production of pie charts showing energy use by sector, of Sankey diagrams depicting the movement of energy from sources of supply through to end use and consumption or forms of input–output accounting like those used in attributing carbon emissions to nation states (Scott and Barrett 2015). It is nonetheless important to remember that producing figures like these depends on abstracting 'energy' from the range of technologies and practices in which it is enmeshed and from the multiple settings and moments in which it is produced, distributed, transformed and used.

Methods of knowing and managing energy that depend on techniques of abstraction, standardisation and equivalence tend to preclude close analysis of how, when and where energy demands are made and reproduced. These time-less, or time-independent representations relate to the tendency to think of energy as a quantifiable resource and as something that can be stored and used at a later date, as is the case with fossil fuels. Such 'purifying' (Latour 2012) moves make it difficult to engage with fluctuations in the timing, duration and sequencing of the various practices that underpin energy demand. These are especially limiting features when thinking about a future in which renewables have a much more significant role than they do today. Developing this idea, any really substantial movement away from fossil fuels almost certainly implies correspondingly substantial shifts away from resource-based views of energy and from related concepts of efficiency and consumption. Put differently, greater reliance on renewables calls for a major overhauling not only of infrastructures and systems of provision, but also of how energy is defined and understood. In effect a more sophisticated account of energy as a feature of the situated and dynamic enactment of social practices is a precondition for comprehending and shaping the timing and the dynamics of supply *and* demand (Shove and Walker 2014).

In exploring these themes, the first part of the chapter discusses the tendency to treat representations of energy as if they had meaning in their own right. The second part considers methods of reconceptualising the place of energy-and-practice within society.

Abstracting Energy

In the span of human history, energy has only recently been isolated and described in the way that it is today. It was not until the 1840s that previously important theories about vital forces gave way to a handful of interlinked ideas which established 'energy' as a common point of reference, enabling further distinctions to be drawn, for instance, between potential, thermal and kinetic forms. The laws of thermodynamics are part of this tradition, as are understandings of how energy is transmitted, converted and 'lost'. Alongside and as part of these theoretical developments, standard units (e.g. the Joule) replaced what were previously localised, variable and situated forms of knowledge about horse power, manpower, candle power, etc.

In the physical and natural sciences, and in policy, energy is now known and discussed in the singular and in terms that are removed from many and varied moments and sites of 'use'. This is reflected in widespread reliance on what have become thoroughly routinized methods of measuring energy and of estimating carbon emissions. Units like Million Tonnes of oil equivalent (Mtoe)[1] are used to represent and summarise different forms of energy provision and consumption. The carbon consequences of a plethora of energy-and-emissions-related activities are rendered comparable in much the same way, being described in standard units of tonnes, or millions of tonnes of CO_2. It is usual to assume that assessments of this kind record and capture relevant trends on a global scale, and in a sense they do. It is nonetheless important to recognise the 'performative' character of such calculations and the forms of averaging and aggregation involved.

By performative I mean that measurement and calculation have an active role in shaping and framing what count as relevant questions and lines of enquiry: they do not simply reflect what is going on in the world. They also constitute and sustain understandings of problems and of potential solutions. Contemporary techniques typically reproduce an understanding of energy as a finite resource: specifically, as *oil* equivalent. As discussed in the next part of the chapter, such representations underpin ideas about efficiency and consumption which percolate through national and international policy agendas.

[1]One tonne of oil equivalent represents the energy content of a metric tonne of crude oil.

Energy Efficiency

Increasing efficiency, for instance of a light bulb or a car engine, depends on knowing, with some precision, how much energy is required to produce a specific result: to deliver a certain amount of light or to enable a standard car to travel a certain distance under certain conditions. Amongst other things, measuring energy input and comparing the outcome depends on defining, stabilising and quantifying 'relevant' features such as levels and qualities of light or aspects of a car's performance. The limits and boundaries that are needed to evaluate efficiency also limit and bound the scope of efficiency-oriented programmes. This works in different ways.

Since measures of efficiency depend on comparing things which purport to deliver the same service, changes in the meaning of 'service' are consequently out of view. In practical terms, this means that if items like fridge freezers increase in size or if they offer additional facilities (ice making, etc.), new protocols are required to ensure that their efficiency is fairly assessed and compared to other equivalent models. Because the focus is on efficiency, not on energy consumption, trajectories of product development are out of scope.

A second related point is that measures of efficiency work with, and thus reproduce particular understandings of what a car is, or of what a freezer should do. Focusing on efficiency tends to obscure the recursive and somewhat longer term co-evolution of technologies and practices. To give a different example, lighting technologies are implicated in constructing and reproducing ideas about what constitutes good or acceptable standards and qualities of light: these qualities are then treated as fixed points of reference when evaluating the relative efficiency of different bulbs and fittings.

Methods of identifying and enhancing efficiency are designed to isolate and abstract the 'energy' from the ongoing conduct (and transformation) of social practices of which more and less 'efficient' technologies are a part. This result is to emphasise technological substitution and equivalence of delivery, forgetting that things like cars, washing machines and heating systems have histories that are themselves bound up with the provision and consumption of energy. As a result, and precisely because they depend on stripping energy out of context, efficiency agendas side-line questions of change which are arguably crucial for any understanding of demand.

In measuring efficiency the meaning of useful work[2] is stabilised and taken for granted: all that counts are the means—i.e. the standardised units of energy—through which this work is done. This separation of means and ends is doubly problematic in that it disguises the extent to which definitions of service (necessarily black boxed in assessments of efficiency) are shaped by the technologies involved. The International Energy Agency's (IEA) suggestion that 'energy efficiency' should be thought of as a 'fuel' alongside others like coal, gas and oil (and

[2]Physicists define energy as 'the means to do useful work'.

measured in Mtoe avoided) compounds these problems, entailing a further level of abstraction by treating energy that is not used (because of efficiency measures) as some kind of virtual supply.

Representing Efficiency as a Fuel

In 2013 the IEA published its first Energy Efficiency Market Report which ‘sits alongside IEA market reports for oil, gas, coal and renewable energy, highlighting its place as a major energy resource’.[3] Producing this assessment required considerable methodological ingenuity, including setting baselines from which to estimate the energy that might have been used had efficiency measures not been introduced,[4] touching on issues of rebound (in which money saved by installing efficiency measures might be used in ways that increase energy consumption elsewhere), and grappling with the complexity of estimating ‘cumulative’ avoided energy.[5] Having established methods of calculating avoided energy, and having assessed the costs of installing energy efficiency measures, the report’s authors go on to calculate what are described as ‘reserves’ of energy efficiency—namely the catalogue of presently cost effective efficiency measures that have yet to be taken. As described, ‘the sum of these opportunities at today’s price levels can be considered to be our “reserves” of avoided energy use. These reserves are analogous to the world’s stated reserves of oil or gas.’ (International Energy Agency 2013: 30).

Estimates of the extent of efficiency-as-fuel are bewilderingly huge. To quote: ‘Between 1974 and 2010, cumulative avoided energy consumption due to energy efficiency in these IEA member countries amounted to over 1350 EJ (32 billion toe).’ (International Energy Agency 2013: 55). Such estimates are also bewildering in the sense that they peg definitions of service (that is of what energy + technologies are expected to deliver in terms of heating, cooling, speed, etc.) to a fixed point (1974) and evaluate efficiency in delivering them, but without acknowledging that expectations of heat, cool, speed, etc. evolve. This is, perhaps, an extreme example but the step-by-step logic of calculating the use of efficiency-as-fuel in units of avoided Mtoe is entirely consistent with dominant methods of conceptualising energy in ways that lift it out of the flow of social practice and out of related patterns of social, cultural and technological change.

[3]http://www.iea.org/publications/freepublications/publication/energy-efficiency-market-report-2013.html.

[4]Examples include better building insulation, more efficient appliances, light bulbs, etc.

[5]Here, the challenge is to quantify the energy not used In the year following the baseline (i.e. when the energy efficiency measure was introduced), and in all subsequent years of the expected lifetime of the ‘measure’—whatever that might be.

Energy Consumption and Energy Consumers

The standardising language of energy goes hand in hand with an also 'flattened' account of consumer behaviour. Representing people as 'energy consumers' rather than as commuters, home owners or chefs prevents further analysis of how patterns of energy consumption follow from the enactment of practices as diverse as those of commuting or having dinner. Instead, all feature as similar if not equivalent instances of energy consumption. The same applies in the field of transport, a sector in which it is usual to compare and quantify passenger kilometres travelled.[6] As with units like Mtoe, any one passenger kilometre travelled is equivalent to the next. Given that it is normal to describe journeys in these terms, policy analyses rarely differentiate between trips to the supermarket or to school: as a result there is often no way of knowing whether diverse mobility dependent practices are evolving in similar or in radically different ways.

Instead, energy consuming behaviours are discussed as if people were also standardised 'units', and as if moments of consumption were, for all practical and analytic purposes, identical. Consistent with this view, initiatives designed to modify energy-related 'behaviour' routinely suppose that trends and patterns are outcomes of a handful of generic behavioural drivers, typically including attitudes towards energy/environment and price (Chatterton 2011).

In all of these, the figure of 'the energy consumer' is multiply removed from the flux of day to day life. Not only is he or she taken to consume 'energy' rather than mobility, heat, entertainment etc., not only is there no historical or situated account of the technologies and practices involved in travelling or keeping warm, there is also no recognition of how such practices change and vary or how they come to be shared across space and time. Instead, increases and decreases in energy use are thought to reflect narrowly defined commitments to energy conservation or carbon reduction (Shove 2010), or the price of fuel.

Different discourses, starting from physics but extending into policy, economics and behavioural science/psychology, reinforce and amplify each other, creating an impression that 'energy' exists as a topic in its own right and that energy is something that people save, consume and waste. Policies grounded in these ideas separate 'energy' from the multiple historically and culturally specific practicalities of demand. They work with a view of energy that is indeed 'oil' equivalent as regards processes of using (up), storage and consistency; they take no account of when consumption happens (units and moments of consumption are equivalent); and they prise 'energy' apart from other forms of consumption, material culture, technology and practice. As such they are quite unsuited to the problems of organising and handling energy which is *not* oil equivalent.

---

[6]https://www.gov.uk/government/uploads/system/uploads/attachment_data/file/489894/tsgb-2015.pdf.

Renewing Ideas

The share of renewable energy in global power generation is expected to rise to over 26% by 2020.[7] At first sight, variable renewables like solar and wind power represent useful, low carbon additions to existing energy sources, and it is in these terms that their actual and potential contribution is generally understood. For example, the IEA reports that 'In 2013, world total primary energy supply (TPES) was 13,555 million tonnes of oil equivalent (Mtoe) of which 13.5%, or 1829 Mtoe, was produced from renewable energy sources'.[8]

Despite this description, renewables are not 'oil equivalent': they are not depleted or stored in the same way, the scale of the 'resource' cannot be estimated in the same terms, and there are distinctive and important variations in the timing and location of harvesting or 'production'. Since there are significant losses involved in converting renewable energy into forms that can be transported over any distance, or stored on any scale there is a distinctive immediacy to the relation between supply and demand.

These features do not prevent analysts from aggregating and averaging renewable energy as if it were fossil fuel (see above). But it is obvious that information about the average annual output of a nation's wind turbines is of limited value for those who are trying to manage and use wind energy on a daily basis. Sometimes turbines produce a lot of power, sometimes not, and to complicate matters, output varies from one location to the next.

As one might expect, efforts are being made to 'tame' renewable energy, whether by slotting it into a world of existing policy, provision and practice or by making it fit established conventions of representation and analysis. In this context it is no wonder that there is so much emphasis on developing energy storage, including batteries and electric vehicles, and in using smart grids to help cope with awkward variations in supply.[9]

This is not the only way to go. Rather than reproducing a fossil fuel mentality, a more ambitious and also more challenging response is to re-conceptualise energy—or to be more precise, to re-conceptualise the relation between energy and social practice—and to do so in terms that are capable of capturing and characterising fluctuations, ranges and variations in the extent and quality of supply and demand across different spatial and temporal registers. Although inspired by the distinctive features of renewables, and especially matters of intermittency and timing, the ideas outlined below entail a more deep-seated revision of the terms in which energy-society relations are understood. Amongst other things, such an approach

[7]http://www.iea.org/aboutus/faqs/renewableenergy/.

[8]http://www.iea.org/publications/freepublications/publication/RENTEXT2015_PARTIIExcerpt.pdf.

[9]Significant fluctuations in supply only count as a problem in a context in which producers and consumers are used to thinking about energy as a uniform, oil equivalent resource, and in which there is an expectation of continuous, uniform and stable supply all year round.

situates the pursuit of efficiency and helps overcome the limitations of a narrow interest in energy consuming behaviour. More fundamentally it involves reversing the tendency to abstract energy or to see it as a separable subject. Instead, energy (supply and demand) is understood within and as part of a more comprehensive analysis of the dynamics of social practice. The following paragraphs give a sense of what such a conceptual renewal might involve, starting with a discussion of the temporal relation between supply and demand.

Renewing Representations of Time and Energy

One of the problems of radically increasing the share of renewable energy is that in a country like the UK, peaks of provision do not correspond to peaks of demand. This begs the question of how and why moments of peak demand come to be as they are, and of whether they might shift. At first sight, these are not questions about energy: rather they are about the range of social practices enacted in society and related aspects of timing, duration, sequencing and synchronisation.

This is not the place to rehearse the ways in which 'time' and especially clock-times have been conceptualised but as Glennie and Thrift explain, time, like energy, 'comprises a number of concepts, devices and practices' (Glennie and Thrift 2009: 9). Amongst other things, this means that understandings of time vary and evolve. Contemporary interpretations of time as a measurable, finite but also abstract resource have a short history, and one that has also revolved around the production of standards, units and notions of equivalence. These parallel histories combine in that methods of defining and managing the problems of matching supply to demand in 'real time' revolve around typically 'detached' understandings both of time and of energy.

Smart grids and smart appliances are, for example designed to influence the timing of energy demand. These and other such strategies suppose that householders and organisations are free to rearrange the standardised currencies of time, energy and money at will, adapting the timing of energy-demanding practices to fit the tariffs of the day. The hope, here, is that these techniques will make it possible to bring the awkward ebb and flow of renewable energy into line. That is, bring it into line with a set of temporal patterns that are themselves outcomes of an historic reliance on fossil fuels.

The fact that energy-time management is a tricky business, and that it is often difficult and sometimes impossible to 'shift' the sequences of daily life is not simply indicative of the fact that social practices are interconnected, coordinated and synchronised. The further point is that energy-time management strategies are designed around a concept of time which is as abstract and as reified as that of energy. These standardised representations have only limited purchase on the flow of daily life in part because the practicalities of timing and scheduling are not in some sense 'outside' the realm of practice, but are outcomes of it (Shove 2009). To put it more directly, social rhythms and meanings and experiences of time reflect and are reproduced through conventions like those of family life, the week and the weekend, and the special characteristics of Friday night (Zerubavel 1982). From this point of view time, like energy, cannot be extracted from the plenum of practice.

To elaborate, arrangements like 'the week-end' or the evening meal are not natural, they do not arise by accident, nor do they develop in ways that are independent from technologies, infrastructures and resources. To give one obvious example, the widespread provision of electric lighting transformed the length and standardisation of the working day. In addition, systems of provision along with infrastructures of power and of transport are typically designed and sized to cope with 'peak' demand. The existence and the persistence of a standard 9 am–5 pm working day is thus multiply woven into present regimes of energy provision.

Efforts to embed renewables into existing systems and societal rhythms currently seek to mimic and maintain forms of energy-society relations that are grounded in an understanding of energy as a temporally stable and reliable resource. Anything else spells trouble. In particular it spells trouble for established metrics and methods of measuring and representing 'energy', none of which make reference to the relational timing of supply and demand. This is curious in that matters of timing are massively important, not for the IEA or national energy agencies, but for companies involved in energy markets.

One response, and one way of bringing the intermittency of renewables into a modified version of normally 'timeless' ways of thinking about energy is to continue working with standardised units of energy but to value and qualify them differently depending on exactly when they are produced and used. A KW at peak time (KWAPT) is thus not the same as a KW outside of peak time (KWOPT). Widespread use of new units like KWAPTs and KWOPs would depend on the production of a standardised, internationally recognised method of assessing the fluctuating and changing relations between energy supplies and patterns of demand. Should such measures exist they would complicate global estimates of energy supply currently represented in temporally 'flat' units of Mtoe, but would share many of the same characteristics. In effect they would fold a standardised time dimension into already familiar processes of averaging, aggregating and managing.

Another more radical alternative would be to develop a thoroughly relational and a thoroughly situated understanding of energy-in-use and of energy-in-time that might actively foster the re-emergence of complexes of social practice and temporal orders more closely attuned to the seasons and the ebb and flow of renewable resources. Should they exist, methods of representing relations between energy-and-the-timing-of-social-practice could not be aggregated or averaged like Mtoe, or like KWAPs and KWOPs. Whilst this has obvious disadvantages, especially given the needs and ambitions of national and international organisations, it would provide an arguably more meaningful representation of the various socio-technical arrangements in and through which energy is, in fact, produced, transformed and used. These suggestions have further consequences (a) for understanding 'efficiency', or to be more precise, for analysing the recursive relation between technologies and practices, and (b) for comprehending the dynamics of energy demand in terms that go beyond limited accounts of energy consumers and their behaviour.

Renewing Ideas About Energy Efficiency

Normal methods of evaluating energy efficiency depend on quite specific forms of abstraction and boundary-making. Are there ways of reconceptualising relations between energy (resources), appliances and practices so as to reveal and perhaps influence their interaction? One way of thinking about this question, and hence of recovering the possibility of conceptualising the dynamics of energy demand, is to reconsider the scope of analysis. For example, rather than stripping energy out of context, and rather than focusing on vehicle efficiency in isolation, it would be possible consider the energy involved in commuting to work. At a minimum, such an exercise would draw attention to issues of distance as well as to modes of transport. Extending the boundary of what is included in a judgement of efficiency—e.g. not the performance of a car engine, but patterns of commuting—would most likely inspire new forms of policy response and investment.

A second option is to work with different terms and units of comparison. As explained above, present methods of assessing efficiency depend on comparing like with like. As such they disguise trends over time. What is needed is a method of highlighting points of non-equivalence and folding these into a more 'rolling' or dynamic mapping not only of how services are provided, but also of how they change.

According to Kris De Decker, the current Citroen C1 does about the same number of miles per gallon (mpg) as a 2CV from the 1950s.[10] Whilst the C1's engine is much more 'efficient' in technical terms, it is used to drive a vehicle that is heavier, that has windows that wind down, and that has all the features one would expect of a car today. The mpg would increase significantly if manufacturers were to put a modern engine inside an old 2CV, but the result would not correspond to what now counts as a 'car'. As already mentioned, focusing on engine efficiency alone, and insisting on comparability, e.g. between the C1 and other similar cars reveals nothing about how the 'yardstick'—in this case the meaning of a car—evolves. Somehow what is required, but also missing, is a means of representing changing relations between energy-and-service that is sensitive to the recursive and dynamic relation between technologies, expectations and practices. It is important not to resort to simplistic notions of function, but it is also liberating to think of how one might compare the energy involved in drying clothes on a washing line as compared with a tumble dryer. The obvious objection that these are not 'the same' is itself part of the story: as I say the challenge is in part one of reintroducing an

[10]http://www.lowtechmagazine.com/2008/06/citroen-2cv.html.

account of difference and of incomparability in order to 'see' the changing roles of energy in society. Whatever else, thought experiments of this kind underline the point that energy is not used 'raw': resources, devices and infrastructures are always interlinked and are, in turn, inseparable from what it is that people do, how they do it, and how this changes.

Renewing Ideas About Energy Consumers

Fossil fuel-based policies and analyses treat energy as a commodity and as a resource. Energy is consequently thought of as something that people 'consume'. Quite different issues come into view if energy is reconceptualised as part of what people do. This is not just a semantic point. The proposition that people do not use energy for its own sake but always and only as part of accomplishing social practices at home, at work or in moving around highlights the need for an account of energy demand as an outcome of social/technical processes, not as an expression of consumer choice.

The significance of reconceptualising 'consumption' in these terms becomes obvious when thinking about what it might mean to match demand to more intermittent and more variable forms of renewable energy supply. As already mentioned, peaks and troughs in demand are outcomes of the collective, not personal, scheduling of different areas of daily life. Whilst it is difficult, but not impossible, to imagine a situation in which working hours were seasonally adjusted, or in which certain activities were commonly re-scheduled depending on the weather (and thus related to the generation of renewable energy), it is quite out of the question to think of such developments as expressions of individual choice.

More ordinarily, and because energy is not dis-embedded or separated from items like cars, freezers or tumble dryers, people do not in any meaningful sense 'consume' energy. In essence this means that any representation of the dynamics of demand is, at the same time, a representation of changing practices. The issue here is that aggregate trends in energy consumption reveal nothing about the detail of exactly which energy-demanding practices are moving, in which direction, at what rate and with what consequences for other interconnected areas of daily life. This argues for methods of analysis and policy-making which set a concern with 'energy consumption' and with the 'energy consumer' aside in order to identify and distinguish between the processes and relationships at play, for example, in relation to driving as opposed to freezing or drying. Rather than trying to isolate a handful of generic factors that propel energy consumption (as if these pertained in all historical contexts, cultures and contexts) such a strategy would favour a range of different enquiries, focusing on how specific conjunctions of technology-and-practice circulate (as is the case with air-conditioning and cooling), or on how food chains have come to rely on a global network of home freezers.

Finally, it is important to recognise that social practices influence each other, forming what have been described as loosely connected bundles or more closely interdependent complexes (Shove et al. 2012). Time use data provides traces of some of these relations. For example, studies of what people are doing at different times of day show that a much higher percentage of French people stop for lunch than is the case in Finland, for example. This collective habit represents a form of societal synchronisation that has knock-on consequences for the timing and scheduling of other activities (Shove 2009).

Energy-dependent practices are linked in various other ways, including via shared reliance on infrastructures of power. Compared with devices and appliances, which are often practice-specific (a toaster, a freezer), infrastructures like an electric grid enable the powering of many practices at once. Figuring out how energy is situated in society is thus a matter of figuring out how resources, appliances and infrastructures work together and what these conjunctions mean not only for the lives of individual practices but also for the emergence and disappearance of different forms of inter-practice connectivity. The terms of such an analysis have yet to be worked out, but it is in any case evident that it is a mistake to think of 'energy' consumption as a category in its own right, or to think of consumers as separate decision-making units.

Renewing Agendas in 'Energy' Research and Policy

This chapter has outlined and polarised two very different ways of knowing energy: one that proceeds by abstracting, the other by embedding.

For the time being, the first approach constitutes what amounts to a dominant paradigm reproduced in research agendas, journal articles, reports and policies around the world. Techniques that enable global assessments of trends and opportunities for decarbonisation consequently depend on (a) stripping 'energy' out of the situations in which it is 'made' and used and (b) conceptualising it as an oil equivalent resource. The understandings that follow feed into compatible, but yet more distanced or abstracted ideas about efficiency, consumption and behaviour. The result is a total package of thinking that hangs together but that is progressively disconnected from what people do. As discussed in other chapters in this volume, renewable energy has the potential to upset the entire apple cart.

So far the response has been to look for ways of overcoming or obliterating the practical and theoretical challenges posed by renewable sources. This is likely to continue. However, there are already signs that the dominant paradigm is starting to creak at the seams. Some utilities are developing business models that depend on the provision of 'energy services' not resources as such. This depends on a much finer grain analysis of what energy is 'for' and on a more direct and obvious involvement in when and how energy is used. Real-time tariffs, designed as means of handling peaks in supply and demand, have new and different roles as the share of variable renewables increases. At some point it may be truly meaningless to try to describe energy without reference to social–temporal rhythms of generation and use.

Likewise, the efficiency agenda, to date a mainstay of energy policy is subject to a number of increasingly powerful critiques. These generally revolve around issues of rebound, suggesting that efficiencies are likely to be counterproductive in the longer run. But the limitations of side-lining more fundamental changes in what energy is used for over time are also coming into view. There are obvious political advantages in promoting efficiency—who could object to the idea of using less to achieve the same?—and equally obvious risks in broaching issues of how expectations change and of whether conventions and practices can or should be steered. But sticking to the seemingly 'safe' ground of efficiency is itself an intervention: effectively stabilising and reproducing specific interpretations of 'need' (e.g. of what makes a car, what a freezer should do). Meanwhile, efforts to modify consumer behaviour (as defined above) putter along, disregarding major developments in the spatial and temporal ordering and organisation of daily life, and instead showing up as marginal (e.g. 5–10%) reductions in energy use, as measured against a fictionally stable benchmark.

Would it really make sense to scrap all talk of Mtoe or Joules and start from scratch? And if so what might an alternative involve? Ironically, some clues are to be found in physicists' definition of energy as the means to do useful work. Work can, of course, be reduced to a small movement, or to a shift in temperature, but there is scope for developing a more historical and sociologically sophisticated interpretation of the concept of 'useful work', and of the practices that constitute it. Emphasising this part of the definition, i.e. the useful work, not the means, reintroduces the possibility of appreciating that energy has no abstract meaning, but is instead always part of some practical undertaking—some form of practice or 'work' broadly defined. Such a move would make it possible to recognise that interpretations of 'use' develop and change alongside the means of provision.

I am not sure I can imagine an international agency of useful work, not least because the practices that constitute 'useful work' have localised and differentiated dynamics, though some aspects may indeed be international. However, it is not too difficult to think about what new style 'energy' policies might entail at other scales. For a start and as indicated by the quotation marks, they would not be confined to departments of energy. Instead, such strategies would recognise that many different areas of policy-making have a hand in configuring and shaping trajectories of practices and practice complexes that matter for the timing of what people do and for the technologies and resources on which those doings depend. In other words, the development of a paradigm anchored in practice depends on taking the 'energy'—as an abstract and reified concept—out of energy policy. Research and policy would no longer focus on identifying opportunities for substitution (e.g. with renewables), efficiency or marginal behaviour change, all geared around present practices and ways of life. Instead, agendas and programmes would form around the challenge of imagining and establishing configurations of technologies, practices and socio-temporal orders that would be compatible with greater reliance on renewables and with reduced demand, accepting that the result might involve the emergence of ways of living that are really very different from those with which we are familiar today.

Acknowledgements This work was supported by the Engineering and Physical Sciences Research Council [grant number EP/K011723/1] as part of the RCUK Energy Programme and by EDF as part of the R&D ECLEER Programme.

References

Chatterton, T. 2011. *An Introduction to Thinking About 'Energy Behaviour': a multi-model approach*. London: Department of Energy and Climate Change.

Glennie, P., and N. Thrift. 2009. *Shaping the Day: A History of Timekeeping in England and Wales 1300–1800*. Oxford: Oxford University Press.

International Energy Agency. 2013. Energy Efficiency Market Report 2013. from https://www.iea.org/publications/freepublications/publication/EEMR2013_free.pdf.

Latour, B. 2012. *We Have Never Been Modern*. Massachusetts: Harvard University Press.

Scott, K., and J. Barrett. 2015. An Integration of Net Imported Emissions into Climate Change Targets. *Environmental Science & Policy* 52: 150–157.

Shove, E. 2009. Everyday Practice and the Production and Consumption of Time. In *Time, Consumption and Everyday Life: Practice, Materiality and Culture*, ed. E. Shove, F. Trentmann, and R. Wilk, 17–35. Oxford: Berg.

Shove, E. 2010. Beyond the ABC: Climate Change Policy and Theories of Social Change. *Environment and Planning A* 42 (6): 1273–1285.

Shove, E., M. Pantzar, et al. 2012. *The Dynamics of Social Practice: Everyday Life and How It Changes*. London: Sage.

Shove, E., and G. Walker. 2014. What Is Energy For? Social Practice and Energy Demand. *Theory Culture and Society* 31 (5): 41–58.

Zerubavel, E. 1982. The Standardization of Time: A Sociohistorical Perspective. *The American Journal of Sociology* 88 (1): 1–23.

Chapter 10
Radical Transitions from Fossil Fuel to Renewables: A Change of Posture

Timothy Allen, Joseph Tainter, Duncan Shaw, Mario Giampietro and Zora Kovacic

Abstract The transition from fossil fuel to renewable resources is more difficult than it at first appears. It is not just a pressing issue of policy and governance; it is a special case of a whole raft of problems that press contemporary society in transition. The trap is that fossil fuel and renewables both are matters of energy in the service of human society, but they are essentially different. The issue invites giving privilege to an engineering level of analysis which is not special except it is regularly chosen by experts. The justification for the privilege of energy as understood by engineers is reification of that level of analysis. Reification in turn leads to an assertion of a situation in material terms, when it is in fact an abstraction. More data do not help if the situation is not material; it is not a data problem. Dominant and recessive genes are not a data problem as conventionally conceived, so the errors coming from reification are commonplace. It has led to 60 years of misconception in the Darwinian new synthesis. The effects of genes do not simply cascade up to phenotype, but instead pass through a hierarchy of physiological processes. Similarly, joules do not simply cascade upward to give sums for fossil fuels and renewables that are equivalent and straightforwardly comparable. The critical complication is the distinction between energy sources versus energy carriers. Embedded in all this are the purposes of energy use. Wheat is an energy source, flour is a carrier, but horse feed uses the source while making cake uses the carrier. At each stage, there are grammars that act as constraints on sources and carriers.

T. Allen (✉)
Department of Botany, University of Wisconsin, Madison, USA
e-mail: tfallen@wisc.edu

J. Tainter
Department of Environment & Society, Utah State University, Logan UT, USA

D. Shaw
Nottingham University Business School, Nottingham, UK

M. Giampietro · Z. Kovacic
Universitat Autonoma de Barcelona (UAB), Institute of Environmental Science and Technology (ICTA), Barcelona, Spain

M. Giampietro
ICREA—Catalan Institution for Research and Advanced Studies, Barcelona, Spain

N. Labanca (ed.), *Complex Systems and Social Practices in Energy Transitions*, Green Energy and Technology, DOI 10.1007/978-3-319-33753-1_10

The language of fossil fuel use is different from renewable energy use. The reference systems for time and energy are simply different. To bring energy systems into equivalent terms, it is crucial that the language of energy capture in the environment be distinguished from language of energy currency inside the system. Energy use is a complex system because it requires more than one level of analysis, with no simple nor necessary translation between levels. Fossil fuel is so fundamentally different from renewable because fossil fuel is simply consumed while renewables must be hugely processed outside the system. These ideas are remarkably general because goods are carriers of service.

Introduction

There is something very different about fossil fuel and its high return on effort (EROI) as opposed to renewable resources with their lower EROI. One face of fossil fuel and renewables would appear to be only different sources of energy. But the difference between them invokes complexity, and that denies the two forms being simple counterparts. Complexity presents difficulty in mapping a transition to renewable energy from fossil fuels because discontinuities deny a simple rescaling. There is an asymmetry between them that puts the respective sorts of resource in a relationship that invokes discrete dissimilarities, which deny a unified treatment. All this is very confusing, because on the face of it they are simply different sources of mechanical force unified through capture and use of energy. Fossil fuel and renewables would seem to be simply materially different, but they are not. They are apples and oranges while both are still fruit. The implication is that they are simply different sorts of energy, with energy as the reference for comparison, but it is not that simple.

This paper moves across several areas of discourse: engineering, energetics, ecology, and the philosophical nuances of epistemology and reification. We see a commonality between business creating goods to satisfy desires and the functionality of postal services, and relate all this to energy offering services to society. We choose examples from complexity as it arises in ecology and genetics. We contrast rate-independent grammars with thermodynamic flux. We contrast energy sources, such as grains of wheat, with energy carriers such as flour. We are conscious that the tensions raised in energy use challenge what society will tolerate politically. And yet this is not a grab bag. In all this, we circle around the issues of energy types situated at different levels in a complex system. Different levels cannot be easily connected and have to be explicitly linked for each example. Simple summation to move across levels is simply delusional. The very diversity of issues raised indicates we are addressing a particularly general problem. All this ties tightly into the tension between using fossil fuel on the one hand and renewable energy on the other hand. The problem here is the usual raft of difficulties and misconceptions surrounding complexity. Complexity is widely misunderstood as being a material issue; it is not, and that is whence comes the difficulty embodied in complexity.

Tainter and Lucas (1983) made the same point for significance not being material. Mind you, once something has been recognized as significant, material findings do allow for some material situations to have more of what it takes to be significant than others, but that is a different discussion that arises only once significance is recognized as not material. Concepts do collide with experience such that we can invoke materiality. Similarly, once we have asserted complexity by asking a question, tooling up to define the situation in material terms actually destroys the intrinsic uncertainty and discontinuity that was in the original conception of complexity. Complexity arises from the point of view that is taken, leaving some significant aspects undefined. The situation degenerates from full blown complexity to mere complicatedness when definitions are imposed. The whole problem appears to be in understanding the role of the observer's cognizance in a complex system. What does the observer do in assessing quality of resources that confuses the situation so that it is not a simple sum of energetics?

Reification

The muddle comes in reification of concepts. Reification is a common problem in biology and ecology. In their recent book, Allen and Hoekstra went back to Arthur Tansley who wrote about reification in ecology as early as 1926.

> One last point, we must always be aware of hypostasizing [reifying] abstractions, that is, giving them an unreal substance, for it is one of the most dangerous and widespread vices through the whole range of philosophical and scientific thought. I mean we must always remain alive to the fact that our scientific concepts are obtained by "abstracting from the continuum of sense experience," to use philosophical jargon, that is by *selecting* certain sets of phenomena from the continuum and putting them together to form a concept which we use as an apparatus to formulate and synthesize thought. This we must continually do, for it is the only way in which we can think, in which science can proceed. What we should not do is treat the concepts so formed as if they represented entities which we could deal with as we should deal, for example, with persons, instead of being, as they are, mere thought apparatuses of strictly limited, though of essential value.[1] Thus a plant community is an essential concept for purposes of the study of vegetation, but is, on the other hand, an aggregation of individual plants which we choose to consider an entity, because we are able to recognize certain uniformities of vegetational structure and behavior within the aggregation by doing so. A climax community is a particular aggregation which lasts, in its main features, and is not replaced by another, for a certain length of time; it is indispensable as a conception, but viewed from another standpoint it is a mere aggregation of plants on some of whose qualities as an aggregation we find it useful to insist… But we must never deceive ourselves into believing that they are anything but abstractions which we make for our own

[1]Footnote (of Tansley): A good example of the hypostatization of an abstraction, exceedingly common 40 years ago, but now happily rare, is the treatment of the process of natural selection as if it were an active sort of *deus ex machina* which always and everywhere modified species and created new ones, as a breeder might do with conscious design. Tansley (1926, pp. 685–6, his emphasis and quotation marks).

> use, partial synthesis of partial validity, never covering *all* the phenomena, but always capable of improvement and modification, preeminently useful because they direct our attention to the means of discovering connections we should otherwise have missed, and thus enable us to penetrate more deeply in the web of natural causation.

In biology, reification comes from the misunderstanding surrounding material systems. It is universal. For example, almost all biologists think that dominance of genes is a material issue. It is in fact a normative judgement not a material fact that is accessible through data. Most problems in biology are not data problems. Once we have decided that a gene is dominant, then there are indeed material happenings that follow, but they do not determine which gene is dominant in the first place; that is a distinction at a higher level of analysis. Dominant versus recessive is a prior compression into the surrounding human values that make the context. Only then can we see how the gene works in a material fashion. The problem for dominance being material is all genes are dominant for the protein for which they code, and are recessive with regard to what all other genes code. Sickle cell anemia is almost always seen as recessive relative to the normal gene for red blood cells, but a wrinkle is that it offers resistance to malaria in the heterozygote. With regard to resistance to malaria, the sickle cell gene is dominant. It is never called dominant because the homozygote is lethal. Dominance carries baggage of "normal," "predominant," or "beneficial." All these are normative values. If the issue is malarial resistance then sickle cell is a dominant gene. It may seem we are raising a special case, but sickle cell is only distinctive in the way it forces an intellectual dilemma. We use sickle cell because it is a powerful pedagogical device in an arena where the conventional mistake is held particularly vehemently.

For over 70 years the new synthesis of Darwinian genetics has held sway with Darwin police making sure heretics are burned, in academic terms. The new synthesis is now in tatters, as physiologists (Noble 2013) have shown that genes do not have the material effect that was given them by orthodoxy. Lamarck's acquired characters are looking less like a failed competitor to Darwinian concepts. The two evolutions look more like an alternative conception that is not competitive, and explains certain common situations. As with dominance, there is some material consequence to genes, but it is always after decisions about privilege have been made. Significance and privilege always reside at the hierarchical level above the discourse. Allen has written extensively on hierarchies (Ahl and Allen 1996; Allen and Starr 1982; Allen and Hoekstra 2015).

Levels of Analysis of Energy

One source of reification is from those who calculate energy and like issues by giving a privileged position to some arbitrary judgement only because it is a commonly accepted benchmark. Almost everyone agrees as to the units, and that is taken as justification to make them real. For instance, some of the authors of this chapter were at an energy meeting in Porto Venere in 2002 where world class exergy experts were insisting on a certain value for exergy of a can of gasoline. Exergy is the amount of work that can be done (it is explicitly not the inverse of entropy). The expert insistence was because the calculation for getting work out of gasoline is based on the assumption that one will burn it all the way to what is mistaken to be the real dead state, down to water and carbon dioxide. And with that assumption in place the experts were right, there was only one valid number for exergy with complete oxidation. But they had forgotten a prior assumption about how the fuel was going to be used. Giampietro mocked the energy engineers by setting his narrative on an island, thus making for a closed system. But in the story, Giampietro had nothing to light the gasoline and anyway burning it to cook food was not the issue. The can was used to kill a rabbit by hurling it at the animal. Clearly, the exergy involved in killing the rabbit was not the same as that involved in burning gasoline.

In the world of physical phenomena, such as energy systems, the tangible properties of the phenomena in question encourage reification of the concepts employed, e.g., the units of energy. In the world of digital services, like web site design and ecommerce, the phenomena of interest are services which are obviously virtual rather than material. And they are more commonly accepted as being subjective (Allen et al. 2013). Services and the systems that produce them are much more malleable than products, they can be customized to fit the needs of a specific individual and they can be produced as and when required. Modern digital services are commonly personalized for specific users because users value a service that fits their specific contexts—their situation, their needs and their timing. It is clear that the relationship is malleable. The exergy involved in killing the rabbit—using the kinetic energy of the can transferred to it by the muscles of the desperate hunter—was not the same as that involved in burning gasoline because of the decision of the different users. This is subjectivity that is based on each user's capabilities for using exergy in different forms.

Beyond sentient decisions of humans, subjectivity matters in biological systems because life plays with the dead state of its fuel. So do social systems that use either fossil fuel or renewable energy. Active manipulation of the dead state is more obvious in society with its sentient humans at work making decisions, but it happens in biology all the time. For instance, yeast given free oxygen can reduce sugar down to the generally accepted physical dead state of water and carbon dioxide. But without free oxygen yeast must ferment the sugar down to carbon dioxide and ethanol as the dead state. The fact that ethanol will still burn is beside the point. As far as yeast is concerned ethanol is the dead state. It all depends on the significance of the fuel for the user. Another favorite of Giampietro is to point out that there is no exergy in a gallon of gasoline if your mode of transportation is a donkey cart.

So the difficulty in calculating energy in a shift from fossil fuels to renewables is that the units of energy are not the same across the divide and so are not additive. The change means a shift in who is using renewables. That changes the calculations, including information as to what these new users would find usable and useful. It is an information issue rather than a material shift. Like any ecommerce web site, energy use suggests products that anticipate what a customer needs.

Reification of Time

Much as the units of energy use and capture, time also is labile and cannot be reified. This adds to the dilemmas of switching from fossil fuel to renewable resources. Time does not progress at the same rate in different phases of ecological progress, and it is open to subjectivity. H.T. Odum worked hard to try and keep time and energy as measurable in certain units for consistency. In his maximum power principle, Odum notes that capital builds in ecological systems that organize to protect the capital. Trees have devices that preserve capital by discouraging fire. Odum makes the distinction between capital and liquid assets. There is a specific phase when capital declines because it is converted to liquid assets. Liquidity matters because it can easily be spent, and always is in short order. In all this Odum moves forward in fixed time units. Sometimes changes are slow, as in accumulation of capital, but at other times capital is quickly converted to liquid assets. But all this is expressed by Odum in the progress of just one set of time units. C.S. Holling has an equivalent scheme, called panarchy, where he plots capital against organization. He slowly reaches a K phase where capital is great and is protected by organization. Holling, unlike Odum, does not distinguish capital from liquid assets; it is all something that can do work under transformation (Fig. 10.1a, b). When Holling's K phase collapses he notes capital suddenly declining. But then in a surprising switch, Holling's capital re-emerges, but in a disorganized state. After the forest dies, a new capital appears in the body of dead trees which decay faster than they grew. They are dead and so have no means to defend their capital. Holling with his dead trees is in fact speaking of ecosystem liquid assets, which are not protected. Quickly Holling's capital-come-asset is used up. Holling uses a Lazy 8 figure where the cycle is slow as it creates capital, but whips around fast from collapse to re-emergence and on to loss of "capital" and organization. So for Odum, time marches on, where for Holling the move to other phases suddenly speeds up or slows down. Time is different for Odum and Holling, even though they are studying the same general pattern in just one material system.

So Odum keeps time moving forward in a way that addresses rate-dependent phenomena of flux and consumption. Holling's system moves forward from one rate-independent event to another. Both are descriptions of the passage of time. We make the distinction between high and low gain (Allen et al. 2001) which is exquisite in distinguishing energy inside versus outside the system. High gain takes in ready-made fuel for great profit. Low gain takes in low quality material and needs to process it to make fuel (like agriculture makes sunshine into something to burn). Fossil fuel and its use is a high gain, rate-dependent description. The system is governed and predicted by limits on the rates of flux. Renewable resource creation takes in low quality resources and has to make it into fuel. Low gain consumption is a matter of applying constraints on flux to increase efficiency. Low quality resource is first forced up a gradient to become more concentrated. Resource is captured and processed to turn it into fuel by the imposition of constraints. Such a scheme is governed by efficiency, and that turns on favored states, which are of

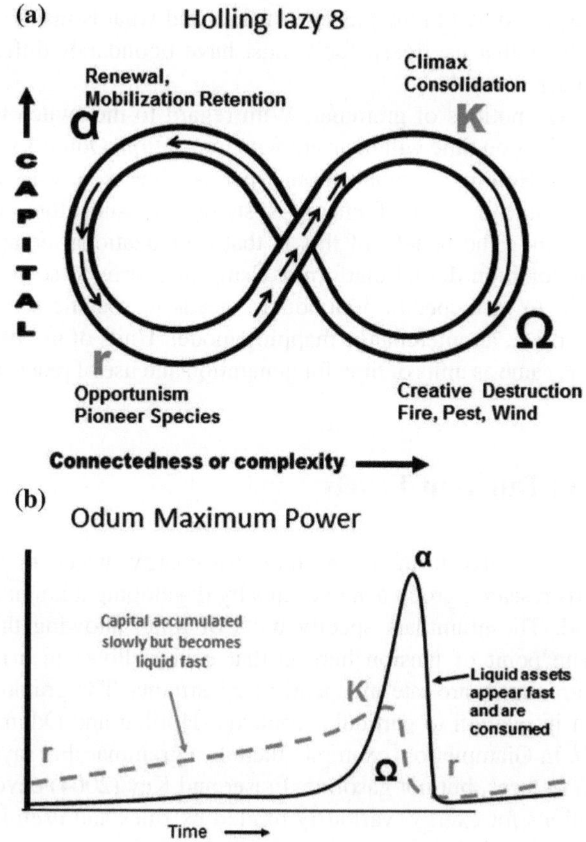

Fig. 10.1 **a** Holling's Lazy 8 narrative, which has come to be known as panarchy (Holling 1986). The Lazy 8 scheme is really a narrative, not a model. Capital builds, is destroyed, and then re-emerges as disorganized capital (did it ever disappear?). The track is from *r* to *K*, to Ω, to α, and back to *r*. At *K*, the system has much capital but is brittle and fragile. As in all fragile systems, collapse to Ω is fast. At α and *K*, capital is high. At α H.T. Odum (1995) would say that capital is converted to liquid assets. For Holling, the distinction is that α is not organized. **b** More like a model than Fig. 10.1**a**'s narrative, Odum's maximum power principle plots capital on one line and liquid assets on the other. The two lines sidestep the inconsistency of panarchy, making it more of a model. In terms of high and low gain, *K* to Ω is the high gain harvesting of standing crop. But the whole cycle for *r* to *r* is a low gain scheme that keeps *r* to *K* growing with maximum power. From Ω–α to *r*–*K* is not just a passive phase of not harvesting; it is part of the low gain strategy of maintaining maximum power. While the panarchy cycle has fast and slow transitions, Odum's scheme moves through time at the same pace, but the passage across the narrow (short time) α peak is where panarchy cycle appears to speed up. Without the Ω and α phases where there is lots of action, both systems would stagnate at *K*. Odum says maximum power always pertains. After Allen and Hoekstra (2015)

course rate-independent. So renewable resource use is best seen as a sequence of rate-independent, but time-dependent events. It is best treated with a narrative rather than models. Fossil fuel is therefore fundamentally different from renewable resource, all the way down to the passage of time and what is inside versus outside the system. Entities that use fossil fuels must have boundaries differently defined from those using renewables.

We invoke here notions of grammar. With regard to the switch to renewables, different perspectives on time will interfere with translation. Odum's sidereal view of time changed to Holling's organizational perspective is a way to translate a time-based view into a view of energy systems that substitutes organizational complexity for time. The benefit of this is that organizational complexity can be mitigated using common digital platforms. Using platforms to solve "dating problems"—like how to fit a specific individual's needs to specific service sources—needs a grammar, i.e., an interchange mapping model. Units of use of fossil fuel are unlikely to be the same as units of time for generation and use of renewable resources.

Grammars at Different Levels

Giampietro has been able to aggregate units (of energy, water, or food) to reach upper levels. His research group achieves this by developing a grammar for energy, water, and food. The grammars specify units of time, allowing them to remain comparable. The point of tension here is that energy flows in a rate-dependent manner while grammars are rate-independent constraints. The grammar gives rules for aggregation in relation to particular contexts (Holling and Odum used alternative grammars). In Giampietro's example, there is a grammar that says donkeys can burn hay down to feces, but not gasoline. Fraser and Kay (2004) developed a set of different definitions for exergy, variously limited by rules and usefulness for given purposes. With all the changes in type and vehicle for energy transfer, regular engineering cannot work with moves over to wind, wave, solar, or thermal and still keep the units straight. All this involves shifts between levels in a hierarchy. If we are working with hierarchies, it might be a good idea to develop a practical theory of hierarchies.

If renewable energies are to substitute for fossil fuels one must identify how particular instances of reification are so damaging. First, what is meant by energy? We distinguish between primary energy sources (such as oil, coal, gas, wind, waves, and solar radiation) and energy carriers (such as gasoline, electricity, and heat). In food, primary energy sources can be understood as wheat, and energy carriers as flour. One kilogram of wheat is not equivalent to one kilogram of flour for two reasons: (1) one kilogram of wheat is of no use in baking a cake, one must first make the flour—the process matters; and (2) processed wheat (flour) is of no use in feeding an animal; other byproducts are used for feed—purpose matters. The usefulness of the units used to measure the quantity of wheat, or of a primary energy source, depends on its end use.

In terms of energy, alternative energy carriers require different production, and different energy carriers have different end uses. In talking about the transition to renewable energy in terms of primary energy sources, these two important distinctions can be missed. Using a gallon of oil to produce gasoline is not the same as producing electricity from burning oil. Using wind and solar radiation to produce electricity does not mean that all end uses are provided (one still needs gasoline for cars, gas for heating, hay for feeding animals, etc.).

The distinction between primary energy sources and energy carriers as well as between different end uses leads to a second issue: how useful is it to measure everything in terms of Joules? The use of the same unit of measurement makes the quantification of energy problematic as it does not provide the conceptual tools needed to distinguish between different forms of energy. At the same time, the use of quantitative measurements gives a false sense of objectivity to the description, inviting reification.

Once the differences in processes and end uses are taken into account, the viability of a transition to renewable primary energy sources becomes most difficult to assess based on the availability of technological solutions. That is, even though it is technically possible to produce electricity using photovoltaic panels, wind mills, water turbines, etc. this information cannot be used to infer that the above energy production processes will satisfy the variety of end uses in society. It may well be incompatible with the resources available (labor force, primary energy inputs, etc.).

The way around reification is translation. Giampietro et al. (2014) have developed grammars to translate units of energy, water, or food, and we have developed grammars to translate between time and organizational complexity. These grammars could be the basis of a digital translation platform that would enable diverse sources and diverse users of renewables and fossil fuels to get, buy-in, plan, act, and operate.

To explain one of the pillars of this platform, an analogy with the delivery of mail is useful. To use Fig. 10.2, one works around the sectors of the figure in a counterclockwise direction. The distinctions are between moving in and out of the environment and system (horizontal movement) on the one hand, and moving from inputs to outputs (vertical movement) on the other hand. Figure 10.2, represents a mail delivery system divided exquisitely into two parts. (1) Actually delivering letters inside the society are postmen playing a role in society. (2) This is in contrast to the context of mail delivery, the society at large served by the mail system. So there is a subtle change in level of analysis: who does the work and how, as opposed to how much work is to be done. In the upper right quadrant of Fig. 10.2 is the total population and the total mail received by that population in one year. That is the context in which the actual delivery of mail by postmen operates. The letters received per capita can be defined as the rate of mail consumption by that population (20 letters per capita per year). Moving to inside the societal system, the upper left quadrant shows the percentage of the population that works as postmen. Outside the system the lower right quadrant shows the percentage of annual flow of letters delivered per day. Finally, the lower left quadrant represents the letters that each postman must deliver per day in order to support the rate of mail consumption of the population. The latter can be defined as the intensity of mail supply (16 letters per capita per day) (after Giampietro et al. 2012).

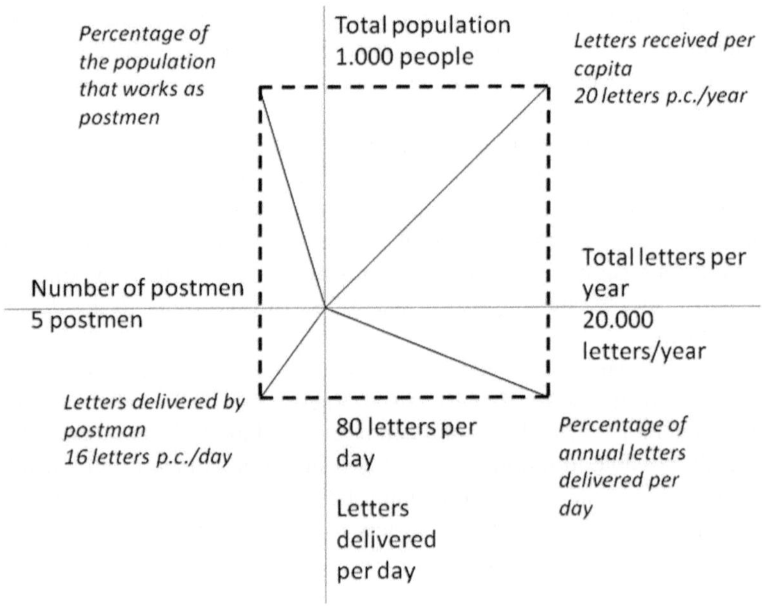

Fig. 10.2 Mail delivery system (after Giampietro et al. 2012)

This representation makes it possible to visualize the forced congruence that there must be between the different quadrants that represent input and output segments. It is forced because once three quadrants are specified the fourth is fixed in a zero sum game. That says, in terms of energetics, in a transition to renewable energy it must all add up. That seems to be obvious enough, but in an elaborate change of energy source, energy vehicles, and uses for energy it is easy to forget this or that translation such that "It does not add up," even if it is mistakenly seen to do so. In the mail system, a change in any of the parameters causes a change in the rest of the system. For example, an increase in the flow of letters per year, would require either that a higher percentage of letters is delivered per day leading to longer working hours, or that the mail delivery system is improved leading to a higher rate of letter delivery per postman per day, or that more postmen are hired leading to a decrease in employment in other economic activities, or to the employment of people from the nonworking population.

Similarly, in the case of energy, one can plot the total population against the total energy throughput—measured in energy carriers in order to refer to end uses (Fig. 10.3). A transition toward renewable energy sources, based on the available technology (a fixed slope of the curve in the lower left quadrant), requires a higher flow of primary energy sources per Joule energy carrier produced. This is due to the fact that fossil fuels have a much higher energy return on energy investment than solar radiation and wind. As a consequence, the slope of the curve in the lower right quadrant would change leading to a lower total energy throughput, which would

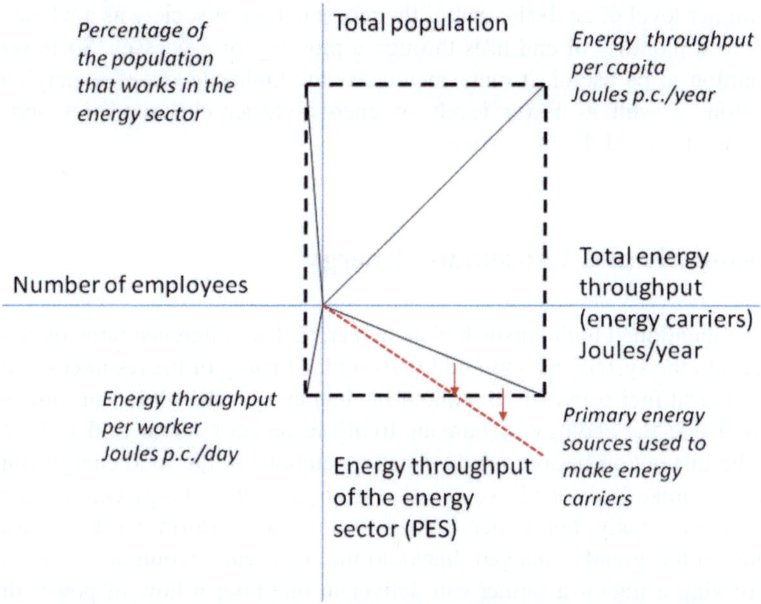

Fig. 10.3 Energy system (after Giampietro et al. 2012). Total population against the total energy throughput—measured in energy carriers in order to refer to end uses

affect food production, manufacturing, and living standards, or would require a higher portion of the human activity to work in the energy sector. This change in proportions would either increase the working hours of those employed in the energy sector, take away from other economic activities, or require the employment of the nonworking population (children, retired people, etc.).

The scenario presented highlights some of the constraints that would be forced by a transition toward renewable energies based on current technologies. This representation is based on the relationships of congruence between the system as a whole and its parts. This is different from quantification of energy flows as an abstract category. In our representation, it becomes possible to distinguish between the energy used by the energy sector to make energy carriers, and the energy available to the rest of society. A finer grain of analysis can be used to distinguish among different end uses in society.

Our representation does not give any indication about which technology should be developed and how, but it does make it possible to assess the possible constraints to the large scale application of available technologies. A further issue is raised as a completely separate concern, scale of analysis. Which scale is the most useful? Reification occurs in the muddling of different levels of analysis in the assessment of renewable energies. The use of renewable primary energy sources refers to a particularly low level of analysis: that of the technical processes used to produce energy carriers. Questions about the transition toward renewable energies refer to a much higher level of analysis: that of the reproduction of society as a whole, which attends to a plurality of end uses through a plurality of processes. So in order for any solution to be useful, it must encompass the higher levels of society's energy production as well as lower levels of energy carrier characteristics and users' capabilities to use different carriers.

Endosomatic and Exosomatic Energy

We have mentioned high versus low gain energy. The difference turns on resources coming into the system, as opposed to raising the quality of the resources within the system. Fossil fuel comes from without the human system, while sunshine is better seen as inside the ecology of humans living in an ecosystem. Lotka (1956) proposed the distinction between endosomatic metabolism (the food energy converted inside the human body) and exosomatic metabolism (the energy converted outside of the human body but under human control to perform work). Exosomatic metabolism has greatly changed thanks to the industrial revolution. "For example, when driving a tractor a farmer can deliver in one hour a flow of power that is a thousand times larger than the endosomatic power delivered in one hour of manual work." (Giampietro et al. 2012).

In relation to renewable energies, the question is whether renewables would make it possible to maintain the same exosomatic metabolic rate as fossil fuels. With the current technology, this is not possible, and it is for strategic not tactical reasons. Tactics would be about technical invention and efficiency. Strategies would have to recognize the fundamental differences between fossil fuels and renewables. Here we again make the distinction between high gain systems that use a ready-made fuel, as opposed to low gain systems that must aggregate and process low grade material into fuel. The process of concentration of the fuel in high gain systems is done by some external system. Crude oil is created by natural processes over many millions of years. Renewable energy is too low quality to use it as a fuel; rather it is a low gain, low quality input that must be processed, often involving extensive transportation. Fossil fuels we burn, but renewable we must make into fuel. Renewable energies have a much lower metabolic rate than fossil fuels (again the steeper slope of the curve in the lower right quadrant of Fig. 10.3). For example, when thinking of biofuels, the suggestion is to go from an exosomatic metabolism (burning oil) to an endosomatic metabolism taking place within the plant (which requires water, fertilizers, time to grow, before it can be harvested, processed, and burned). Fossil fuels are focused, while biofuels are extensive so that gathering is a much larger part of the enterprise.

In his conclusions in *Drilling Down*, Tainter expresses these arguments in similar terms, emphasizing how renewable energy will have huge environmental impacts. Tidal energy:

> could mean treasured coastlines would need to be engineered to become industrial environments. Renewable energy that gives the same power per person as we enjoy today would not be free of environmental damage. Indeed, in the large land areas that it would require, renewable energy would cause more environmental damage than that caused by our use of fossil fuel.

Tainter argues extensively that money and energy are in a sense the same thing (Tainter and Patzak 2012). Much as concentrated heat from burning coal can be dissipated in doing work in a steam engine, gold in treasuries can do work by being degraded into copper coin, and then dissipated in the payroll given to a diffuse body of individual workers. The second law of thermodynamics applies to energy and gold. Furthermore, the accumulation of energy can also accumulate treasure. Excess agricultural production can be stored in a transformation into precious metals; miners are fed on surplus food. Tainter suggests that a mere 5% decline in energy availability in the transition would bring about financial distress as large as the great recession of 2008 (Tainter and Patzak 2012). To put it in perspective he reports street lights turned off, roads returning to gravel, teachers laid off, and:

> Britain is eliminating whole agencies of government, and planning to implement the most drastic curtailment of public services since World War II. All this is happening at a time when energy is still abundant and relatively inexpensive.

All of industrial society will be reorganized. This is not just a change of energy sources. Cities in decline are already reorganizing so that the rich return to the city centers. The poor are moved to the margins. In Europe, the rich never left.

The central problem here is as much cultural as anything. Even in economic recovery the less educated in the United States are angry and supporting an overt loss of reason and civility reminiscent of Germany in the early 1930s. And times are not yet hard. The kudos of quantification in science and engineering allows scientists to "prove it!" Able to prove it, experts feel obliged to do so. That has two effects. First, it slows down the progress achieved by experts. Second, it narrows scope so it applies to the proof. It is not that the conventional approach is illogical; it is that it is cast at such a local scale as to have limited utility in addressing the pressing questions of our energy future. Some changes will involve quantitative differences. But complexity invoked by the pressing questions means we must use qualitative approaches as they become necessary.

The General Condition

We can generalize our discussion to show how ubiquitous are the points we make. There is a similarity to Vargo and Lusch's (2004, 2008) argument that products are just service carriers, i.e., products are vehicles to achieve "what is needed to be done" for users (2004). In the world of ecommerce where products and services are sold on web sites the problem is called "personalization." Personalization is based on understanding each service user's individual context, which is the reason for the modern corporate thirst for customer data. Customers increasingly expect to consume product and services, especially digital services, that fit their exact needs. This applies even if customers do not know their own needs themselves. "Google Now," a product of Google, is a good example of an advice service that can give advice before you even ask because it has enough data to understand your individual context.

The problem of fitting the characteristics of specific energy carriers to specific energy users at specific times and in specific situations is a *translation* problem. It is a mapping and interchange problem. In ecommerce, multinational firms have many different country web sites and if a change is made on one then it is common for all others to need changing. But human translators, who are capable of translating cultural concepts not just words, usually speak only a *pair* of languages. So any translation project—a document or a whole web site—requires a specific configuration of translators and their *language pairs*. And projects can last minutes or weeks. Shaw and Holland (2010) have explained how this boils down to a real-time matching problem, like a dating web site, and how digital platforms can facilitate it. Transitions in energy is not just a matter of energetics, it is a huge matching and translation problem.

Carriers are what users use; they do not generally use sources. What matters are the characteristics of energy use that fit a user's needs. This is a translation problem that could be eased by a digital platform. That platform uses information describing the characteristics of energy carriers *and* the contexts of energy users at two levels: (a) fit specific sources to users and to (b) aggregate all these specific matches to higher levels of matching, for insights about policy. At the heart of such a digital platform are the ideas that we have described including energy aggregation grammars, time to organization mapping and values forced by fixed sums. We note that Figs. 10.2 and 10.3 are implementations in just two dimensions of interest.

We warn against reification because it canalizes conception and so action to a few and often just one level. In our more open view, we give several methods for translation (using grammars) and a method for seeing the big picture at different system levels. The grammars translate between carriers' characteristics and a user's need/abilities. The method of Figs. 10.2 and 10.3 helps one to see the implications of this. Together, these methods can be made into a technically simple and highly scalable web platform to get buy-in, spread the word, and support decision-making.

References

Ahl, V., and T.F.H. Allen. 1996. *Hierarchy Theory, a Vision Vocabulary and Epistemology*. New York: Columbia University Press.

Allen, T.F.H., and T.W. Hoekstra. 2015. *Toward a Unified Ecology*, 2nd ed. New York: Columbia University Press.

Allen, T.F.H., D.R. Shaw, P.C. Allen, and J. Spohrer. 2013. Insights into the relationship between products and services coming from biology. *Systems Research and Behavioral Science* 30: 570–579.

Allen, T.F.H., and T.B. Starr. 1982. *Hierarchy: Perspectives for Ecological Complexity*. Chicago: Chicago University Press.

Allen, T.F.H., J.A. Tainter, J.C. Pires, and T.W. Hoekstra. 2001. Dragnet ecology—"Just the facts Maam": The privilege of science in a post-modern world. *BioScience* 51: 475–485.

Fraser, R.A., and J.J. Kay. 2004. Chapter 9. Exergy analysis of ecosystems: Establishing a role for thermal remote sensing. In: *Thermal Remote Sensing in Land Surface Processing*, ed. D.A. Quattrochi and J.C. Luvall. Boca Raton Florida, USA: CRC Press. http://dx.doi.org/10.1201/9780203502174-c10.

Giampietro, M., R. Aspinall, J. Ramos-Martin, and S. Bukkens. 2014. *Resource accounting for sustainability: The nexus between energy, food, water and land use*. New York: Routledge.

Giampietro, M., K. Mayumi, and A.H. Sorman. 2012. *The metabolic pattern of society: Where economists fall short*. New York: Routledge.

Holling, C.S. 1986. The resilience of terrestrial ecosystems: Local surprise and global change. In *Sustainable Development of the Biosphere*, ed. William C. Clark, and R.E. Munn. Cambridge: Cambridge University Press.

Lotka, A.J. 1956. *Elements of Mathematical Biology*. New York: Dover.

Noble, D. 2013. Physiology is rocking the foundations of evolutionary biology. *Experimental Physiology* 98 (8): 1235–1243.

Odum, H.T. 1995. Self-organization and maximum empower. In *Maximum Power: The Ideas and Applications of H. T. Odum*, ed. C.A.S. Hall. Boulder: University Press of Colorado.

Shaw, D.R., and C.P. Holland. 2010. Strategy, networks and systems in the global translation services market. *The Journal of Strategic Information Systems* 19 (4): 242–256.

Tainter, J.A., and G. Lucas. 1983. The epistemology of the significance concept. *American Antiquity* 48: 707–719.

Tainter, J., and T. Patzak. 2012. *Drilling Down: The Gulf Oil Debacle and Our Energy Dilemma*. New York: Copernicus Books.

Tansley, A. 1926. Succession: The concept and its value. In *Proceedings of the International Congress of Plant Sciences*, Vol. 1, pp. 677–686. Ithaca, New York, 16–23 August. (Published in 1929, Menasha, Wis.: George Banta).

Vargo, S.L., and R.F. Lusch. 2004. Evolving to a new dominant logic for marketing. *Journal of Marketing* 68 (1): 1–17.

Vargo, S.L., and R.F. Lusch. 2008. Service-dominant logic: Continuing the evolution. *Journal of the Academy of Marketing Science* 36 (1): 1–10.

Chapter 11
An Analysis of Everyday Life Activities and Their Consequences for Energy Use

Jenny Palm and Kajsa Ellegård

Abstract In this chapter, we discuss the need for deeper knowledge about the relation between people's daily activities and their electricity use and how to increase our knowledge through time use surveys and the visualization of aggregate activity patterns. To understand people's energy consumption and how to improve energy efficiency or reduce demand during certain peak hours requires an understanding of households' daily activity patterns. The activity patterns can be revealed when people keep time diaries, from which we analyze where, when, and for how long specific energy-related activities occur. In this chapter, we discuss how energy consumption varies in the course of the day and differs between people in different age groups. This has implications for how individuals should be approached and indicates that policies and advice should differ when directed to people in different life stages. By utilizing many time diaries from a population we can analyze differences in aggregate activity patterns. In Sweden, women, for example, use more electricity for activities related to cooking and household care than men do, which makes them the most relevant target group when it comes to giving feedback on how much electricity an appliance uses or on alternative ways of doing certain activities. Time diaries and visualization tools can also be useful as a reflective tool for the households when discussing their members' various daily activities in relation to energy consumption. This can be used by energy advisors when targeting individual energy behavior.

Introduction

Developing a 100% renewable energy system requires substantial changes in the structure of both supply and demand. The smart grid has appeared as a solution, where one idea is that demand is actively managed to fit energy supply over time.

J. Palm (✉) · K. Ellegård
Department of Thematic Studies, Technology and Social Change,
Linköping University, 581 83 Linköping, Sweden
e-mail: jenny.palm@liu.se

N. Labanca (ed.), *Complex Systems and Social Practices in Energy Transitions*,
Green Energy and Technology, DOI 10.1007/978-3-319-33753-1_11

With the smart grid follows a vision where the grid can deliver low carbon electricity more efficiently and reliably, while it enables consumers to manage and reduce energy use and minimize costs to the benefit of all (Lunde et al. 2015; Nyborg 2015). The concept of flexible consumers has been well recognized within this discourse.

The idea of a flexible consumer relates to and relies on individuals' daily choices and household routines, i.e., what they do in their everyday lives. As a consequence, the development of policy means or business models targeting people's energy consumption requires an understanding of how energy consumption is related to a household's daily activity patterns. Energy is an important resource used for performing a multitude of activities that form people's everyday rhythm. To support improved energy efficiency at household level, it is necessary to understand this rhythm of people's everyday life, which preferably can be done in terms of the timing and duration of energy-related activities within an individual's full daily activity sequences. More knowledge is needed about the basic dynamics and temporalities of demand to develop an informed demand-side policy (Walker 2014). In this chapter, we offer time-geographic concepts and tools to gain such insights.

The individual's activity sequence in daily life reveals when, where and for how long the electricity consumption is generated by her daily activities. One important implication is that each individual must be regarded as an indivisible whole, which implies that activities cannot be altered or moved without influencing other activities of importance for the individual. From individual time diaries, we will analyze and learn about where, when, and for how long specific energy-related activities occur in the whole activity sequence of individuals. We will discuss how energy consumption varies in the course of the day and differs between people in different age groups. This has implications for how individuals should be approached and indicates that policies and advice should differ when directed to people in different life stages.

By aggregating activity sequences from the time diaries written by many individuals we can reveal differences in aggregate activity patterns in larger groups, for example, men and women, people in and outside the labor force, single households and families, adults, and children.

In this chapter, we start by contextualizing the need for deeper knowledge about the relation between people's daily activities and their electricity use. Then we present data and methods. Thereafter, we show the activity sequences of an individual, to explain the principle of the visualizations, and then we visualize the aggregate activity pattern of a larger population. Following that, we show how electricity use is derived from the aggregate activity patterns. Finally, we present our conclusions on the relation between daily activities and electricity use.

Activities, Time, and Everyday Energy Consumption in a Wider Context

The major tool the EU and governments use to reduce household energy consumption is information. Information and education are also often regarded as basic and necessary preconditions for achieving more efficient energy use (Gyberg and Palm 2009; Palm 2010). The use of information touches on the moral aspects of household members' energy behavior, rather than strictly economic ones, as they will need to become more aware of and involved in their energy consumption (Oikonomou et al. 2009). People will then need to start reflecting on their energy consumption, and data collected from time diaries are suitable for creating tailored information.

We know from social science research that the use of energy is interwoven in everyday life, with its routines, meanings, social dynamics, and technical infrastructure (Lutzenhiser 1992; Wilhite et al. 2000; Wilhite 2005; Bartiaux 2008; Bartiaux and Salmón 2014; Palm and Ellegård 2011; Ellegård and Palm 2011; Gram-Hanssen 2010; Shove 2003). Energy consumption is embedded in cultural processes (Aune 1998; Stephenson et al. 2010). Shove (2003) emphasizes the importance of understanding consumption, technology, and social change from the perspective of "invisible practices". When it comes to energy systems today it is more of a fact that few consumers are interested in the delivery of energy per se, but rather in the functions and conveniences it can provide. Energy is required for such needs as preserving and preparing food, supplying heat and light, and maintaining health and sanitation. How energy is produced and distributed is not of prime interest from this point of view, since the important purpose is how to fulfill those needs. When for example developing and introducing sustainable solutions, these need to be integrated in a way that they "provide and sustain what people take to be normal services" (Shove 2003, p. 198). Consequently, what people do in their everyday life, their activities, is an important starting point for understanding energy use in a social context.

The understanding of energy demand and how it relates to a sustainable everyday life requires different approaches to analyze and explain change. A common approach has been to model and forecast household energy consumption, rather than to understand the practices that give rise to energy demand (see Anable et al. 2014 for a more thorough discussion on this). If change in energy demands follows changes in social practices, then we need a better understanding of the complexity and dynamics of these practices (Walker 2014).

In this contribution, we will put people's activity sequences in focus for the analysis, and we use concepts and visualizations developed within the time-geographic approach. Time-geography emphasizes a bottom-up perspective and one basic ontological assumption is that the individual is an indivisible whole, which implies that it is not sufficient to regard the average time use for activities per individual. According to Ellegård and Palm (2011), understanding of energy use at household level needs to be framed by all activities that are the basis for everyday life, in which habits are embedded. This understanding is complex since flexible work schedules and school hours, as well as increase in mobility, create new types of activity patterns and routines that have implications for energy use. However, the result of these will differ considerably depending on into which activity sequence these new activities are interwoven.

The time-geographical approach is based on the human impact on the ecological balance on Earth and was developed by Hägerstrand (1985), Hägerstrand et al. (2009), Hägerstrand and Lenntorp (1993), Hägerstrand (1970). In time-geography the individual's indivisibility is emphasized, and the individual is regarded as continuous every day from birth to death (Hägerstrand et al. 2009).

Time-geography assumes that time may serve as a measuring device for all existence and thereby things that seemingly are not connected still coexist in time and space and can be described and analyzed together because of their time-space coexistence. From the early days of time-geography, the importance of individuals' daily activities to analyze and understand society was emphasized (Mårtensson 1974) which includes how people can arrange their daily projects depending on the location of the home and their opportunities to transport themselves (Lenntorp 1977; Ellegård et al. 1977).

Time-geography is a contextual approach, where the everyday activity context is defined as the sequence of activities that the individual has performed in the course of the day. Then, the individuals' projects which they pursue by performing activities relate to their household context. A household's project is defined as a set of activities distributed among and performed by the household members to achieve a goal they have agreed upon, e.g., feeding the family, maintaining the home and belongings or raising children. A short-term project is for example preparing a specific dinner or buying a new washing machine. Projects are fulfilled by household members performing the necessary activities. For example, preparing dinner for the family comprises activities, such as reading a recipe, finding the ingredients, cooking the dish, setting the table, serving the dinner, and eating it (Ellegård and Palm 2015).

Practice theory has in recent years become a common approach to apply when discussing consumption and everyday life (Reckwitz 2002; Gram-Hanssen 2010; Shove and Spurling 2013; Warde 2014; Welch and Warde 2014; Shove 2003; Røpke 2009; Strengers 2012). Time and space was not discussed as much to start with, but has become more in focus within practice theory too (Shove et al. 2009). In their study Røpke and Christensen (2012) combined time-geography and practice theory to analyze energy impacts of ICT. In this study, they also discuss how practice theory and time-geography relate to each other and we will repeat some of the main points here (for the full comparison see Røpke and Christensen 2012).

In practice theory it is possible to identify clusters of activities where coordination and mutual dependency makes it possible to see them as an entity, as a practice. A practice is also recognized across time and space and an important assumption in practice theory is that a practice is reproduced over time. In a practice theory perspective people are engaged in practices, such as cooking, eating, work, sleep, etc. A time-geographic project consists of specific activities to achieve a goal set up by one or more individuals, while in practice theory this is expressed as a series of practices needed to complete an intention. Røpke and Christensen (2012) describe this as a metapractice to which several subpractices relate.

Practices may be related to each other without being bound together through projects. Pantzar and Shove (2010) distinguish between bundles and complexes of practices. A bundle of practices is recognized by the coexistence of two or more practices that are minimally related by for example being co-located. A complex of practices is practices that are closely related and mutually dependent (Røpke and Christensen 2012).

Elementary in time-geography is that everyday life develops in time and space and that every individual's movements can be described by a path, or trajectory, in time and space. A basic everyday life challenge is also to participate in projects within the constraints set by time and space and where each individual's activity sequence is coupled with the activity sequences of other individuals. Most multi-person households usually have some kind of division of labor. Perhaps one member always performs all the activities in a certain household project or all household members cooperate and alternate the responsibility for performing certain activities over time (for further discussion on this see Ellegård and Palm 2015; Isaksson and Ellegård 2015). This rhythm of everyday life changes however over time. It changes because the composition and social context of households change, and because the household's project goals change.

In this perspective Garabuau-Moussaoui's (2011) interesting historical study shows how individuals in France, at each stage of life, build a specific relationship with energy and energy efficiency. She shows that people have different relations to energy in different stages in life and emphasizes the importance of developing generational policies and tools. Each life stage has its own behavior according to Garabuau-Moussaoui (2011), with its own possibilities and restrictions to change energy consumption patterns.

This conclusion is interesting and calls for a deeper investigation. We will therefore use time-diary data from the Swedish population from a study undertaken by Statistics Sweden in 2010/11, to compare the temporal distribution of activities by individuals belonging to different generations and consider weekdays and weekend days. But before we do that, we will first describe the methods used.

Method: Time Use Survey and Visualization

In everyday life many activities are performed by routine and without any deeper reflection. Therefore, it can be hard to collect reliable data on when and for how long activities are performed. Collecting diaries of individual time use is one way to deal with this problem. Time diaries reveal what activities are performed, when they are performed and for how long. However, time diaries do not usually include information on motives, so information of that kind needs to be collected through other methods.

In the type of time diary commonly used in national time use surveys, each participant of the study writes what activities she is occupied with during a whole day, divided into 10-min intervals, with whom, where and if they use a computer. Other time-diary methods let the participant define the starting time of an activity which at the same time serves as the stop time of the previous activity. This diary method is used in studies aiming at a dialogue between a researcher and the participants (see, for example, Orban 2013). The data from any kind of time diary can be used to visualize the activity sequences of individuals and the activity patterns at aggregate levels, and at all levels the rhythms of everyday life are revealed. In addition, they show the social context in which the individual performs various activities, and finally the geographical context, which shows where the activities are performed.

Time can be defined and analyzed in different ways (see Hellgren 2015 for a more thorough discussion on this). Here, the time used for a specific activity is defined as the duration of the uninterrupted time period when an individual carries out the activity. This definition excludes for example peoples' experiences and their perception of time. Another way to collect data on people's time use is to ask respondents to retroactively estimate the average time they spent on an activity during a day or a week. This so-called stylist method has a bias to over- or underestimates actual time used for an activity. This method does not consider the context an activity is carried out in, but the interest is on singular activities (Hellgren 2015).

Another way to collect data on time use is experiential sampling methods where the respondents report what activity they are doing and answer supplemental questions when they are alerted at randomized points in time. This method yields what an individual is doing at these randomized times and what happens in between is left out. Thereby the individual's activity sequence is lost and the rhythm of everyday life is not identified (Hellgren 2015).

There are various ways to do large-scale data collection via time diaries (Hellgren 2015). One is the American Time Use Survey (ATUS) used by the US Bureau of Labor Statistics. Another is the Harmonised European Time Use Study (HETUS) developed by Szalai (1972) and Harvey (1993). Still another way is a low-budget time use survey diary with a limited number of predetermined activities (see, for example, Skolverket 2013). Researchers make efforts to develop methods for time-use data collection by means of ICT devices, and (Minnen et al. 2014) developed methods to collect time diaries over the Internet. The use of smart phones with apps and GPS devices is reported and discussed by many, for example, Shaw et al. (2008), Silm and Ahas (2014), and Shoval et al. (2014). However, GPS data will give information about travel between and stays at various places but no information about what activities people perform. A smart phone app for collecting activity data from time diaries is in its tryout phase (Vrotsou et al. 2014).

In this chapter, we use data from Statistics Sweden's time use survey 2010/11 which follows the HETUS guidelines (Eurostat 2009). These guidelines recommend that data be collected during 2 days, a weekday and a weekend day from every participant. This is due to cost aspects and that it should not be too much effort for the participants to fill out the diaries. One negative aspect of this choice is that data are just collected from two singular days and there might be substantial differences in activities at an individual level between days. Also, projects (various activities that relate to the same goal) that persist over a longer time will not be covered since just some of their constituent activities are performed during the diary day (Hellgren 2015). Statistics Sweden has however added a question on how well a day represents an average day to ensure validity.

In the Statistics Sweden time use survey 2010/11, 3244 respondents yielded a total of 6477 diaries covering 3233 weekend days and 3244 weekdays. The age span was from 15 to 84 years. The study included individuals in households of various sizes and living in various regions of Sweden. The data can be used to capture the unique and complex everyday life activity pattern of each individual, as well as to demonstrate that there are basic patterns common to many people (the aggregate activity pattern) and that people share a similar rhythm of everyday life with others.

The Swedish time use survey of 2010/11 was initially coded by Statistics Sweden according to the HETUS guidelines with its five main activity categories. This coding is recoded to fit the activity categories of VISUAL-TimePAcTS by Hellgren (2015). The recoding translated the activities into a code scheme which has seven main categories: care for oneself, care for others, household care, recreation/reflection, transportation, prepare/procure food and work/school (Ellegård and Nordell 1997; Ellegård 2006).

The combination of people's uniqueness and similarities is revealed by using the visualization software VISUAL-TimePAcTS[1] (Ellegård and Vrotsou 2006; Vrotsou 2010). The software was developed to analyze, at different levels of aggregation, people's everyday activity context from their diaries (Ellegård and Vrotsou 2006; Ellegård and Cooper 2004). Later, this software was refined to estimate, in real time, the electricity used by people's utilization of electric appliances when fulfilling their activities. The opportunity to estimate electricity use is based on a model developed by Widén (2010). This refined software, VISUAL-TimePAcTS/energy use, tracks the relationship between activities, appliance use, and electricity consumption. The software may handle several levels of aggregation, i.e., individual, household, group, and population, though the individual is in all cases the basis for analysis.

The software visualizes individuals' everyday activity sequence in a systematic and standardized way. Hence, the software helps visualize people's activities performed as a sequence in the course of the day, from 4 am to 4 am the next morning, just as they are written in the diaries.

[1] VISUAL-TimePAcTS: VISUAL = visualization, P = place, Ac = activity, T = technology, S = social companionship; time is, of course, time.

Results: Visualization of Activities in Everyday Life

The rhythm of everyday life describes the dynamic of repetition that permeates everyday life and provides temporal structures that organize the social world (Walker 2014). The rhythm can be observed on an aggregate level but it is made up by the many activities individuals are doing in the course of the day. The rhythm to a large extent relates to the timing of individuals sleep and meals and of them being out of home for work and education activities.

In Fig. 11.1 we show the activity sequence of one time-diarist, a woman, to visualize how her day (using a 24-h clock) is created by showing the activities performed in sequence (Fig. 11.1a) and what activities she performed indoors when at home (Fig. 11.1b). A time-geographic assumption is, as already mentioned, that activities are performed sequentially by each indivisible individual during the day, and that movements are regarded as activities among other activities. The visualization of the activities performed by the woman in Fig. 11.1 shows the duration of each activity, what activity precedes each activity and what activity follows each activity. The purpose of showing this example is twofold: first, to show how the activity sequence is built up by activities sequentially performed by the indivisible individual, and second, to help the reader understand the principle that is behind the aggregated activity patterns of many individuals to be shown in the next section.

The visualization shows the seven main categories with different colors. However, the diaries are much more detailed than that. The software can handle five levels of detail, but not display them in different colors because of blurring nuances. Therefore, to handle this problem, all sublevel categories within each of the seven main activity categories can be detected by the "activity separator lines" in the figure.

The woman's day is structured around her physiological needs for sleep and food and her work time. Work time serves as a constraint to when she has to get up from sleeping in the morning, have breakfast and when it is possible for her to have dinner in the afternoon. Her work time takes into account the need for food, so she has a lunch break at noon. The evening activities are less steered by authorities, but maybe she is following a TV series and therefore she might be steered by the timetable.

In the time-diary material from 2010/11 it is only possible to follow the daily activity sequence of individuals with no social relation to each other, so, for example, the household level cannot be investigated in that dataset. In another, older time-diary material (from 1996) time diaries were written by individuals who live together in households. Then it is possible to use the diary information to investigate who does what activities (division of labor) and when the different household members are in the home (Ellegård and Palm 2015; Isaksson and Ellegård 2015). There are many advantages to collecting time diaries from all members in the households. Then it is possible to identify constraints for an individual due to other members in the household doing an activity that needs the other household members' attention or due to competition over appliances.

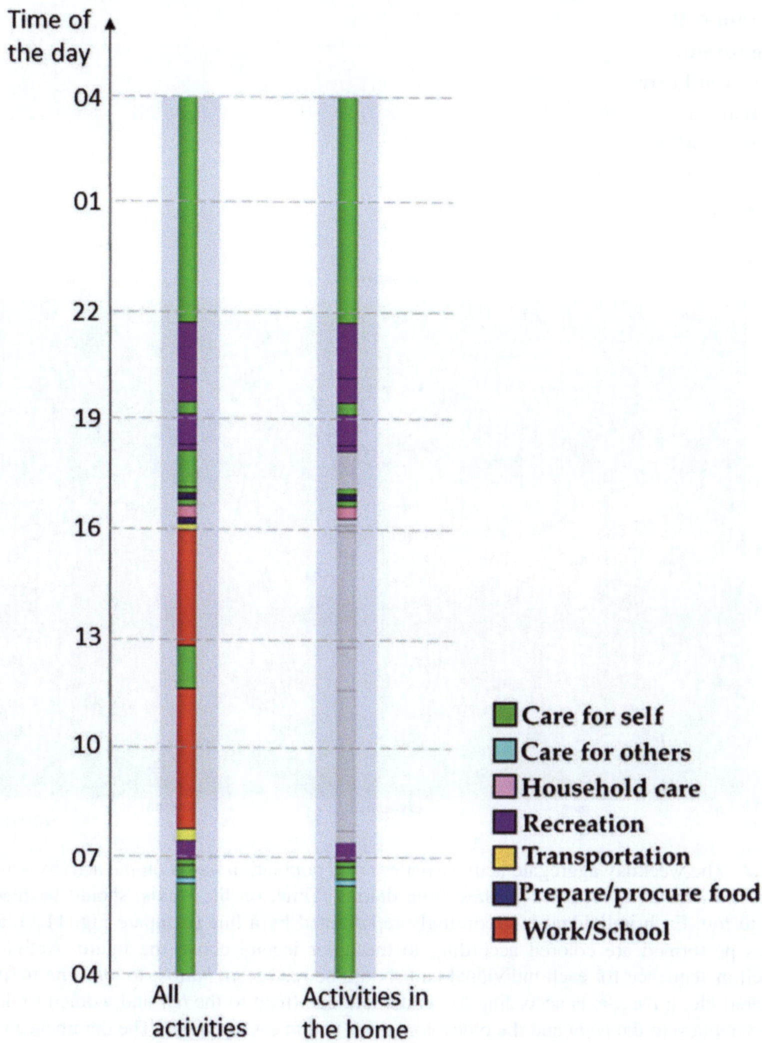

Fig. 11.1 A weekday activity pattern of one woman. In VISUAL-TimePAcTS it is possible to follow her activities during the day. In the figure to the *left* it is shown that she sleeps until 6:30, when she wakes up and takes care of others. She does her hygiene and then she has breakfast at 7:00. At 7:10 she reads the newspaper. At 7:40 she takes her moped to work. From 8:00 to 11:50 she works. Between 11:50 and 13:00 she has lunch and then she works again until 16:10 when she takes her moped to the supermarket. She does some shopping. At 16:30 she is at home cleaning until 16:50 when she does her hygiene. At 17:00 she cooks and she eats at 17:10. After dinner, she has a cup of coffee. At 18:20 she starts watching TV. At 19:20 she has supper. At 19:39 she sits at her computer. At 20:20 she watches TV again. At 22:00 she goes to sleep. In the figure to the *right* her being indoors at home is shown. She is in her home primarily when sleeping, cooking, and performing reflection/recreation activities in the evening

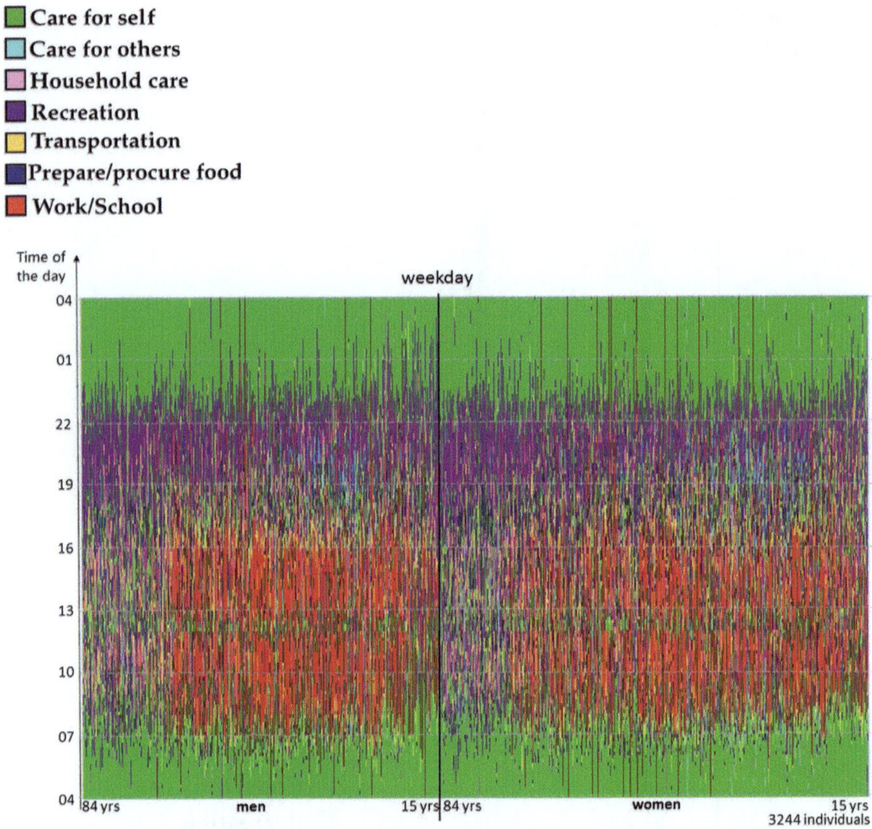

Fig. 11.2 The weekday aggregate activity pattern of a population based on the activity sequences of 3244 individuals as recorded in their time diaries. Time, on the *y*-axis, should be read from *bottom* to *top*. Each individual is accordingly represented by a line (compare Fig. 11.1), and the activities performed are colored according to the color legend above the figure. Activities are presented in sequence for each individual and should be read from *bottom* to *top*. The individuals are ordered along the *x*-axis according to gender and age (men to the *left* and women to the *right* and the youngest to the *right* and the oldest to the *left* within each gender). The dominance of sleep during night-time hours (*green*, care for oneself) and work/school activities (*red*) during daytime hours is evident. Travel to work/school in the morning and after these activities in the afternoon is indicated by the yellow parts of the line. In the evening, reflection/recreation activities (*dark lilac*) dominate, the most significant of them being watching TV

The main drawback is that it is difficult to get all members of many households to keep time diaries.

Figure 11.2 presents a visualization of all activity sequences performed by the individuals in the Swedish time use survey from 2010/11 on weekdays. The time is running vertical on the *y*-axis from 4 am at the bottom to 4 am of the following day at the top. Each individual is represented by her activity sequence (according to the principle presented in Fig. 11.1a for one individual) and they are placed along the

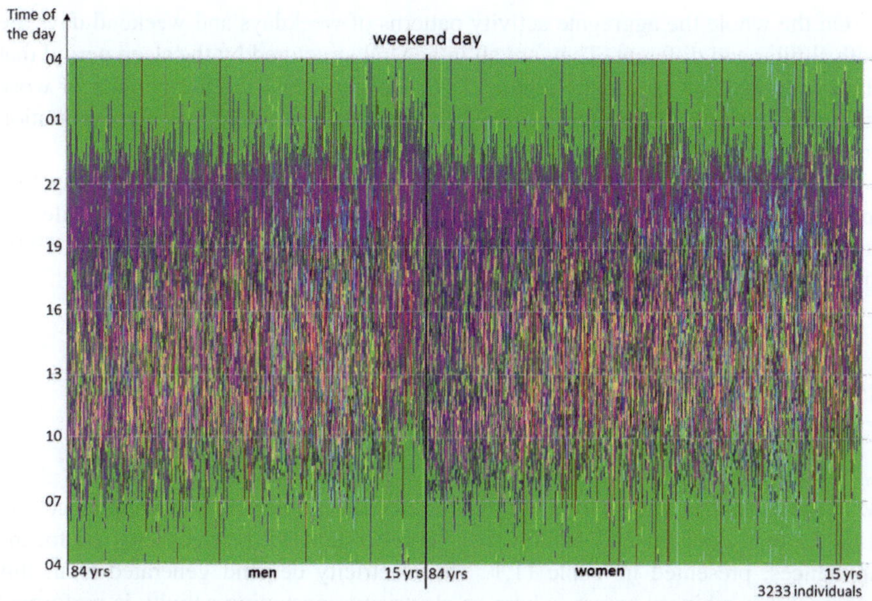

Fig. 11.3 The weekend day aggregate activity pattern of a population based on the activity sequences of 3233 individuals as recorded in their time diaries. Each individual's activity sequence is visualized with the same principle as in Fig. 11.1. There is still a dominance of sleep during night-time hours (*green*, care for oneself). The day is however now more scattered and it is more unusual to have one activity that lasts for several hours. In the evening, reflection/recreation activities (*dark lilac*) dominate, the most significant of them being watching TV

x-axis, each activity sequence beside the other, ordered by gender and age. Men are displayed to the left and women to the right. The youngest individuals are placed to the right and the oldest to the left.

This aggregate activity pattern shows some common structures: sleep dominates early mornings, nights, and late evenings. The oldest people spend more time for sleep and meals in the morning than do people working or going to school. There is a very distinct structure of the day given by the lunch break for all people in the active population. Also, a pattern that covers all people is the reflection/recreation activities (primarily watching TV) in the evenings.

Figure 11.3 presents a visualization of all activity sequences performed by all individuals on a weekend day. The seven activity categories are colored according to the same legend as in Figs. 11.1 and 11.2. Weekend days have a very different aggregate activity pattern from weekdays since only a few individuals are subject to constraints from work and school schedules. Instead, the youngest individuals sleep longer in the morning. Just like weekdays, the evenings are dominated by reflection/recreation activities (mostly watching TV). There are also more household care activities and more time-consuming travels undertaken on weekend days. Among retirees, weekdays and weekend days look very much the same.

On the whole the aggregate activity patterns of weekdays and weekend days are both similar and different. They are all in general structured by the sleep period that starts around 22:00 and ends around 7:00. During weekdays the periods of work structure the day and for most people free time starts around 17:00. Transportation connects home-based activities and work and school activities.

During weekends most people use the whole day for activities according to a more irregular pattern. There is however still a rhythm that shapes everyday life and structures it. Watching television is e.g. mainly done in the evening between 19:00 and 21:00, including during weekends.

Visualization of Everyday Activities and Electricity Consumption

Many households' daily activities consume electricity since electrical appliances are used for their performance. Widén et al. (2009) estimated the electricity use for the appliances, presented in Table 11.1. The electricity demand generated from this approach has a binary nature, where an electricity-consuming activity is performed and then uses the full amount of estimated electricity or is not performed at all and in such cases uses no electricity. The modeled synthetic load profiles were validated with measurement of electricity use, and the generated load profiles were found to be representative (Widén et al. 2009).

When we use these estimated parameters in the software we add an electricity load profile to the aggregate activity pattern, and thereby visualize the electricity consumption pattern during a weekday, in Fig. 11.4, and a weekend, in Fig. 11.5.

Figures 11.4 and 11.5 visualize activities performed on weekdays and weekend days respectively and how much electricity these activities require when appliances are used. We got a load profile for this population that can be used for information to inspire people to action leading to a change of the load profile. Interesting to note is how similar the load curves are when the electricity demand peaks are achieved

Table 11.1 Appliances and the parameters, power, and runtime developed by Widén et al. (2009) and used in the software

	Power (W)	Runtime (min)
Cold appliances	100	–
Audio	100	–
Computer	100	–
TV	200	–
Cleaning	1000	–
Ironing	1000	–
Drying	1650	90
Washing	490	130
Dishwashing	430	160
Cooking	1500	–

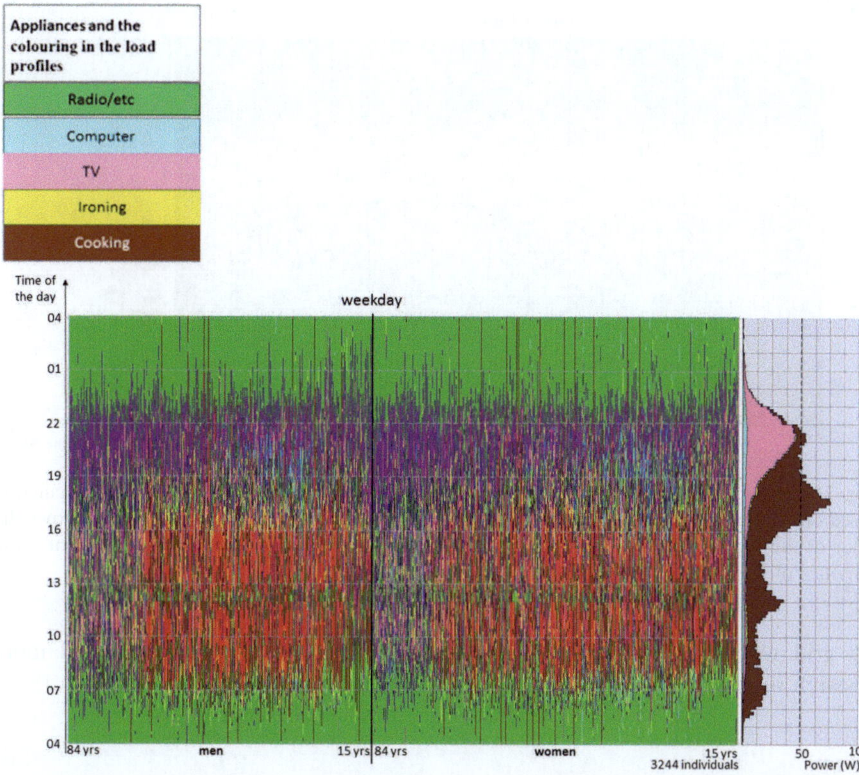

Fig. 11.4 The load profiles of a weekday from the aggregate activity pattern of a population based on the activity sequences of 3244 individuals. The colors in the load profiles follow the colors of the legend above. The load profile yields the estimated averaged of these activities of the sample ($N = 3244$)

around lunch time and in the evening, both during weekdays and weekends. This visualization clearly shows the connection between activities and electricity consumption and that electricity consumption is a consequence of activities performed and not a demand than can be understood as something isolated. From this it is possible to start elaborating about possible changes in load curves about the possibilities to change activities that consume electricity. Activities such as going to work/school and sleeping are quite inflexible and need to be done at certain times. This also creates a restriction concerning which activities can be moved in time. Activities are in general much more flexible during weekends than during weekdays. The determination of which activities are perceived by households as flexible needs further investigation however, using other methods such as interviews or surveys. The load curve can also be changed by appliances storing energy for later use by the appliance itself or by appliances storing energy that can be delivered to

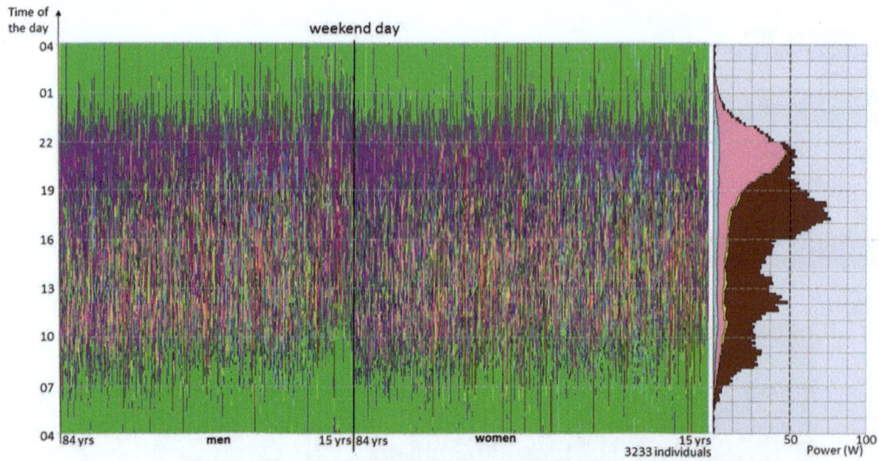

Fig. 11.5 The load profiles of a weekend day from the aggregate activity pattern of a population based on the activity sequences of 3233 individuals. The coloring in the load profiles follows the color of the legend Fig. 11.4 and again the load profile yields the estimated averaged of these activities of the sample ($N = 3233$)

the grid when there is high demand. These kinds of solutions also need more research in relation to how the households perceive and would accept them.

Everyday Life Activities in Different Generations

As noted above, Garabuau-Moussaoui (2011) finds in her study in France that different stages of life had specific connotations for energy and energy efficiency. From earlier studies we also know that choice of housing, i.e., detached houses or apartment buildings, are not only related to household income and household type but also to the phase in the life cycle a person is in. In Sweden, many households in which parents live together and have children living at home, live in detached houses. Single parents with children usually live in a rental apartment in a block of flats. Households of immigrants usually live in rental apartments. Young single persons often live in rental apartments, while older single people live in rental apartments or housing associations. During their lifetime Swedes usually change housing according to specific events in life, e.g., when young they move into their own home, when they have children they move to a bigger home, when they are aging or when families are reconstituted by divorce or new cohabitation they move to a smaller home (Lindén 2011).

With this in mind we checked what the activity patterns and load curves looked like in different age spans. Below we have categorized the individuals according to their age. The first figure shows those who are between 55 and 84 years old,

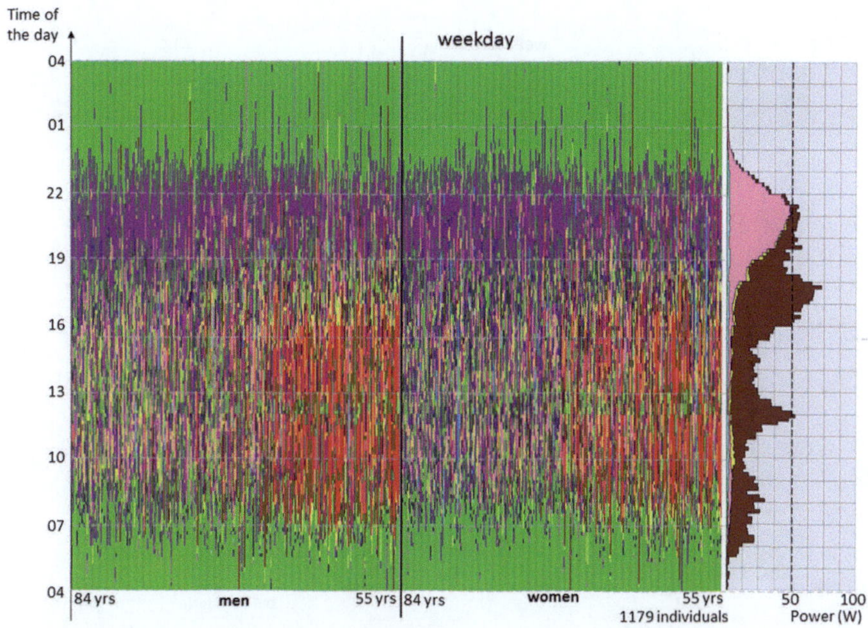

Fig. 11.6 The weekday aggregate activity pattern and the load profile of the 1179 oldest people in the surveyed population, born between 1927 and 1955

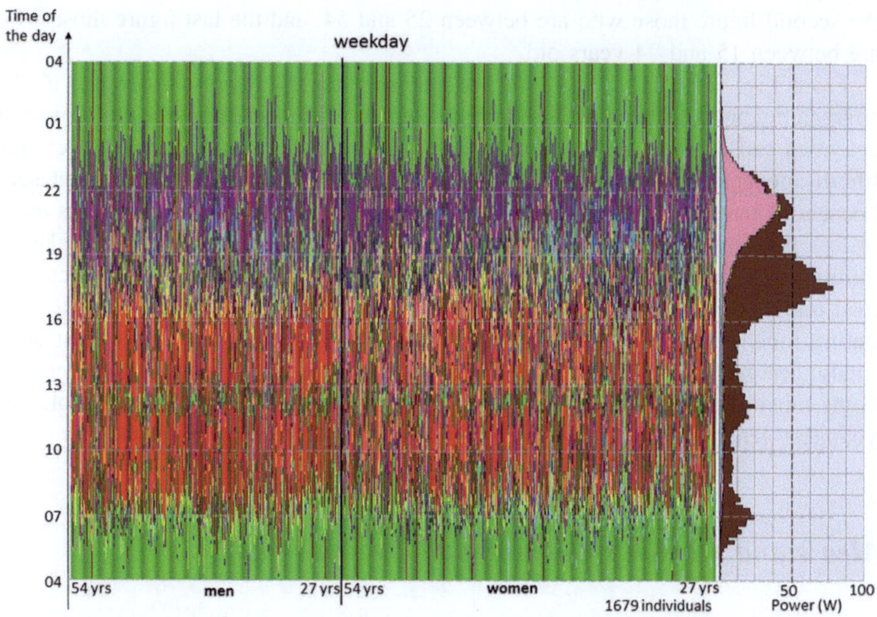

Fig. 11.7 The weekday aggregate activity pattern and the load profile of the 1679 surveyed individuals in middle age, born between 1956 and 1985

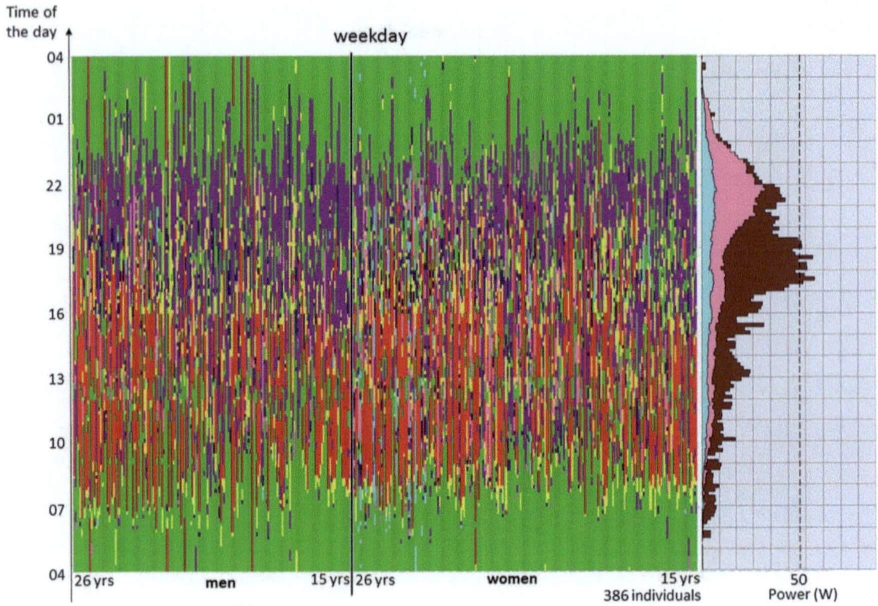

Fig. 11.8 The weekday aggregate activity pattern and the load profile of the 386 surveyed the youngest individuals, born between 1984 and 1996

the second figure those who are between 25 and 54, and the last figure those who are between 15 and 24 years old.

From Figs. 11.6, 11.7 and 11.8 we can see that the biggest difference in the aggregate activity pattern occurs with retirement. The restrictions in the timing of activities caused by work schedules that until retirement have structured everyday life disappear and the days become filled with activities, such as care for oneself, household care, and recreation/reflections. Still, the electricity load curve has more or less the same shape over the generations. People in the oldest group wake up during a longer time-span in the morning, which slightly flattens their electricity load and moves it to later in the morning. But otherwise they also have a peak during lunch time and in the evening. The oldest generation also uses most electricity, according to the estimated load profile. This is mainly due to them being at home most of the day, while the other generations are at workplace or school. The youngest generation uses more electricity for computers all through the day.

Men's and Women's Cooking and TV Activities

It is also interesting to analyze if there are gender differences. The aggregate activity pattern shown by VISUAL-TimePAcTS/energy use visualizes differences in men's

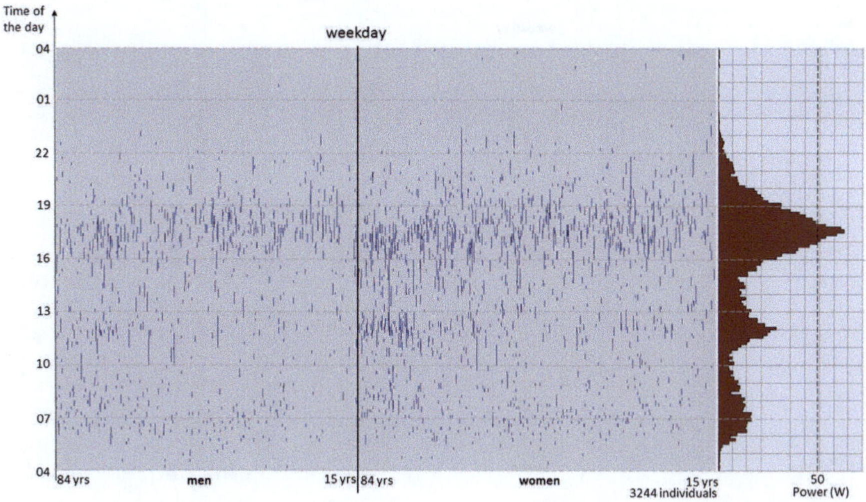

Fig. 11.9 Cooking activities during a weekday for the whole population, where the men are to the *left* and the women to *right*

and women's aggregate activity patterns with focus on specific activities. In Fig. 11.9 we have visualized the cooking activity of the whole population during a weekday. The men's cooking activity is shown to the left and the women's to the right.

Figure 11.9 shows that women perform cooking activities both more frequently and during longer periods than men. A deeper analysis of this activity shows that the men participate in preparing food, while the women bear the main responsibility for this activity. The women use more electricity than men in relation to this activity. In earlier studies, we have also shown that it is the women that consume most electricity in the home, simply because they perform more activities in the homes that require electricity (Ellegård and Palm 2015) and much of this electricity is used for the good of the entire household.

The electricity used for watching TV is visualized below. Figure 11.10 shows the women's activities, while Fig. 11.11 shows the men's activities.

When it comes to electricity used for watching TV, the differences between men and women are erased and both groups report more or less the same time use for this activity. It is however interesting to note that most women stop watching TV at 10 pm, while the men's stop time is more fluid. But for both groups the electricity consumption peaks between 9 and 10 pm.

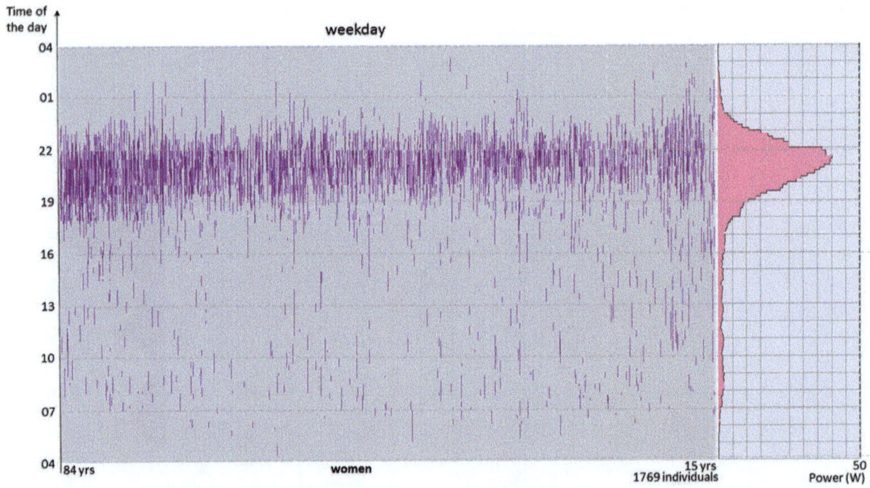

Fig. 11.10 Women's TV activity during a weekday, with the load curve profile to the *left*

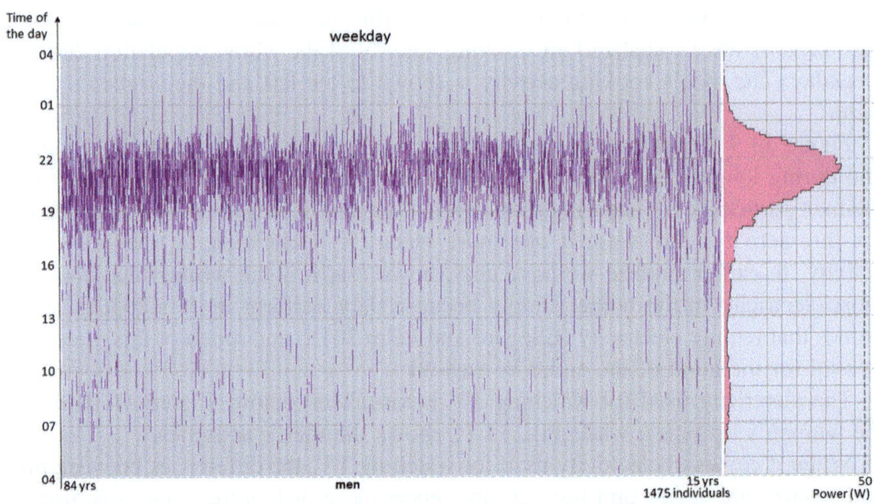

Fig. 11.11 Men's TV activity during a weekday, with the load curve profile to the *left*

Conclusions

Public policy on more efficient energy consumption and on climate change mitigation needs more informed knowledge about everyday life activities in households, since these activities give rise to household energy consumption. To analyze and understand energy consumption in relation to people's activity patterns is one way to increased knowledge, both on how people live their everyday lives and how the household sector can contribute to energy efficiency targets.

It is very important to relate information to the household members' own activity patterns and their rhythms of everyday life. By utilizing many time diaries from a population we can, as shown, discover differences in aggregate activity patterns, for example, between generations and between men and women. In general, women use more electricity for activities related to cooking and household care than men do, which makes them the most relevant target group when it comes to giving feedback on how much electricity an appliance uses or on alternative ways of doing certain activities.

Time diaries and visualization tools such as the VISUAL-TimePAcTS/energy use software are useful for learning more about everyday activities that need energy. They are useful for researchers, policymakers and others that want to be informed about energy-related activities in households or want to reach out to household members with tailored information about energy efficiency. But they are also useful for households as a reflective tool when discussing their members' various daily activities in relation to energy consumption. Then, the method can provide direct feedback to the household members and the information is valid since it emanates from their own reported activities. It is then possible to begin discussions on how activities can be changed and energy consumed without losing values and routines that the households believe contribute to maintaining their good life. This is a tool for energy advisors trying to target individuals' energy behavior.

Household energy use is however a collective rather than an individualized process. In the future, it is important to collect time diaries from everyone in one household to also be able to analyze the links between members of households and the energy use resulting from this. This linking of members is increasingly being recognized, and households will be a key unit for analysis in the future.

Recommendations for Policy Makers

Relevant and efficient policy promoting renewables and more efficient energy consumption needs informed knowledge about individuals' everyday life activities. These activities are part of the projects people want to realize and the activities give rise to household electricity consumption. It is simply not efficient to base energy policymaking on an average individual taken out of his/her context.

One way to increase the understanding of everyday life activities is to take the perspective of the households by analyzing the time diaries kept by the household members that describe their day. By starting from an analysis of people's everyday activities in their sequential appearance knowledge is gained of whom to target with policy instruments. Tailored information can use knowledge from time use surveys to find out who in the household does what activity and with whom to communicate.

Time diaries are equally useful for energy advisors and others who give advice to households. A time diary can be used as a reflective tool when discussing the family members' daily lives together in relation to energy consumption. The method is useful to provide direct feedback to households about their own reported activities. On this basis, it is possible to begin discussing how activities can be changed without affecting the values and routines that households believe help maintain their good life.

Household energy use is a collective rather than individualized process. To collect time diaries all members in a household should be prioritized in the future. It is analytically quite simple to handle single individuals' time at home, but much more complicated when a whole household is to be taken into consideration. This means that the links between members are as important to understand as individual activity performance per se if the goal is to reduce household energy consumption.

References

Anable, J., B. Anderson, E. Shove, and J. Torriti. 2014. *Categories, Concepts and Units: Representing Energy Demand in and Through Time*. Working paper 3. Lancaster University, Lancaster: DEMAND center.

Aune, M. 1998. Nøktern eller nyttende: Energiforbruk og hverdagsliv i norske husholdninger. PhD, Senter for teknologi og samfund. Norges teknisk-naturvidenskapelige universitet, Trondheim, Norway.

Bartiaux, F. 2008. Does environmental information overcome practice compartmentalisation and change consumers' behaviours? *Journal of Cleaner Production* 16 (11): 1170–1180. doi:10.1016/j.jclepro.2007.08.013.

Bartiaux, F., and L.R. Salmón. 2014. Family dynamics and social practice theories: An investigation of daily practices related to food, mobility, energy consumption, and tourism. *Nature and Culture* 9 (2): 204–224. doi:10.3167/nc.2014.090206.

Ellegård, K. 2006. The power of categorisation in the study of everyday life. *Journal of Occupational Science* 13 (1): 37–48.

Ellegård, K., and M. Cooper. 2004. Complexity in daily life—A 3D-visualization showing activity patterns in their contexts. *Electronic International Journal of Time Use Research* 1 (1): 37–59.

Ellegård, K., and K. Nordell. 1997. *Att byta vanmakt mot egenmakt: [självreflektion och förändringsarbete i rehabiliteringsprocesser]: [metodbok]*. Stockholm: Johansson & Skyttmo.

Ellegård, K., and J. Palm. 2011. Visualizing energy consumption activities as a tool for making everyday life more sustainable. *Applied Energy* 88 (5): 1920–1926. doi:10.1016/j.apenergy.2010.11.019.

Ellegård, K., and J. Palm. 2015. Who is behaving? Consequences for energy policy of concept confusion. *Energies* 8 (8): 7618.

Ellegård, K., and K. Vrotsou. 2006. Capturing patterns of everyday life: Presentation of the visualization method VISUAL-TimePAcTS. In *International Association for Time-Use Research Conference*, Copenhagen, Denmark. Danish National Institute of Social Research.

Ellegård, K., T. Hägerstrand, and B. Lenntorp. 1977. Activity organization and the generation of daily travel: Two future alternatives. *Economic Geography*: 126–152.

Eurostat. 2009. *Harmonised European Time Use Surveys 2008 Guidelines*. Luxembourg: Office for Official Publications of the European Communities.

Garabuau-Moussaoui, I. 2011. Energy-related logics of action throughout the ages in France: Historical milestones, stages of life and intergenerational transmissions. *Energy Efficiency* 4 (4): 493–509. doi:10.1007/s12053-011-9124-8.

Gram-Hanssen, K. 2010. Standby consumption in households analyzed with a practice theory approach. *Journal of Industrial Ecology* 14 (1): 150–165. doi:10.1111/j.1530-9290.2009.00194.x.

Gyberg, P., and J. Palm. 2009. Influencing households' energy behaviour-how is this done and on what premises? *Energy Policy* 37 (7): 2807–2813. doi:10.1016/j.enpol.2009.03.043.

Hägerstrand, T. 1970. What about people in regional science? *Papers in Regional Science* 24 (1): 7–24.

Hägerstrand, T. 1985. Time-geography: Focus on the corporeality of man, society, and environment. In *The Science and Praxis of Complexity*, ed. S. Aida, 193–216. Tokyo: The United Nations University.

Hägerstrand, T., and B. Lenntorp. 1993. Region och miljö–sammanfattning av ett projekt om ekologiska perspektiv på den rumsliga närings-och bosättningsstrukturen. Eng Region and environment–a summary of a project about ecological perspectives on the spatial structure industry and settlements) NordREFO 5: 229–237.

Hägerstrand, T., K. Ellegård, U. Svedin, and B. Lenntorp. 2009. *Tillvaroväven*. Stockholm: Forskningsrådet Formas.

Harvey, A.S. 1993. Guidelines for time use data collection. *Social Indicators Research* 30 (2–3): 197–228. doi:10.1007/BF01078728.

Hellgren, M. 2015. *Energy Use as a Consequence of Everyday Life*. Linköping: Linköping University.

Isaksson, C., and K. Ellegård. 2015. Dividing or sharing? A time-geographical examination of eating, labour, and energy consumption in Sweden. *Energy Research and Social Science* 10: 180–191. doi:10.1016/j.erss.2015.06.014.

Lenntorp, B. 1977. Paths in space-time environments: A time-geographic study of movement possibilities of individuals. *Environment and Planning A* 9 (8): 961–972.

Lindén, A.-L. 2011. Vardagens aktörer och deras hushåll. In *Sammanvävt. Det goda livet i vardagsforskningen. En vänbok till Kajsa Ellegård*, ed. J. Palm, and E. Wihlborg, 28–39. Linköping, Sweden: Linköping University.

Lunde, M., I. Røpke, and E. Heiskanen. 2015. Smart grid: Hope or hype? *Energy Efficiency*. doi:10.1007/s12053-015-9385-8.

Lutzenhiser, L. 1992. A cultural model of household energy consumption. *Energy* 17 (1): 47–60. doi:10.1016/0360-5442(92)90032-U.

Mårtensson, S. 1974. Drag i hushållens levnadsvillkor i Ortsbundna levnadsvillkor. In *Bilagedel till Orter i regional samverkan*, vol. SOU 1974:2. ed. 1974:2 S, Appendix 2. Bilagedel.

Minnen, J., I. Glorieux, T.P. Van Tienoven, S. Daniels, D. Weenas, J. Deyaert, S. Van Den Bogaert, and S. Rymenants. 2014. Modular online time use survey (MOTUS)-translating an existing method in the 21st century. *Electronic International Journal of Time Use Research* 11 (1): 73–93. doi:10.13085/eIJTUR.11.1.73-93.

Nyborg, S. 2015. *Smart Homes in Transition: Investigating the Role of Households in the Development of Smart Grids in Denmark*. PhD, Aalborg University, Denmark.

Oikonomou, V., F. Becchis, L. Steg, and D. Russolillo. 2009. Energy saving and energy efficiency concepts for policy making. *Energy Policy* 37 (11): 4787–4796. doi:10.1016/j.enpol.2009.06.035

Orban, K. 2013. The process of change in patterns of daily occupations among parents of children with obesity: Time use, family characteristics and factors related to change. Diss., Lund University, Lund.

Palm, J. 2010. The public-private divide in household behavior: How far into home can energy guidance reach? *Energy Policy* 38 (6): 2858–2864. doi:10.1016/j.enpol.2010.01.018.

Palm, J., and K. Ellegård. 2011. Visualizing energy consumption activities as a tool for developing effective policy. *International Journal of Consumer Studies* 35 (2): 171–179. doi:10.1111/j.1470-6431.2010.00974.x.

Pantzar, M., and E. Shove. 2010. Temporal rhythms as outcomes of social practices: A speculative discussion. *Ethnologia Europaea* 40 (1): 19–29.

Reckwitz, A. 2002. Toward a theory of social practices a development in culturalist theorizing. *European Journal of Social Theory* 5 (2): 243–263.

Røpke, I., and T.H. Christensen. 2012. Energy impacts of ICT—Insights from an everyday life perspective. *Telematics and Informatics* 29 (4): 348–361. doi:10.1016/j.tele.2012.02.001.

Shaw, S.L., H. Yu, and L.S. Bombom. 2008. A space-time GIS approach to exploring large individual-based spatiotemporal datasets. *Transactions in GIS* 12 (4): 425–441. doi:10.1111/j.1467-9671.2008.01114.x.

Shoval, N., M.P. Kwan, K.H. Reinau, and H. Harder. 2014. The shoemaker's son always goes barefoot: Implementations of GPS and other tracking technologies for geographic research. *Geoforum* 51: 1–5. doi:10.1016/j.geoforum.2013.09.016.

Shove, E. 2003. *Comfort, cleanliness and convenience: The social organization of normality*. Berg Oxford.

Shove, E., F. Trentmann, and R. Wilk. 2009. *Time, consumption and everyday life: Practice, materiality and culture*. Berg.

Shove, E., and N. Spurling. 2013. *Sustainable practices: Social theory and climate change*, 1–14. doi:10.4324/9780203071052

Silm, S., and R. Ahas. 2014. The temporal variation of ethnic segregation in a city: Evidence from a mobile phone use dataset. *Social Science Research* 47: 30–43. doi:10.1016/j.ssresearch.2014.03.011.

Skolverket. 2013. *Lärarnas yrkesvardag: En nationell kartläggning av grundskollärares tidsanvändning*. Stockholm: Skolverket.

Stephenson, J., B. Barton, G. Carrington, D. Gnoth, R. Lawson, and P. Thorsnes. 2010. Energy cultures: A framework for understanding energy behaviours. *Energy Policy* 38 (10): 6120–6129. doi:10.1016/j.enpol.2010.05.069.

Strengers, Y. 2012. Peak electricity demand and social practice theories: Reframing the role of change agents in the energy sector. *Energy Policy* 44: 226–234. doi:10.1016/j.enpol.2012.01.046.

Szalai, A. 1972. The use of time: Daily activities of urban and suburban populations in twelve countries. *Publications of the European Coordination Centre for Research and Documentation in the Social Sciences*. The Hague: Mouton.

Vrotsou, K. 2010. *Everyday mining: Exploring sequences in event-based data*. Department of Science and Technology, Linköping University, Norrköping.

Vrotsou, K., M. Bergqvist, M. Cooper, and K. Ellegård. 2014. PODD: A portable diary data collection system. In: Proceedings of the 2014 international working conference on advanced visual interfaces, 2014. ACM: 381–382.

Walker, G. 2014. The dynamics of energy demand: Change, rhythm and synchronicity. *Energy Research and Social Science* 1: 49–55. doi:10.1016/j.erss.2014.03.012.

Warde, A. 2014. After taste: Culture, consumption and theories of practice. *Journal of Consumer Culture* 14 (3): 279–303. doi:10.1177/1469540514547828.

Welch, D., and A. Warde. 2014. *Theories of practice and sustainable consumption*. In Handbook of Research on Sustainable Consumption, 84–100. doi:10.4337/9781783471270.00013.

Wilhite, H. 2005. Why energy needs anthropology. *Anthropology Today* 21 (3): 1–2.

Wilhite, H., E. Shove, L. Lutzenhiser, and W. Kempton. 2000. *The legacy of twenty years of energy demand management: We know more about individual behaviour but next to nothing about demand.* In Society, behaviour, and climate change mitigation, 109–126. Springer.

Chapter 12
Making Energy Grids Smart. The Transition of Sociotechnical Apparatuses Towards a New Ontology

Dario Padovan and Osman Arrobbio

> *For when asceticism was carried out of monastic cells into everyday life, and began to dominate worldly morality, it did its part in building the tremendous cosmos of the modern economic order. This order is now bound to the technical and economic conditions of machine production which today determine the lives of all individuals who are born into this mechanism, not only those directly concerned with economic acquisition, with irresistible force. Perhaps it will so determine them until the last ton of fossilized coal is burnt.*
>
> Weber (2005: 123)

Abstract The analysis of the assemblages and the functioning of conventional energy grids is the starting point of any process of smartness. Even if smarter elements already exist in energy grids, a full transition towards smartness is still far away. To investigate the starting conditions of a claimed process towards smartness, we realized an investigation in the city of Turin exploring the socio-technical development of its district heating network. The social elements it is composed of have been the object of an empirical investigation, based on 38 interviews and 3 focus groups and aimed at depicting its features from the various perspectives of the many roles that are played in it, from the professionals of the energy utility to the end users. We use two main perspectives. The first one is to conceive energy grids as technological zones, in which metering standards, communication infrastructures, and social evaluation assemble. The second one is to conceive energy grids as apparatuses or dispositives in which asymmetric lines of power, knowledge, information, decision-making, intensity and artefacts, constitute the ontology of the grid itself. An apparatus is an assemblage or a hybrid of technical and social elements, which has the strategic function to respond to an urgency. Foucault refers to the apparatus as a device consisting of a series of parts arranged in a way so that they influence the scope. This device exerts a normative effect on its "environment" because it introduces certain dispositions. In their effectiveness, energy networks

D. Padovan · O. Arrobbio (✉)
Department of Culture, Politics and Society, University of Turin,
Lungo Dora Siena 110/A, 10153 Turin, Italy
e-mail: osman.arrobbio@unito.it

N. Labanca (ed.), *Complex Systems and Social Practices in Energy Transitions*,
Green Energy and Technology, DOI 10.1007/978-3-319-33753-1_12

are apparatuses made of variable and disparate assemblages of natural, technical, and social elements, a continuous process fostering differences and repetitions. Based on our outcomes, we can consider thermal grids as a kind of complex system or network of agents in which energy power circulates in a way very similar to the circulation of social power.

Introduction

In this chapter, we describe the assemblages and functioning of conventional energy grids at the beginning of the smartness process. This exercise is useful because it makes possible to pinpoint obstacles, barriers, resistances, conflicts, differences, necessities in the process of energy grids' democratization and aligning. Usually, the description of an energy smart grid consists of a list of properties that the grid needs to get to be called "smart". Thus, smart grids are tools that can make imaginable the management of "direct interaction and communication among consumers, households or companies, other grid users and energy suppliers" (European Commission 2011). A smart grid gives smart information, allows for savings, allows for good and real-time information, connects providers and users. Yet, what is still lacking in the claim for smart grid is an ontological dimension of both energy and grid. In our idea, it is not enough to enunciate an amount of technical characteristics that should mark the grid and its smartness. What we are trying to do is to provide a deeper and more complex frame for the energy smart grid implementation.

To accomplish this task, we use two main perspectives. The first one is to conceive energy grids as technological zones, in which metering standards, communication infrastructures, and socio-technical evaluation assemble. The second one is to conceive energy grids as apparatuses in which asymmetric lines of power, knowledge, information, decision-making, intensity and artefacts, constitute the ontology of the grid itself. This "irreducible inequality", this transcendental injustice, which marks the grid—likely any grid or network of relations—is what the smartness has to reduce but also to convert in new qualitative characteristics of the grid itself and of its components. A smart grid that wants to align or flatten the original disparities making itself more effective must change by actualizing its creative potential. Insofar as an apparatus such as an energy grid is constituted by heterogeneous components such as corporate actors, people and devices, its ordering is always unstable and challenged by the mutating conditions of the environment. However, despite the fluctuating orders capable of entering into communication, everything that happens and everything that appears into the grid is correlated with orders of differences: differences of level, temperature, pressure, tension, potential, intensity. These differences, when aligned, produce new configurations between agents of the grid. This is what policymakers and technical makers have to take in mind when they foster the smartness of the grid. These new alignments are what allows the smartness of the grid.

In the first section, we depict the present discourses aimed to foster smart grids. In the second, we illustrate the characteristics of conventional energy grids how we knew them during our investigation. In the third section, we enter the description of a thermal grid, more specifically, the thermal grid constituting the district heating system of the city of Turin in Italy. This grid has been the object of an empirical investigation, based on 38 interviews and 3 focus groups, aimed at depicting its features from the various perspectives of the many roles that are played in it, from the professionals of the energy utility to the end users. Some of the socio-technical features of the district heating network are thus described in that section, while the conceptual framework of this work mainly derives from, and was mainly tested against, the results of our empirical investigation. In the fourth section, we resume the transitional perspective towards the smartness. In the fifth, we introduce the problematic of technological zones by pinpointing three different configurations that might support the transition towards smart thermal grids: metrological zones, infrastructural zones, zones of interoperability. In the sixth, we introduce the concept of apparatus, trying to using it to understand the very nature of energy grids. In the seventh paragraph, we underline the distribution and formation of asymmetries of power inside the grid. In the eighth, we claim for a deep change in energy apparatuses and we provide some advice for policy and technical makers.

Energy Smart Grids

To manage the transition to a more sustainable energy system based on fluctuating and asymmetric energy production and consumption, a new highly complex, self-balancing energy system called 'smart grid' has been designed. Though elements of smartness already occur in many parts of existing grids, the difference between today's grid and a smart grid of the future is mainly the grid's capability to handle more complexity than today in an efficient and effective way (European Commission 2011). The European Commission described smart grids as "energy networks that can automatically monitor energy flows and adjust to changes in energy supply and demand accordingly".[1] Smart grids are regarded as an upgraded energy network to which two-way digital communication between supplier and consumer, intelligent metering and monitoring systems have been added. Combining information on energy demand and supply can allow grid operators to better plan the integration of renewable energy into the grid and balance their networks. Smart grids also open up the possibility for consumers who produce their own energy to respond to prices and sell excess to the grid. In a few words, the smart grid is a process of defining and developing intelligent control technologies to control and coordinate flexible consumption in order to maintain over time a

[1]http://ec.europa.eu/energy/en/topics/markets-and-consumers/smart-grids-and-meters [Last accessed: 07-11-2016].

balance between production and consumption in the overall energy system. It should open up possibilities for consumers to directly control and manage their individual consumption patterns, providing, in turn, strong incentives for efficient energy use if combined with time-dependent energy prices. Improved and more targeted management of the grid translates into a grid that is more secure and cheaper to operate. Smart grids should be the backbone of the future decarbonised power system. They can enable the integration of vast amounts of both on-shore and off-shore renewable energy and electric vehicles while maintaining availability for conventional power generation and power system adequacy.

The development of smart grid visions and solutions are influenced by many different interests, ideas and actors, which contribute to a high degree of complexity within the field. However, despite the plethora of R&D and demonstration projects, only little has been achieved in terms of actually realizing the smart grid visions fully. The smart grid system is still much in the making, and there is still a gap between the ideas of the future system and the practical realization of these ideas (Gram-Hanssen 2009). In order to get an effective transition towards smart grids, important aspects that are so far considered merely technological have to be managed, faced and, where possible, overtaken. To understand the particular form smart grids requires, detailed empirical and historical analyses are needed. Moreover, the evolution of these technical configurations is not predetermined. The particular circumstances of their development are of considerable significance, and these particular circumstances depend on the construction of a whole series of inter-relations which all agents or stakeholders are involved in.

Conventional Thermal Grids

Our chapter gives an understanding of the current functioning and potential evolution of conventional thermal energy grids. As said before, elements of smartness already exist in many parts of existing grids. But these elements have to be integrated, harmonized and pushed at work. Conventionally, thermal grids convey energy by using water as a carrier. Water, hot or cold, is conveyed through underground hubs, which then distribute water throughout different buildings' thermal plants or boilers and then among final users. This is the reason why they are often seen as composing the district heating system. Thermal energy grids are technically different from electric smart grids, mainly regarding final users. For example, from the point of view of metering, given the current infrastructure in Europe, it is easier to provide feedback on electricity consumption than on thermal energy consumption (EEA 2013). Residential thermal energy consumption (as well as gas consumption in the case of domestic autonomous consumption) is determined principally by structural dwelling characteristics, while electricity consumption varies more directly with household composition and social standing—and thus may be more responsive to behaviour change programmes (Brounen et al. 2012). Heat consumption practices, when enacted through thermal grids, i.e.

through centralized heating systems, differ for some other aspects from the electricity consumption practices, notably from those which are not aimed at providing warmness or coldness. First, heat consumption practices are marked by a temporal dimension. Indeed, unlike what happens with electric appliances (again, excluding those which are not designed to modify temperatures), the required energy service is not obtained, nor it ceases, instantaneously. As it is well known, radiators only gradually gain and lose heat. Moreover, switching radiators on or off is only effective when enacted within the building heating time. Second, differently from electricity consumption, which is paid only based on singular household consumption, the apportionment of heating costs is based on different criteria. Where heat costs allocators are not put in place, they are apportioned only based on the cubic meters of the apartments. Where heating costs allocators have been installed, heating costs are partly apportioned based on cubic meters and partly based on heat consumption. It is in the last few years that, as the installation of thermostatic valves and heat costs allocators became mandatory, the costs apportionment procedures have been moving from the former to the latter. Under the former system of apportionment, thermal energy consumption is characterized by a process of compensation. Differences in thermal energy efficiencies of apartments and in thermal energy consumption behaviours among final users do not directly translate into differences in heating expenses. People who spend more time at home (e.g. homemakers, the elderly, sick persons, part/full-time unemployed) thus benefit from an advantageous price per unit of thermal energy they directly enjoy. The same happens to people living in the coldest sections of buildings. It means that a sort of "thermal equity" issue is (was) in some way addressed or implied. Thermostatic valves and heat costs allocators are thus pulling towards a more individualized way of consumption that erodes the former compensation process described here.

Based on our investigation we can say that conventional thermal grids are not only a set of technical devices aimed at the provision of warmness or coldness, but are a more complex arrangement of technical objects, practices, rules, and aims regulating, driving and compensating the actions performed by agents. In our case, the thermal energy grid regulates and performs the comfort condition in relation to two aspects: (a) by determining and deciding prices, conditions of use and provision of thermal energy and (b) by providing people with some tools in order to freely and autonomously control the energy apparatus. This latter aspect, at least in our investigation concerning the district heating system of Turin, is very limitedly addressed. Temperature, schedule, time, starting and stopping, are often not controlled by the final user. Compared with electric energy grids, thermal grids are so unmanageable by the final users that we got the idea that they are still victims of a centralized and untouchable power. This condition generates an asymmetry of power that is at the core of socio-technical apparatuses developed in the context of modern society and that is often managed and controlled by corporate actors. Before starting the smartness processes, agencies, public authorities, providers and final users have to be aware of the complex arrays of relations and configurations that make possible a smart energy grid. Here we provide some possible interpretation of smart grid development in order to foster an interplay among technical and social agents and agencies to reach a real smartness.

The Turin District Heating System

Despite all the existing literature and pilot projects on smart energy grids, their actual implementation seems to be far from giving effective results. As revealed in our case study, i.e. the district heating system of a medium size city, the implementation of the smart grid is not happening quickly enough. The reasons why this case study looks interesting are varied:

- it relates to an energy system under transition: from the actual situation to a virtual or desired one;
- it shows some of the difficulties characterizing the process. For instance, the problematic interactions between the agents of the grid;
- it shows some of the frictions that final users have with technical apparatuses.

As already mentioned, the Turin district heating system has been the object of an empirical investigation based on 38 interviews and 3 focus groups aimed at defining its features from the various perspectives of the many roles that are played in it, from the professionals of the energy utility to the end users. Some of the socio-technical features of the district heating network are thus described in this section, while the conceptual framework described in the following mainly derives from, and was mainly tested against, the results of our empirical investigation.

The process of smartness implementation is not at its first stages. It can be said that it had already started with the remote monitoring of substations as well as of other heating network components. The district heating network in Turin is quite huge, supplying 56 millions of cubic meters to 550,000 inhabitants, out of a population of almost 900,000. At present, the point of saturation, given the current infrastructure, has been almost achieved. Not differing from other district heating systems, Turin's processes trying to implement smartness in thermal grids has to be assessed against its capacity of leading to consumption peak shaving. Being the most part of the delivered thermal energy produced by means of CHP (Combined Heat and Power), shaving the peak would mean that the number of hours in which the integrative HOBs (Heating Only Boilers) have to be used should be reduced, leading to a more efficient heat delivery with reference to the primary energy used. Such an outcome would also be positive for the metropolitan area as a whole. Indeed, it would decrease the pollutant emissions, at the same time allowing more cubic meters to be connected to the network.

There are two ways in which more and/or enriched information and data could be used to achieve this outcome. Firstly, a more reliable model of the functioning of the network could be used by the heating company for peak smoothing purposes, meaning for optimization interventions (invisibly) carried out by the heating company itself. Secondly, this information could be used to foster changes in the way the other actors act. In order to understand which information and data could serve to this aim, let us have a look at the features of the actors of the socio-technical system represented by this district heating network by selecting two main categories of buildings connected to this network: private (residential) buildings and public buildings (owned by the Municipality).

Private Buildings

The management of private buildings is run by professional building administrators. They manage the great majority of apartment buildings in Turin. It means that they need to be annually and formally appointed as building administrators by the residents' assembly. With reference to heating aspects, building administrators receive and pay (to the supplier) the heating bills, receive any heating reports, communicate with the district heating company which is the desired heating time schedule and set point for the building, answer to complaints coming from residents, and so on. Building administrators also act as mediators between the district heating company and final users.

Residents' heating practices are much thinner. Indeed, many final users have almost completely delegated the heating practices to the building administrators. As a result, they only have vague or wrong ideas about which is the heating system of the building as well as about the building heating time schedule. Moreover, they usually only receive an annual report containing the heating costs as they were apportioned among apartments according to different methods (cubic meters alone or in conjunction with metering performed through heat cost allocators, where available). In sum, residents have, by default, very few possibilities to make comparisons with previous years, self-metering being one of them. Understanding if any improvement or worsening in heat expenses is attributable to their behaviours, to other residents' behaviours, to errors, to insulation measures, to tariff changes, to supplier's policies, to a mild or harsh winter, etc., is even more difficult. On the other side, they are in charge of the management of the thermal comfort in their apartments, even if they have very few (or dependent on income and on homeownership) ways to reduce heat consumption.

Public Buildings

With a few exceptions, the management of thermal issues is only in a limited way part of the tasks of public buildings managers. They do not receive any heating bill, nor do they receive any heating report. They are not given guidance or objectives to reduce thermal energy consumption either. Seen from a public building managers' perspective, thermal issues are thus essentially related to guaranteeing the thermal comfort for workers and employees. Indeed, the management and setting of the thermal systems, as well as the decisions related to refurbishments and maintenance, are left to an external company managing all energy aspects for all public buildings owned by the Municipality. Neither the managers of the public buildings, nor the employees using them, are asked on a regular basis, or they are not asked at all, to monitor and steer the heating practices as they manifest in these buildings. Nonetheless, this does not prevent some employees to adopt in their workplace some of the elements constituting the heating practices as carried out in their private buildings (thermometers, complaints and clothing). As we are going to show in the following sections, public buildings are the least flexible and the interactions between actors and technical devices are problematic.

Thermal Grids Transition: Interpretations, Visions, Dynamics

Notwithstanding their differences, smart thermal grids are claimed to play, as smart electric grids do, an important role in future smart cities by ensuring a reliable and affordable heating and cooling supply to various customers with low-carbon and renewable energy carriers like waste heat, waste-to-energy, solar thermal, biomass and geothermal energy. Smart thermal grids allow adapting to changing conditions in supply and demand in the short, medium and long-term, and facilitate participation of final users, for instance, by allowing supplying heating or cooling back to the network. To do so, they need to be spatially integrated in the complete urban energy system and to interact with other urban infrastructures, such as networks for electricity, sewage, waste, ICT (Lund et al. 2012, 2014). By optimising the combination of technologies and enabling a maximum exploitation of available local energy resources through cascade usage, smart thermal grids can contribute to improving the efficiency of urban heating and cooling, while increasing the cost efficiency and increasing the security of supply at a local level. The scale of smart thermal grids can range from neighbourhood-level systems to citywide applications, depending on heating and cooling demand and urban context.

The technical elements of smart thermal grids cover thermal generation systems like small-scale low-carbon heating and cooling systems, combined heat and power systems (CHP), thermal storage technologies and innovative network improvements. Network-integrated sensors and smart heat meters allow for more effective and efficient use of the separate components when supported by overarching energy management (Schmidt et al. 2013; Lund et al. 2014). This so-called 4th Generation District Heating concept is interesting, but it lacks some key elements, the main ones being the final users involvement in the idea of smartness and, consequently, a broader and smarter design of the grid that includes these users.

The design of the 4th Generation District Heating concept is enlightening because it clearly shows that the technological dimension is widely overriding the future scenario. Here, institutional planning policies, social motivations, economic incentives, costs/benefits perspectives, are compressed and superposed to the technical architecture. Moreover, the juxtaposition of sectors and parts of the grid and the imaging of their amalgamation is not enough if the processes, in virtue of which they integrate, are not decomposed and analysed in their effective dynamics, oppositions, and tensions.

Technological Zones

Thermal energy grids are situated socio-technical systems that combine hard technical infrastructures and devices with expectations of ordinary and pre-established actions and behaviours from both distributors and final users. In this sense, their working needs repetitive interactions among all human agents and technical devices involved and locally composing the grids. Often thermal grids do not need to be large to be conventionally efficient. Their spatial dimensions are often coincident with the urban dimension, covering large sectors of urban settlements but not embracing entire regions. The situatedness of thermal grids is also understandable when looking at its implementation. Grids are carriers of energy (electric or thermal as in the case of district heating), better they are carriers of energy intensity and their performance is based on an ontology of difference. The energy intensity conducted by water coming from the provider's station is very different from the energy intensity arriving at the final user location, and then to the station again. Generally, the water leaving a heating station is around 100–120 °C, it reaches buildings boilers at around 80–90 °C and it comes back at the heating central station around 60 °C. This intensity can be measured by different standards such as Joule/second/m^2 or also in more trivial terms of initial (power plants) and final (end users) temperature. However, we would like to suggest a revision of the classical thermodynamic interpretation of energy flowing into grids, as well as of conventional engineers' interpretation of grids.

Embracing an idea coming from Gilles Deleuze's philosophy, we understand intensity as the force that determines difference and produces repetition (Deleuze 1994; Crockett 2013). Every phenomenon is marked by differences. For Deleuze, "Every phenomenon refers to an inequality by which it is conditioned. Every diversity and every change refers to a difference, which is its sufficient reason. Everything that happens and everything that appears is correlated with orders of differences: differences of level, temperature, pressure, tension, potential, difference of intensity" (Deleuze 1994, p. 222). What is interesting here is the fact that the different agents of the grid are connected via difference (of energy/power obviously, but also of status, role, control, income). The difference in intensity conveyed by the grid, and the way it is flattened, is very crucial for the efficiency of the grid itself. Energy intensity dissipates along the grid, producing entropy. The process, "from more to less differentiated, from a productive to a reduced difference, and ultimately to a cancelled difference" (Deleuze 1994, p. 223), is immanent to the grid's dynamic (will the grid never come to deliver customized temperatures?). The smartness process, *how is until now designed*, strengthens the process of annulment of difference, without thinking that new qualities emerge from it. This is the reason why fourth generation smart thermal grids are designed to cut down the water's temperature they are carrying and to shave daily, weekly, seasonal peaks. Actions and applied devices are implemented in different zones of the grid to smooth this difference in intensity. These devices and operators vary from smart metering to home appliances, from dynamic tariffs to network management, focusing on different technological and social areas as shown by JRC studies (Mengolini and Vasiljeska 2013). Thus, thermal grids, as we collected information on its establishing, functioning and deploying, are areas where differences and asymmetries between the agents of the grids are often present, but where these ones tend to be shaved. Energy grids can thus be viewed as technological zones that work as operators or "differentiators" aimed to reduce differences, transforming intensity into extensity, as said by Deleuze (1994, p. 223).

> We know only forms of energy which are already localized and distributed in extensity, or extensities already qualified by forms of energy. Energetics defined a particular energy by the combination of two factors, one intensive and one extensive (for example, force and distance for linear energy, surface tension and surface area for surface energy, pressure and volume for volume energy, height and weight for gravitational energy, temperature and entropy for thermal energy ...). It turns out that, in experience, intensio (intension) is inseparable from an extensio (extension) which relates it to the extensum (extensity). In these conditions, intensity itself is subordinated to the qualities, which fill extensity (primary physical qualities or qualitas, and secondary perceptible qualities or quale). In short, we know intensity only as already developed within an extensity, and as covered over by qualities.

In this perspective, a technological zone can be understood as an extended space where differences and intensity are reduced thanks to standardized techniques, procedures and forms. As suggested by Barry (2006), such technological zones take broadly one or a mix of three forms: (1) metrological zones associated with the development of common forms of measurement; (2) infrastructural zones associated with the creation of common connection standards; and (3) zones of qualification that come into being when objects and practices are assessed according to common standards and criteria. Smart grids visions such as those developed by different European Commission DGs and agencies such as JRCs are made of varying combinations of these traits. An analytical approach to such technological zones that forge thermal grids is required in order to pinpoint hotspots where intervening to trigger a grid transition.

Metrological Zones

At the core of a smart grid there is a metrological zone based on smart metering. Intelligent metering is usually an inherent part of smart grids, forming a common form of measurement. Without a homogeneous metrological zone where power metering is standardized in order to make all agents aware of their contribution to the grid functioning, we find no smartness. When coupled with smart metering systems, smart grids reach consumers and suppliers by providing information on real-time consumption. This process is called feedback. Feedback is claimed to be a strong condition for the grid's smartness (Pullinger et al. 2014). With smart meters, consumers can adapt—in time and volume—their energy usage possibly avoiding different energy peaks throughout the day, the week, the month, the season, and so on, fitting different prices for saving money by consuming more energy in lower price periods. Smart metering systems support devices that give feedback aimed at encouraging behaviour changes, specifically to reduce energy demand and spending on energy. Detailed standards specify the minimum technical capabilities of the smart meters and feedback devices. It means that the adoption of smart metering systems should raise specific functionalities for the final users, providing readings from the meter to the customer and to equipment that he/she may have installed, and updating these readings frequently enough to allow the information to be used to achieve energy savings. Moreover, the establishment of a smart metrological zone allows all customers to possess and control meter data and to transmit these data to in-house devices. Finally, a smart metering allows the provision of messages or other information to the users from the energy supplier.

All aspects here evoked have the goal to reduce energy consumption that in terms of thermal energy provision and consumption means that the indoor comfort have to be managed in a more smart and flexible way, adapting to the variability of daily life circumstances, such as outdoor weather conditions, people time-use patterns, services provision's patterns in the case of public buildings. The development of common measurement standards and practices that make information comparable between different locations, agents, and final users aims clearly to change the so-called thermal behaviour, or in other words to change the final users pursuing of thermal comfort. The assumption behind a smart metrological zone is that energy consumption behaviours can be altered by reminders on energy consumption data provided by ICTs devices, and that consequently behaviour can be monitored and changed where needed (Cakici and Bylund 2014).

However, research on feedback information effects illustrates its own limits for fostering behavioural change. For Hargreaves et al. (2010), householders interaction with feedback is marked by different and contrasting aspects. For one side, overtime, smart energy devices gradually become 'backgrounded' within normal household routines and practices, increasing the householders' knowledge of and confidence about the amount of energy they consume. For the other side, beyond a certain level and for a wide variety of reasons, these devices do not necessarily encourage or motivate householders to reduce their levels of consumption. Once equipped with new knowledge and expertise about their levels of energy consumption, household practices may become harder to change as householders realize the limits to their energy saving potential and become frustrated by the absence of wider policy and market support.

In the course of our investigation, we did not notice of feedback devices used by users. We found that conventional energy grids are far from the use of these devices, and that they completely miss metrological zones and common modes of measurement. Final users, being private flat owners or public services workers or costumers, are blind about their consumption. Often, also institutions get not comprehensive data about their consumption. Moreover, we think that detailed, different and adjunctive indicators are needed to meet smartness, even though it is plausible to think that people do not want to pay too much time to the reading of data provided by these indicators. In any case, indicators of energy mix, energy time using, primary and secondary energy consumption, energy prices variability, CO_2 emissions, and of the making of communities that practice the distributed energy production, have to be considered and deployed. In short, the deployment of smart metrological standards requires in depth adjustments of the conventional and existing metering systems in order to meet the needs of different stakeholders and make all agents aware of their reciprocal and ontologically differential contribution to the grid.

Infrastructural Zones (Connection and Communication Standards and Feedback)

The development of common connection standards makes it possible to integrate systems of production, distribution, and communication, as well as to exclude consumers and producers who do not conform to the standard. Connection standards establish infrastructural zones that have a critical importance in the development of smart energy grids governed by information and communication technologies. In this case, it allows remote reading of meter registers by metering operators and by third parties. Moreover, these functionalities allow on-demand frequent regular readings by the meter operator. The provision of meter reading information by the supplier to the customer is thus very crucial. This would include regular readings of peak demands where the tariff is based on these; ability of linking several meters (electric, gas, water, etc.) into a single smart Meter System in order to facilitate communications; data storage within the meter; correct billing, both on a regular basis or on demand (say on the change of occupier or energy supplier). In this perspective, smart energy grids imply infrastructural zones associated with the creation of common connection standards. Infrastructural zones are areas of interoperability among different agents. It means that the thermal system must be monitored using sensors, collecting data, crossing data, performing algorithms, building platforms, enabling feedback processes. These infrastructural zones serve to make social practices of heating and cooling possible and possibly less disordered or redundant compared to what they already are. Infrastructural zones also serve to reduce the power disparity and differential among agents that is an ontological condition of energy grids. These have a critical importance in the development of information and communication throughout the grid, in order to facilitate agents' exchange of information on conditions, settings, socio-technical arrangements, and final users' practices performances.

Zones of Qualification and Improvement

Smart energy grids imply at the end the existence of a zone of assessment, in which evaluations related to grid quality and to its capacity to generate comfort while saving energy are performed. The decoupling between comfort and energy consumption is the core of smart energy grids goals and it has to be detected by using metering, devices and some comfort indicators. The development of common regulatory or quality standards has become critical to the governing of energy. Such standards govern the quality of practices enabled thanks to energy, which may exist within a particular domain. Necessarily, such standards depend on the development of various technical devices, which make it possible to assess and compare the qualities of technical devices and practices performed. However, we may speak of the existence of a zone of qualification when the technical devices allow for practices that meet common criteria, such as environmental standards. Here the role of final users such as households becomes the core component in the smart grid. The role of dynamic users that support the energy system by e.g. being flexible in their consumption and able to produce autonomously warmness or coldness, thus helping the grid to face peaks of consumption is now increasingly acknowledged (Nyborg and Røpke 2011; Strengers 2012). The changing role of final users in the transition and functioning of energy grids to smartness is year-by-year revaluated, not only at the academic level, but also in the more important documents of EU and other European bodies (European Environmental Agency 2013).

The problem is that, the way for facing the human scope in energy grids is mainly psychological or behavioural, what has been termed by Elizabeth Shove the ABC syndrome (Shove 2010). Our exploration of conventional thermal grids towards smartness confirms that this is the main vision shared by designers, engineers, and administrators. Behind this approach is the idea that individuals are fully rational beings and that they should be aware of what they are spending, consuming, and dissipating not only in monetary terms, but also in thermodynamics terms. As it has been demonstrated by several studies, not only money is for people a volatile and sometime invisible object, but also energy is difficulty understandable in its nature and ontology. Energy flows are in many ways invisible to residential energy consumers. This makes energy management and conservation practices both difficult and unusual. The more modern energy systems provide increasingly invisible means of meeting demands for heating and cooling. Warm water that flows seamlessly and silently into homes meeting our demand of comfort makes it without any notable trace of their presence (Ehrhardt-Martinez et al. 2010; see also Schwartz et al. 2013). The only way to get an account of energy use is the practices that people perform thanks to energy, such as heating. Household's everyday practices are indicators of how much energy is consumed and dissipated, the involuntary way to make energy visible. In our case of district heating, shaving the peak loads and avoid primary energy consumption is a consequence of comfort practices performed by households' agents, whereas technologies and nature of households' engagement and a new displacement of time for heating play a very crucial role. This time shift in governing energy production, distribution, and consumption, plays an important role in European energy policies. Energy saving is also the expected outcome of the evolution of this reincorporation of a time-based perspective into technological zone.

The introduction of these new perspectives, such as the temporal ones, poses the question of the useful function of economic incentives to drive the behavioural change. The idea that everything can be obtained given the right incentive is an appealing idea. However, it is not enlightening enough. What is becoming clearer is that a rational and informational view of energy users is not enough to foster change in energy system and related social practices. An approach based on theory of social practice to address this problematic is far more useful. Practice theory contributes to understanding how thermal energy is used and how this changes over time by focusing not on energy per se, but on the everyday routines of space heating through which energy is used, and on the roles of technologies, material environment, skills, rules and habits in constraining or enabling change. Practice theory is providing insights into the likely effectiveness of feedback devices and the forms of feedback they provide (Shove and Walker 2014; Schatzki 2011; Pullinger et al. 2014).

Sociotechnical Apparatuses

The technological zones as previously described are mainly technology-oriented. It is not wrong to depict energy grids in terms of technical standardization but this seems to exclude something else. Here we broaden the Foucauldian perspective suggested by Barry embracing the very interesting concept of dispositive or apparatus forged by Michel Foucault along all its oeuvre (see Agamben 2009; Raffnsøe 2008; Bussolini 2010). An apparatus is "a thoroughly heterogeneous set consisting of discourses, institutions, architectural forms, regulatory decisions, laws, administrative measures, scientific statements, philosophical, moral, and philanthropic propositions—in short, the said as much as the unsaid. Such are the elements of the apparatus" (Foucault 1980, 194). The apparatus itself is the network that can be established between these elements, but it is also an assemblage or a hybrid of technical and social elements, which has the strategic function in a given moment to respond to an urgency. Foucault refers to the apparatus as a device consisting of a series of parts arranged in a way so that they influence the scope. An apparatus indicates an arrangement that exerts a normative effect on its "environment" because it introduces certain dispositions.

According to Foucault, there are two important moments in the apparatus's genesis. A first moment is that oriented to a prevalent strategic objective.

In a second step, the apparatus as such is constituted and enabled to continue in existence insofar as it is the site of a double process. On the one hand, there is a process of "functional overdetermination", because each effect—positive or negative, intentional or unintentional—enters into resonance or contradiction with the others and thereby calls for a readjustment or a reworking of the heterogeneous elements that surface at various points. On the other hand, there is a perpetual process of "strategic elaboration" that allows the apparatus to establish and reproduce different fields of power relations (Foucault 1980, 195). Being its nature essentially strategic and goal-oriented or teleological, it implies a certain manipulation of relations of forces, a rational and concrete intervention in the relations of forces, either to develop them in a particular direction, or to block them, to stabilize them, to utilize them. Finally, an apparatus is also always linked to certain limits of knowledge that arise from it and, to an equal degree, condition it. In short, an energy grid is a set of strategies of the relations of forces supporting, and supported by certain types of knowledge.

Foucault applies his concept of apparatus to asylums, prisons, schools, factories, and hospitals, as apparatuses of disciplining and transformation of practices. In our view, it appears reasonable to apply the concept of apparatus, as depicted here, to energy grids. Norms are thus developed and inscribed in the case of energy grids into a play of power, aimed to overcome resistances, or to change inertial habits, or again to orient future choices. Data standardization and collection is crucial to monitor the functioning of the energy grid, to drive it towards more efficient ways to provide and use energy, and to discipline agents of the grid for more appropriate behaviour, as for example the harmonization of demand and supply. Infrastructures provide the architectural frame in which power and prescriptions flow. Moreover, in the case of the energy grid, "functional overdetermination" refers to the interactivity between effects of constructive or destructive interaction/interference that might create a need to adjust or rework the connections between elements. A perpetual process of "strategic elaboration" happens whereas the strategic objective is the reduction of energy dissipation alongside the grid favoured by different changes of agents' practices. This energy grid transition is not peaceful or irenic, but constellated by more or less critical contradictions that ask for perpetual adjustments and strategic elaboration. It is a process that never ends. This holds for example the interest of provider to supply increasing energy (and not reducing it) or the aspiration of the final user to freely use the desired amount of energy without constraints, or again the right of a final user to exercise a quasi-total control on his/her piece of apparatus.

What we discovered is that our final users would take place inside the apparatus, cooperating in it, sharing the power circulating in it. The problem is that they cannot do it because they are off-grid, separated from the apparatus or deprived of their potential or virtual agency to act on it. Moreover, when they are incorporated into the grid, they fight with the grid's devices, that resist any intervention and intrusion. Final users expect to be active grid supporters and not only passive objects of grid, aiming to drive and sway technological improvement dynamics. They also are not really persuaded that "dynamic pricing" should manage their enrolment in the grid as anticipated by the EU and enclosed in the Commission Staff Working Document SWD 442 (2013). We got insights also from the energy provider, underlining that they are very disappointed with other grid's agents behaviour. They think that energy managers and facility managers are an obstacle to innovation, that people do not understand their proposals such as to provide thermal energy during the night or to shave energy peak early in the morning when they start to warm buildings. Therefore, they show a clear distrust in potential behavioural changes, and human agents are the problem not the solution.

A more general question arises regarding the role of technical devices and artefacts in the evolution of the apparatus. As already pointed out, Foucault uses this term to designate a configuration or arrangement of elements and forces, practices and discourses, power and knowledge, that is both strategic and technical. He mentions material arrangements as part of the apparatus, but he does not pay much interest in developing this, as it would deserve. As observed by Karen Barad, he does not provide a satisfying and articulate explanation of the precise nature of the relationship between discursive practices and material phenomena. The dynamic and agential conception of materiality that takes account of the materialization of all bodies (nonhuman as well as human) and that makes possible a genealogy of the practices, is not examined by Foucault (Barad 1998, 2007). He only alludes to the ways in which technical apparatuses provide intimate, pervasive, and profound reconfiguring of the practices performed by agents, and that this reconfiguring is often unstable and unfixed. The definition of apparatus provided by Deleuze sounds more fitting our idea of energy grid, underlining the disconnected and rather precarious character of such ensemble of heterogeneous elements.

> But what is a dispositif? In the first instance it is a tangle, a multilinear ensemble. It is composed of lines, each having a different nature. And the lines in the apparatus do not outline or surround systems which are each homogeneous in their own right, object, subject, language, and so on, but follow directions, trace balances which are always off balance, now drawing together and then distancing themselves from one another. Each line is broken and subject to changes in direction, bifurcating and forked, and subject to drifting. Visible objects, affirmations which can be formulated, forces exercised and subjects in position are like vectors and tensors. Thus the three major aspects which Foucault successively distinguishes, Knowledge, Power and Subjectivity are by no means contours given once and for all, but series of variables which supplant one another (Deleuze 1992, 159).

Our issue of energy grid might be a clear example in which a satisfactory transformation of practices should be understood only in the light of new assemblages of technology and human activity. The notion of apparatus, readjusted and moved towards a consistent materiality where the inseparability of objects and subjects is acknowledged, can give energy grid a different interpretation, allowing the pinpoint of a surface where to attach a strategy of transition. In short, a conventional energy grid is an apparatus in which humans act as depending from devices driven by incorporated knowledge and language. A smart energy grid is an apparatus in which devices and humans try to communicate to adapt to new conditions.

Asymmetries of Energy and Power

In their working, thermal grids bring and convey both energy power for heating and social power in forms of rules, norms, and dispositions. They apply on subjects, but in doing it they also change the current state of affairs. Our investigation rises up the problematic of the flows and links between energy and power, the problem of how energy is appropriated, transformed, converted, distributed, used and disposed and the way in which these processes change the actual configurations. The agents of those processes that all contribute to the building and functioning of the grid, and how their nexuses and relationships work out, become a matter of investigation. How is the "power of power" maintained, conditioned and disputed by coalitions of agents, dominant and resistant, performing different but interlinked social practices from which these emerge? (Mitchell 2011). This asks for analysis of how power flows through complex systems, how it supports and makes existing positive and negative feedback loops between production and consumption of energy, how technical devices, knowledge, enunciations, build up energy machines, regimes, apparatuses, that make society likely. Social forms, as living systems, depend upon flows of energy maintaining their systemic viability far from thermodynamic equilibrium (Smil 2010). Since only the simplest forms of energy may be harnessed without infrastructures, energy resources are always mediated through socio-technical systems (Smil 2010, p. 12; quoted in Tyfield 2014, p. 61). Keeping different forms, energy is central to a social system's metabolic reproduction (Padovan et al. 2015; Padovan 2015a, b).

In their effectiveness, energy networks are analogous to social networks, been made of the same substance: a variable and disparate assemblage of natural, technical, and social elements, a continuous process fostering differences and repetitions. Based on our outcomes, we can consider thermal grids as a kind of complex system or network of agents in which energy/power circulates. This power and the way in which it works have very great similarities with social power. As stated by Bertrand Russel, "The fundamental concept in social science is Power in the sense in which Energy is the fundamental concept in physics. Like energy, power has many forms, such as wealth, influence, communication. No one of these can be regarded as subordinate to any other, and there is no form from which the others are derivative" (Russell 2004, Or. ed. 1938 p. 4). As in the social networks in which power flows reproduce asymmetries and differences (but also negating them), in these technical energy networks energy flows reproduce asymmetries and dissimilarities. The analogy can go further whereas we pinpoint dynamics of energy/power circulation, disciplining, and control: how is the grid governed? Who benefits in terms of energy provision, consumption and comfort? Is the smart energy grid a dispositive that assures a win-win mechanism? Our investigation tries to give some answer to these questions, not looking at thermal grids as a vertical apparatus going from the centre to the periphery, but understanding energy/power circulation by looking at its extremities, at its outer limits where it becomes capillary (for this perspective see Foucault 2003). For instance, we discovered continuous attempts made by final users to understand how much they are consuming, how to save energy, how to regulate temperature, how to intervene on devices, how to make the apparatus more flexible, how to manage a common thermal comfort in public spaces. Our goal, similar to the Foucault one, has been to analyse energy/power regulation at the point where it is invested in real and effective practices, where it relates directly and immediately to what we might call its object, its target, its field of application, or, in other words, the places where it produces its real effects. So the question is this: what happens in the continuous and uninterrupted processes that go throughout the grid making energy circulate and investing bodies, directing gestures, and regulating forms of behaviour? In other words, rather than asking ourselves who simply rules or governs the grid, we should try to discover how multiple bodies, forces, energies, matters, desires, thoughts, are gradually, progressively, actually and materially constituted as subjects in the making of the thermal grid, or to grasp the material agency that uses energy. For instance, we realized that conventional grid leaves agents in a state of blindness regarding the heating system functioning. On the other hand, the deployment of smart grids implies a process of subjectivation whereas agents are invested by a twofold dynamic of freedom and individual responsibility. While water flows through pipelines, the grid conveys also data, prescriptions, rules, advises for users, disciplining and regulating their practices. The study of the multiple peripheral bodies, the bodies that are constituted as subjects by power effects in the frame of thermal grid, enlightens the way in which it acts and is enacted by different users. In our investigation, we noticed also that agents can bend, in some cases, the grid towards their own goals, or can refuse at all the regulating power conveyed by it. Forms of adaptation, rejection, manipulation, constellate the grid along its entire length, becoming often sources of controversies and conflicts mainly in buildings where different tenants experience different intensities and performances of the grid, or in different areas where grid shows some malfunction. Energy/power handling is not a homogeneous phenomenon; it is marked by different levels of intensity and power. Energy/power is something that circulates, or rather is something that functions only when it is part of a chain. It is never localized here or there, it is never in the hands of someone, and it is never appropriated in the way that a commodity can be appropriated. Power functions. Power is exercised through networks, and individuals do not simply and passively receive the energy circulating in those networks; they are in a position to both submit to and exercise this power. They are never the inert or consenting targets of power; they are always its relays. In other words, power passes not only through the pipelines but also through the users. It is not passively applied to them, but it asks for a process of subjectivation.

Thermal (but also electric) grids are complex systems of connection of different agents equipped with different agency and different power of influence and intervention on consumption and environmental impact. It is in some way self-evident the fact that big energy providers and final users are very a-symmetrical in the influence on energy management. At the theoretical level, we can see two big categories of agents: natural and corporate. Natural persons are obviously those people that do not have any legal definition, the final users of the energy. They can only make contracts and agreements with providers and managers. The other category is the one of corporate actors that range from extractors, refiners of oil and gas, sellers and intermediaries placed at the core of the big energy market, national and local providers, energy managers of other corporate actors such as cities and firms. All of them are corporate actors in the sense that they act as fictional persons, law rules their actions, and finally they get an internal structure composed of positions rather than persons, a structure in which persons are merely occupants of positions (Coleman 1974, 1982).

We can define corporate actors as those organized actors, which participate directly in (policy-oriented) decision-making, which are formal organizations, have a real constitution and a real membership, purport to represent the interests of their membership, but often have been challenged for misrepresenting these interests by both internal and external critics. Their decisions result in the establishment, maintenance and transformation of rule regimes (Flam 1990). The main consequence of the corporate actors' agency is an asymmetry of the relations in the energy grid. The relationship between the two actors (natural personas and corporate actors) is asymmetric in the types of parties they involve, but are asymmetric —often extremely so—in two other respects as well: in the relative sizes of the two parties, and in the numbers of alternative transaction partners on each side of the relation. One main consequence of the asymmetry is that a corporate actor nearly always controls most of the conditions surrounding the relation. The result is that two parties beginning with nominally equal rights in a relation, but coming to it with vastly different resources, end with very different actual rights in the relation. This asymmetry of rights taking place in the evolution of a system of relations is one of the main reasons that make people distrust regarding generally processes of socio-technical innovation at the point to refuse them. Distrust, disappointment, discontent are conditions that shape people when they realize the asymmetries of power in which they are involved. This is the case of such top/down strategies of smart grid deployment, whereas actions of involvement are not contemplated or are merely claimed.

Conclusions: Transitional Apparatuses as New Frame for Policymakers

Because of path-dependency mechanisms deployed by the development of fossil fuel conventional energy grids, the transition towards smart energy grids must start from them. A counterapparatus, far more suitable and acceptable for current purposes of energy transition of those existing nowadays, can be built only on already existing infrastructures. There is not alternative to begin from this constraining stage of thermal grid evolution. Consequently, we need to know how conventional grids work and where their potential for change is. As technological zones they are rather rigid, linear, inelastic and thus useful only to a certain extent. In the case of district heating the situation is even worse in the sense that the rigidity and path dependency of co-generation apparatuses is very strong: likely it will be very tough to emancipate this energy provision from its fossil fuel primary source. Moreover, the socio-technical vision of grids transition is considerably naïf: the list of stuff that "should be done" is not enough to ensure a successful transition.

The notion of apparatus or dispositive seems to us more useful to adopt strategies of transition. This notion is similar to concepts such as assemblage (DeLanda 2006) and arrangement (Schatzki 2011, 2015), which outline a relational system for dissimilar elements and practices. Apparatuses, assemblages, and arrangements are concepts that often overlap, and that at the empirical level can operate symbiotically to explain the forging and emerging of practices such as energy production, distribution, usage, and dissipation. All in all, that of apparatus seems to us a more intense, dynamic, and agential concept than the other ones. The co-evolution of varying lines and strata of practices, techniques, discourses, and singularities establishes it. An apparatus is more concerned with its security and functional certainty than an always virtual and a never fully actualized assemblage. Moreover, it is purpose-oriented in the sense that an apparatus organizes people, artefacts, enunciations, and things according to functions, statuses and relations of agents involved in it (see Schatzki 2015 when it describes Deleuze "regimes of power"). Finally, it denotes large systems of real life, such as energy systems with a time-space relevant dimension, which are incessantly changing. Regarding energy grids, it is undoubtable that they are greatly concerned with their security and continuity in time and space, being them an indispensable support for the societal reproduction. They aim towards clear purposes, are spatially deployed and, finally, they are under an incessant process of change depending on the practices performed within them. This tension for change is what distinguishes an apparatus from other kind of socio-technical configurations such as arrangements or assemblages.

Apparatuses are made of lines, which show continuous variations. The features of nonlinear change, emergent properties, spontaneous self-organization, fractal becoming, and so on are perceived to represent not the abnormal conditions of existence of physical, chemical, biological and even socio-historical processes but rather their 'normal' conditions of existence (Ansell-Pearson 1997). The fact that human agents always belong to apparatuses and act within them, interacting with their lines of functioning, means that apparatuses exercise a certain power on them but also that these agents can change them by performing their own practices or fighting against them, as said by Agamben (2009). In other words, apparatuses are agents of change aimed to secure in this way their own continuity and the immortality of the society where they act (Garfinkel 1988). Each apparatus shows lines of breakage and fracture. Sometimes these are situated at the level of powers; at other times at the level of knowledges; other times more at the level of structures of practical action. More generally, it should be said that the lines of subjectivation indicate fissures and fractures. Change depends on the content of the apparatus, and each apparatus deserves its own diagnostic, its own archaeology. Moreover, an apparatus creates a propensity for certain types of events, a trend that some things "happen". The application of this concept to an energy grid opens up the possibility of its change towards the smartness. Can an apparatus become smart, or flat or democratic or equal or differentiated in its functions and provisions? Might an apparatus such as an energy thermal grid be designed and managed in order to generate insensible but enduring changes in the agents' performance? Or to be flexible enough to change in virtue of agents' performance?

An apparatus can change whereas it gives visibility to variable creativity arising out of itself. What counts is the newness of the regime in which a new perspective arises (Deleuze 1992). The newness of an apparatus in relation to what is going before is what one could call "actuality". The new is the current that we are fostering. Each apparatus is thus defined in terms of its newness content and its creativity content, this marking at the same time its ability to transform itself, or indeed to break down in favour of a future apparatus, unless it concentrates its strength along its hard, more rigid, or more solid lines. Apparatuses are composed of lines of visibility and enunciation, lines of force, lines of subjectivation, lines of splitting, breakage, fracture, all of which criss-cross and mingle together, some lines reproducing or giving rise to others, by means of variations or even changes in the way they are grouped. Important consequences arise whereas these lines are more or less rigid and unyielding when they try to orchestrate new configurations and bundles of practices along the grid. An apparatus has in some way the capacity to capture, orient, determine, intercept, model, control, or secure the gestures, behaviours, opinions, or discourses of living agents. An energy grid is an apparatus aimed to capture bodies making them subjects, in the sense of subjectification. As said by Agamben (2009), a subject is that which results from the relation and, so to speak, from the relentless fight between living beings and apparatuses.

Our scrutiny of conventional thermal grids suggests recommendations to policy makers and technical designers to be taken in mind when they decide to build up smart grids.

- Conventional thermal grids offer—because they already are a web of reciprocal actions and of functional interdependencies—the ground on which smartness can be built upon. It means that constraints and opportunities are already established, and they have to be in some way forced up to reach desirable goals of smartness.
- The transition process towards the smartness is often, if not always, seen as a simple addition of different technical operations. From our point of view, these operations are too naïf, socially inappropriate, and driven by a mechanic and linear causality. We suggest thinking in terms of apparatus, or in terms of circularity and co-evolution.
- Different and contrasting aspects mark the deployment of feedback devices. On one side smart energy devices can gradually become 'backgrounded' within routines and practices, increasing the agent's information about the amount of energy they consume. On the other side, beyond a certain level these devices do not encourage householders or customers to reduce their levels of consumption.

They realize the limits to their energy saving potential and become frustrated by the absence of wider policy and market support.

- Smart grids are aimed to shave peaks. This goal can be reached by orchestrating new patterns of differences and repetitions in the heating practices performed by actors.
- A thermal grid bears at the same time energy, power, heat, information, rules, codes, data, and so on. The fact that human practices and technical systems do not act in synchrony is often a matter of communication, interpretation and reciprocal interpenetration.
- The intensity deployed in a thermal grid is fated to disappear in favour of extensity. However, it does not mean that difference inside the grid cancels out. The extension process produces new orders of differences and individualization that inhabit the depth of the reality. A smart thermal grid with all its devices pulls for new singularities and attracts new differences and asymmetries.
- The deployment of change depends on the struggle between agents and apparatuses. It means that in the fostering of smart thermal grids a process of singularization occurs, which implies a harmonization of interests and practices among the different agents of the grid.
- The process of smartness lies into technological areas where metrological, infrastructural and qualification aspects have to be met synchronically, in order to give start to a new becoming involving all socio-technical aspects and the way they communicate and foster feedback.
- The system and its history and dynamics—which is the combination of the intensity that drives the system and the extensity that it exhibits—produces at the end novel patterns depending on its own entropy. In this vision of non-equilibrium thermodynamics, socio-technical systems evolve in an unpredictable way.
- In this perspective, we can say that physical, technological and social agents of the grids are caught into a process of individualization and singularization that opens up new energy arrangements, new intensity/extensity configurations and new forms of socio-technical organization.
- Having described grids as apparatuses we can also say that they are aimed to endorse disciplined actors in order to secure energy provision. However, the way in which people are pushed to behave in way to make energy use more efficient is also unpredictable.

References

Agamben, G. 2009. What is an Apparatus? Translated by David Kishik and Stefan Pedatella. Stanford University Press.

Ansell-Pearson, K. 1997. Deleuze Outside/Outside Deleuze: On the Difference Engineer. In *Deleuze and Philosophy: The Difference Engineer,* ed. Ansell-Pearson K. London and New York: Routledge.

Barad, K. 2007. *Meeting the Universe Halfway: Quantum Physics and the Entanglement of Matter and Meaning*. Durham: Duke University Press.
Barad, K. 1998. Getting Real: Technoscientific Practices and the Materialization of Reality. *Differences* 10 (2):87–128.
Barry, A. 2006. Technological Zones. *European Journal of Social Theory* 9 (2): 239–253.
Brounen, D., N. Kok, and J.M. Quigley. 2012. Residential Energy Use and Conservation: Economics and Demographics. *European Economic Review* 56 (2012): 931–945.
Bussolini, J. 2010. What is a Dispositive? *Foucault Studies* (10):85–107.
Cakici, B., and M. Bylund. 2014. Changing Behaviour to Save Energy: ICT-Based Surveillance for a Low-Carbon Economy in the Seventh Framework Programme. In *Proceedings of the 2014 Conference ICT for Sustainability*, vol. 2, 165–170.
Coleman, J. 1974. *Power and Structure of Society*. New York: Norton.
Coleman, J. 1982. *The Asymmetric Society*. Syracuse University Press.
Crockett, C. 2013. *Deleuze Beyond Badiou. Ontology, Multiplicity and Event*. New York: Columbia University Press.
DeLanda, M. 2006. *A New Philosophy of Society*. London: Continuum.
Deleuze, G. 1992. What is a Dispositif? In *Michel Foucault Philosopher*, ed. J. Timothy, 159–168. Armstrong. New York: Routledge.
Deleuze G. 1994. *Difference and Repetition*. New York: Columbia University Press (or. ed. 1968).
Ehrhardt-Martinez, K., K. Donnelly, and J.A. "Skip" Laitner. 2010. *Advanced Metering Initiatives and Residential Feedback Programs: A Meta-Review for Household Electricity-Saving Opportunities*. Washington, DC: ACEEE.
European Commission. 2011. "Definition, expected services, functionalities and benefits of smart grids", Accompanying documents to communication from the commission to the European Parliament, the Council, the European Economic and Social Committee and the Committee of the Regions, "smart grids: from innovation to deployment", com (2011) 202 final Brussels, 12.4.2011 sec (2011) 463 final.
European Environment Agency. 2013. *Achieving energy efficiency through behaviour change: What does it take?* EEA Technical report No 5/2013.
Flam, H. 1990. Corporate Actors: Definition, Genesis, and Interaction, Max-Planck-Institut für Gesellschaftsforschung, MPIFG Discussion Paper 90/11.
Foucault M. 1980. *Power/Knowledge: Selected Interviews and Other Writings, 1972–1977*, 194–196. New York: Pantheon Books.
Foucault M. 2003. *Society Must Be Defended*, Lectures at the Collège de France, 1975–76, Picador, New York.
Garfinkel, H. 1998. Evidence for Locally Produced, Naturally Accountable Phenomena of Order, Logic, Reason, Meaning, Method, etc. in and as of the Essential Quiddity of Immortal Ordinary Society (I of IV): An Announcement of Studies, *Sociological Theory* 6 (1):103–109.
Gram-Hanssen, K. 2009. Standby consumption in households analysed with a practice theory approach. *Journal of Industrial Ecology* 14: 150–165.
Hargreaves, T., M. Nye, and J. Burgess. 2010. Making Energy Visible: A Qualitative Field Study of How Householders Interact with Feedback from smart Energy Monitors. *Energy Policy* 38 (10): 6111–6119.
Lund, H., et al. 2012. From Electricity Smart Grids to Smart Energy Systems. A Market Operation Based Approach and Understanding. *Energy* 42: 96–102.
Lund, H., S. Werner, R. Wiltshire, S. Svendsen, J.E. Thorsen, F. Hvelplund, and B.V. Mathiesen. 2014. 4th Generation District Heating (4GDH). Integrating Smart Thermal Grids into Future Sustainable Energy Systems. *Energy* 68: 1–11.
Mengolini A., and J. Vasiljeska. 2013. The Social Dimension of Smart Grids. Consumer, Community, Society. European Commission, Joint Research Centre, Institute for Energy and Transport, Report EUR 26161 EN.
Mitchell, T. 2011. *Carbon Democracy*. London: Verso.
Nyborg S., and I. Røpke. 2011. Energy impacts of the smart home—Conflicting visions, ECEEE 2011 SUMMER STUDY. Energy efficiency first: The foundation of a low-carbon society.

Padovan, D. 2015a. Assembling Societal Metabolism and Social Practices: The Dynamics of Sustainable and Unsustainable Reproduction. In *The Consumer in Society. A Tribute to Eivind Sto*, ed. Pal Strandbakken, and Jukka Gronow, 335–362. Oslo: Abstrakt Forlag AS.

Padovan, D. 2015b. Metabolic Exchanges and Practices of Regulation. The Assemblage of Environment and Society in Early Social Sciences. *Ecological Informatics*, 6–17.

Padovan, D., F. Martini, and A.K. Cerutti. 2015. Social Practices of Ordinary Consumption: An Introduction to Household Metabolism. *Journal of Socialomics* 4 (2): 1–11.

Pullinger, M., H. Lovell, and J. Webb. 2014. Influencing Household Energy Practices: A Critical Review of UK Smart Metering Standards and Commercial Feedback Devices. *Technology Analysis & Strategic Management* 26 (10): 1144–1162.

Raffnsøe, Sverre. 2008. Qu'est-ce qu'un dispositif? L'analytique sociale de Michel Foucault, Symposium (Canadian Journal of Continental Philosophy/Revue canadienne de philosophie continentale): vol. 12, pp. 44–66.

Russell, B. 2004. *Power. A New Social Analysis*. London & New York: Routledge (or. ed. 1938).

Schatzki, T. 2011. Where the Action Is (On Large Social Phenomena Such as Sociotechnical Regimes), Sustainable Practices Research Group, Working Paper 1, University of Lancaster.

Schatzki, T. 2015. Spaces of Practices and of Large Social Phenomena. EspacesTemps.net, Works, 24.03.2015.

Schmidt, R.-R., N. Fevrier, and P. Dumas. 2013. *Key to Innovation Integrated Solution. Smart Thermal Grids*. European Commission, Smart Cities and Communities.

Schwartz, T., G. Stevens, L. Ramirez, and V. Wulf. 2013. Uncovering Practices of Making Energy Consumption Accountable: A Phenomenological Inquiry. *ACM Transactions on Computer-Human Interaction* 20 (2), Article 12 (May 2013).

Shove, E. 2010. Beyond the ABC: Climate Change Policy and Theories of Social Change. *Environment and Planning* 42: 1273–1285.

Shove, E., and G. Walker. 2014. What Is Energy For? Social Practice and Energy Demand. *Theory, Culture & Society* 31 (5):41–58 (September 2014).

Smil, V. 2010. *Energy Transitions*. Santa Barbara, CA: Praeger.

Strengers, Y. 2012. Peak Electricity Demand and Social Practice Theories: Reframing the Role of Change Agents in the Energy Sector. *Energy Policy* 44: 226–234.

Tyfield, D. 2014. 'King Coal is Dead! Long Live the King!': The Paradoxes of Coal's Resurgence in the Emergence of Global Low-Carbon Societies. *Theory, Culture & Society* 31 (5):59–81).

Weber, M. 2005. *The Protestant Ethic and the Spirit of Capitalism*. London & New York: Routledge.

Chapter 13
Grid Dependencies and Change Capacities: People and Demand Response Under Renewables

Mithra Moezzi

Abstract Much of everyday activity in highly technologically developed societies involves electricity from a centralized grid. This is most evident during blackouts—at which point the availability of many routine forms of information, communication, light, money, and other connectors are quickly depleted. The expectation of perfect electricity has accompanied an evolution of social practices that absolutely require a working electricity system, while practices that escape that system become abandoned or antiquated. By definition, during supply shortages, societies adapt. In less-developed countries, especially those experienced with unreliable power, and with less-dense ties to the grid, there is established capacity to cope, including substituting non-electricity for electricity, and adjusting the timing of activities. In areas that expect perfect electricity, and rarely experience failures, however, reliance on electricity is higher and coping is more fragile. Drawing on social practice theories and history of technology, this chapter explores examples in the evolution of the grid dependence and develops a concept of sociotechnical resilience. Sociotechnical resilience refers to the degree to which basic activities can be decoupled from the grid, and how they do so. This resilience obviously matters in the case of blackouts and severe supply restrictions, but it also speaks to flexibility within "portfolios" of practices in terms of their synchronization with electricity supply. Demand flexibility is expected to become increasingly important in future scenarios where electricity supply has evolved to include much higher penetrations of renewables. To date, most of the debate on how this flexibility will occur has focused on "demand response," particularly through individual end-user behaviors, and well as through isolated and largely private backup systems to provide temporary power. Focusing instead on sociotechnical resilience broadens the scope of flexibility by looking at people, technologies, and adaptation in a more connected and intricate combination. In addition to the power markets and generation capacity markets that already exist, there is thus a need to recognize, maintain, and further develop the sociotechnical capacity to do without electricity. This possibility is rarely included within the usual boundaries of debates about the renewables and the

M. Moezzi (✉)
Portland State University, Portland, OR 97207, USA
e-mail: mithra@pdx.edu

N. Labanca (ed.), *Complex Systems and Social Practices in Energy Transitions*,
Green Energy and Technology, DOI 10.1007/978-3-319-33753-1_13

grid, or balancing supply and demand. To illustrate, the chapter provides examples from supply disruptions in both more-developed and less-developed countries, explores how policies, language, technology design, and the public sphere might better recognize and build this sociotechnical capacity.

Introduction

Planning for a lower carbon future and hedging against risks of fossil fuel dependence, countries and jurisdictions increasingly envision a future where electricity generation is dominated by renewable resources. A departure from fossil-fuel-centered generation means that the easily dispatched, predictable, storable, and energy-dense nature of fossil fuels that industrialized portions of the world have grown to depend on would be replaced by power sources of a different nature. Renewables-based power is more dependent on uncontrollable environmental conditions, and more variable across time in both short and long terms, than fossil fuels (Labanca et al. 2015; Powells et al. 2014; Stephens et al. 2013). Wind changes from moment to moment; droughts reduce the availability of hydropower and biomass; sunlight is not always available; floods, earthquakes, and high winds can destroy infrastructure for renewables and fossil fuels alike (Michaelowa et al. 2010).

Nuclear power, continued use of fossil fuels, robust portfolios of renewable energy sources, improvements in storage, grid management, and various other supply system technical strategies and innovations can be applied to manage some of the challenges of renewables variability and uncertainty (Lund et al. 2015). Most discussions of a future renewables-based grid, however, include demand response as an important part of that future, usually with an argument that the need for demand response will be greater than in largely fossil-fuel-based electricity systems (e.g., Aghaei and Alizadeh 2013; Denholm 2015; Roscoe and Alt 2010).

References to the "smart grid" then often step into the policy narrative as a solution that would help make this new order workable by efficiently providing price signals, enabling direct load control, and by improved supply and distribution management (Tricoire 2015; Skjølsvold et al. 2015). With limited exceptions (e.g., Higginson 2014; Strengers 2011), the public discussion of demand response in a future renewables-centric grid has been fairly narrow and schematic relative to the scope of social change and responsiveness that seems envisioned. In particular, there has been little serious discussion of what might be required from energy users to provide this higher level of demand response, the limitations and consequences of this response, nor of the larger changes in technologies and practices that might increase demand response capacity or lower the need for short-term demand response through longer term shifts in demand. Rather, price seems to take the default role of an efficient, unbounded, and fairly unproblematic device to unlock potential.

This chapter is not intended to dismiss the tools and insights of conventional demand response. Rather it begins with the recognition that research in load management and demand response have had little direct truck with people, what they do or could do and why, but instead generally includes them implicitly as end users who participate, or not, in generating demand or demand reduction. Demand response is framed as an almost entirely economic mechanism, to be supported by relatively minor adjustments to end-use technologies (e.g., price-responsive smart controls). From this perspective, demand response can be scaled by gaining more participants into the fold of the demand response mechanism and by increasing the financial incentives for reducing demand. There is, however, little reason to assume that current levels of demand response are simply scalable at any level, and little discussion of what price mechanisms might miss, such as equity (Darby 2012), societally reasonable allocation of electricity under conditions of scarcity, or whole new ways of doing things for which electricity is not crucial. In envisioning the future of demand response, a richer dialog about the capacity for, and costs of, energy users voluntarily "changing their load shapes"—that is, what they do and how and when they do it—is needed. These less-spoken alternatives are relevant not only for developing short-term capacity for reducing electricity use when needed (narrowly, demand response) but also for considering the possibilities of reshaping demand over the long term.

Compared to the demand response questions of past decades, as noted above, high penetrations of renewables will likely need a higher degree of flexibility in demand, including both long-term shifts in demand and shorter term "responsiveness." The aim of this chapter is to help expand the current "demand response thinking" that now fills the discursive slot on how the variability and uncertainty of renewables will be integrated into future electricity systems. In particular, the chapter aims to open a fuller appreciation of what activities provide, or could provide, demand response (or some of the benefits of demand response, but from a conceptually different viewpoint), and to begin to ask questions about the consequences and requirements for this expanded capacity in terms of technology design, policies, social structures, and expectations, as well as about the roles of people in producing and enduring this capacity. Paschen and Ison (2014: 1083) note that visions of change, as encoded in "adaptation" and "innovation," often end up "maintaining existing production systems—including … systems of knowledge production, therefore traditional framings and knowledge paradigms that restrict the meanings and implementations of adaptation." Their comment provides an entrée for identifying these paradigms and some of their limitations.

Approach

This section provides an abbreviated preview of the argument to be elaborated in the remainder of the chapter. Table 13.1 compares a more conventional and narrow notion of demand response and demand management to "sociotechnical capacity" viewpoint that sets its sights on broader strategies for shifting activity across time and lowering dependence on electricity. For example, in conventional demand response, people are seen to react to prices or to social calls for curtailment by offering precise actions, generally at the device level, e.g., to reduce air conditioning, to charge batteries off-peak, etc. The broader view moves beyond devices to attend to why activities happen at particular times and in particular ways and what is needed to create affordances for change. Opportunities may be afforded by provision systems that can function without electricity (the Capacity row in the table) or by non-grid electricity (Realm of Analysis row) and that can better manage costs (Load Management Technologies and Priority During Scarcity rows). Elements in the leftmost column overlap and interact with one another.

Lutzenhiser (2014) argues that the vocabularies and politics of conventional energy efficiency debates can quietly limit what can be seen, what counts, and how it can be debated. Table 13.2 outlines some of these conventions as they appear in demand response and smart grid literature, and comments on their potential consequences in restricting the view of what is at play in a renewables-centric future.

The experiences of a smart grid and a largely renewables-based supply are in the future. To explore the expansions hinted at in Tables 13.1 and 13.2, this chapter turns attention to blackouts, load shedding, and power shortages, the exact events that demand response is usually intended to prevent. While drawing from examples in developing countries, the geographical focus for interpretation is the United States, the sector focus is generally residential, and the intent is exploratory: to help open existing technology-centered vocabulary and concepts of the energy efficiency rubric beyond their normal limits, connecting them to a larger framework that includes practices, practitioners, and the possibility of a more historically based evolutionary view.

Dependence and Disruptions

Practice theorists describe practices as "recruiting" practitioners (Shove et al. 2012). Electrical grids, as infrastructures, have done a remarkable job of recruiting interconnected bundles of practices, practitioners, and their devices to rely on these grids. Mundanely, electricity is often called "the lifeblood of modern society" (e.g., NAS 2016). A huge proportion of daily social activities circulate based on the availability of a somewhat fragile centrally planned electricity network. This network is so crucial to everyday life in some areas that power shortages cause panic and the suspension of many normal activities, resulting in high economic and other costs. The point is not just that centrally planned electricity systems are "brittle" (Lovins and Lovins 2001; Smith 2003) but that extreme dependence on this system also creates vulnerabilities.

In evolutionary terms, some of this dependence stems from interests in creating demand and in shaping it to economically suit supply, a dynamic present from the beginning of the grid electricity industry (Deumling 2004; Hausman and Neufeld

Table 13.1 Conceptual shifts to move from narrower to broader notions of demand management

Element	Narrower mode: demand response	Broader mode: sociotechnical capacity
People	Individuals prepared to reduce their energy use in response to price signals (as economic actors) or for non-economic reasons	Practices: configurations of practitioners (people), materials/technologies, and actions
Organizations	Organisms, almost the same as individuals	Contain individuals, but do not function as individuals
Energy services, energy service technologies	Supplies services for pre-existing needs and wants	Expand the notion of demand management beyond simple notions of choice to consider systems of provision and the technical and social creation of needs and wants
Load management technologies	Design for economic efficiency of the utility	Design to manage social costs
Capacity	Primarily offered by a combination of dispatchable generation and limited sets of discrete short-term conservation actions	Sociotechnical capacity: an enlarged notion of capacity emphasizing fostering technical and social abilities to deploy alternative practices that reduce electricity use without dramatically curtailing activities
Realm of analysis	Electricity and the grid as an all-encompassing framework	Incorporate non-electricity (e.g., natural gas, solar thermal cooling, passive design, manual management) and non-grid-tied electricity into the analytical framework; include what's outside the grid as well as what is in it
Prioritization during scarcity	Decided on the basis of ability to pay and on agreements of essential services (e.g., hospitals)	Prioritization cannot be solved by prices alone
Load shape flexibility	Debates about influencing long-term load shapes relatively hidden from view; focus is on short term	Consider the long-term construction of load shapes and how policies and technologies affect electricity rhythms of days and seasons

Table 13.2 Oddities, ironies, catch-22s of renewables in smart grid imaginaries

Capacity and conservation compete. Demand response depends on creating conservation capacity—the ability to get users to voluntarily curtail or shift load—that can be dispatched in the short term. Energy savings that persist in the long term, whether from traditional energy efficiency or by other means, do not provide this conservation capacity. Focus on short-term capacity makes it difficult to recognize the potential contributions of a variety of approaches to a renewables-centric future (e.g., solar thermal, better passive cooling, more robust design) that could provide flexibility via providing non-electricity-based energy services alternatives that shift long-term demand as well as support short-term demand response when required. Here conservation capacity is distinguished from energy storage and energy services storage
Promising complex strategies may be difficult to see. Pressures to track and prove the successes of demand management mean that complex strategies are difficult to promote or evaluate. Instead simple results (e.g., from pilot experiments based on volunteers) may be used to create invalid inferences, and accountability rules may delegitimate certain actors—such as "free riders," who conserve in the absence of an intervention—or ignore other complex responses, placing them outside the frame
Demand management paradigm is based on old models of electricity and grid management needs. Experience to date has been based on creating economically valuable load shapes for fossil fuel or nuclear-centered portfolios
Smartness creates electricity dependence and increases load. The "smart grid" means that more things depend on electricity, with potential domino effects. Smartness builds load and capacity requirements, and increases societal sensitivity to power loss. Sensitivity can manifest in tiny hassles (e.g., inability to open the clothes washer door if a wash cycle is aborted in an outage) as well as larger consequences
"Perfect is the energy of the good". Insistence on having or expecting a perfect grid can exacerbate social and technical vulnerabilities to power outage, since the main "solution" imagined is that perfect grid, rather than ameliorating problems when there is not enough power
Boosterism. Many stakeholders in the "smart grid" are incented to emphasize attractive scenarios, and to render reservations as unimportant or simply solvable. This boosterism makes it difficult to actually cope with problems that might occur
Limitations of technologically centered policy framings. The tendency to express future imaginaries in simple terms renders some elements magical and difficult to question (e.g., in the case of the smart grid, the power of prices, the idea that the problem is to "engage" customers). Jasanoff and Kim (2013) note, "a well-known feature of the American sociotechnical imagination is that technology's benefits are seen as unbounded while risks are framed as limited and manageable." This creates metaphoric lock-in (e.g., what prices exactly, and what about the things that prices can't or won't do?) that restrict the terms of debate. The "people problem" of demand response is usually framed as one of getting people to sign on to rates and to understand, respect, and adequately respond to price signals, rather than taking the world beyond these signals seriously. Adding questions such as "Why?","Why not?", and "What alternatives are available?" can open up a broader arena within which to reason and plan
Variability, heterogeneity, knock-on effects, and other realities are left out. Imaginaries may often rest on convenient stereotypes rather than attending to the diversity and heterogeneity of energy users and usage in the real world

1984; Nye 1990). Contemporary goals of electrification (substituting electricity for other energy sources), such as electric vehicles and the conversion of natural gas end uses to electricity, as well the “smart grid” in general (Powells et al. 2014; Stephens et al. 2013) are also aligned with these interests. Couched in the context of climate change mitigation, environmental improvement, system efficiency, reliability, modernity, etc., the self-interested nature of these arguments is sometimes forgotten. In sanctioning what appears to be a more controllable form of energy supply, these electricity-centric trajectories also create dependencies that can limit societal flexibility and resilience, as well as increase it.

Electricity grids have been described as mammoth infrastructures that tie technology together with social practices, generally through the mediation of appliances (Shove 2015; Skjølsvold et al. 2015) and devices. People, technologies, and activities are entangled, mutually shape and limit each other, and constantly co-evolve (Rupp 2016). Visions and actualizations of a smart grid and a corresponding “internet of things” create a denser network of dependent connections, which at the same time become less tangible and visible. This smart grid involves a “physical, technological and social reconfiguration of the entire energy supply system” (Ballo 2015) as well as adding different types of load (Labanca et al. 2015; Stephens et al. 2013).

Blackouts in the US

Power blackouts have been examined by anthropologists (Rupp 2016), sociologists (Matthewman and Byrd 2014), historians (Nye 2013; Trentmann 2009), and others (Deverell 2009) as moments of rupture of the normal fabric of society. While dependence on grid-tied electricity is obvious, blackouts can make the details of this dependence much clearer both in terms of technical ability (e.g., will this phone work in a power outage?), and interlinked systems providing communications, goods, environmental control, etc. They can also reveal insights about the nature of social connections including both those operating when power is available as well as latent types of social interactions that are activated or become more evident in these liminal periods (Rupp 2016); some of these insights are discussed later in this section.

In the US, blackouts are not common, and usually do not last long. One report estimates an average of 214 min of blackout every 9 months (Amin 2011; Apt et al. 2006), which amounts to a rate of less than 0.1% of time. That is a higher rate than in other countries such as the UK, France, and Japan (Apt et al. 2006), but still quite rare. Usually outages are short enough that they are annoyances or inconveniences that can be coped with, with relatively few people or entities sustaining major losses. In 2003, however, the Northeast blackout affected over 50 million people in Canada and the US. This remains the largest North American blackout to date. Some customers only lost power for several hours, but others were without power for more than two days, and some faced rotating blackouts for 2 weeks (NERC 2004).

In the US, large-scale blackouts are interpreted first as extremely costly economic events.[1] For example, a US federal report estimated that costs from weather-related electricity outages alone ranged from $25 to $70 billion annually, with costs expected to get worse due to increasing severity of weather-related events (White House 2013). In the aftermath of the Northeast Blackout, newspaper and television coverage often referred to the US electricity grid as being "third-world." Though technically this was not a reasonable comparison, it signaled a substantial national shame reflex. This became paired with a renewed emphasis on avoiding blackouts altogether and creating a "perfect grid" providing flawless power, rather than making use of lower quality power when necessary and managing blackouts when they do happen (Marnay and Bailey 2004). The flip side of the expectation of a perfect grid is high dependence on this perfection and potentially high societal vulnerability in the case of power failures.

There are, of course, safeguards in place. There are federal emergency and security analyses and procedures; property, building, and enterprise managers often have plans and sometimes conduct drills for dealing with local power emergencies; and isolated groups of "preppers" take steps to protect their families and communities and talk about these steps with others (e.g., Luther 2014). Overall there is surprisingly little public discussion of how the various costs of blackouts might be reduced, despite the relevance to climate change and societal adaptation and resilience. In the meantime, elements of the technological fabric of everyday life change rapidly, and the dependencies that come along with these changes can be difficult to see without empirical evidence.

Reports by the International Energy Agency and others outline how communities have responded to reduce electricity usage in mid-term and longer term power emergencies, such as during the California "electricity crisis" of 2002 and the Fukushima Daiichi nuclear plant shutdown in 2011 (IEA 2005; Kimura and Nishio 2013; Pasquier 2011). The 2005 report estimated energy savings of between 0.5 and 20% in response to electricity shortfalls, depending on the country. Other authors (Nye 2013; Rupp 2016; Trentmann 2009) have analyzed social responses to blackouts, as discussed further below. What appears to be missing, however, is a detailed account of the technological, infrastructural, and social responses to blackouts insofar as the functionality of basic systems—transportation, water available and quality, sewage management, basic heat and light, communications—and the effects of dysfunctionalities are concerned.

Failure stories that report on these problems can expose individuals, organizations, or technologies to blame and legal repercussions, which likely inhibit the circulation of these stories. In addition, during power outages, the ability of individuals to control the environment on their own is radically changed, since instant

[1]Another common stream of the effects of outages covers accidents or conditions affecting health or mortality (e.g., Anderson and Bell 2012).

power, movement, and information are suddenly much less available. Accordingly, there is more need to cooperate to get things done, though this cooperation requires a different set of skills and rules than those in electrified environment.

Analyzing blackouts in New York City between 1965 and 2012, Rupp (2016) describes an increasing "atomization" of society, wherein each individual circulates largely on his or her own, or at least has the ability to do so. Communication devices connect people as nodes of a social system in a way that is quite different with respect to space, time, and the qualities of sociality compared to those requiring face-to-face interactions. Worldviews and the organization of society echo (and are echoed by) the electricity system, though it would be incorrect to assume a simple formula by which the ubiquity of individual communication devices creates higher levels of individualism from a psychological perspective. In any case, this organization has obvious implications for the electricity system, among them the fact that individualization can increase energy demands because certain energy services become less shared.

According to Rupp, in the case of New York, atomization is fostered by the availability of electricity-mediated technologies that override traditional geographic space as well as the limitations of natural cycles of light and dark. Technically, one can communicate with nearly anyone at any time. Blackouts, Rupp argues, halt these abilities, "resulting in a contraction of space and distances traveled, a slowing of ever-accelerating time, and a thickening of social relations from ethereal, disembodied, mediated communications to communication that is immediate." Thus blackouts create a space that can reignite older forms of sociality, and less atomized relationships, by temporarily evacuating the "normal" electronically based communications infrastructure. Trentmann (2009) also emphasizes the importance of sociality and sharing that arise during power outages. Rupp suggests that sociality is a dynamic matrix where humans and nonhumans continually interact and affect each other. Forms of sociality are certainly emergent based on the forms of the available networks (be they electronically mediated or in physical space), but these forms are presumably path-dependent. That is, interactions and expectations will depend on past experiences and skills.

Load Shedding and Blackouts Where They Are Common

The discussion of power shortages above concerns situations where blackouts are unexpected and uncommon. But what about where they are common and expected? Up until very recently, energy supply has been primarily based on renewable energy sources. Currently many less-industrialized nations currently use high percentages of renewables. Considering electricity itself, 80% of electricity produced in Sub-Saharan Africa is largely fossil fuel based, but in some countries most electricity is hydropower, e.g., 97% in Cameroon (Williamson et al. 2009). According to the World Energy Outlook (2015), in 2013, an estimated 1.2 billion people did not have access to electricity and 2.7 billion relied on traditional biomass for cooking. Even when grid electricity is available, there can be frequent power shortages (Burlando 2010; Kavishe 2015; Nkwetta et al. 2007; Trovalla and Trovalla 2015). According to *The Economist*, in South Africa, people check the reports of Eskom (the national power utility) like weather forecasts to be prepared for probable load shedding (Economist 2015). The Twitter hashtag #loadshedding is active for Eskom as well as other utilities in African and Asia. In short, people in countries with unreliable power expect power outages and organize their activities and expectations to prepare for them.

In providing this analogy, there is no intent of celebrating blackouts (see Onishi 2015), or of exaggerating parallels between countries that are relatively unindustrialized and have poor power infrastructures with those that are highly industrialized and expect cheap plentiful power to be always available. Presumably, in countries where power outages are common, there are greater skills in improvising solutions and in relying on various kinds of social and technical backup systems, due to the frequent need to do so, relative to places that rarely need these skills and systems.[2] In academic and internet-available popular literature, however, there is little detailed information about how this coping, adaptation, and management take place; rather, it is clear that it does take place. At least speculatively, however, a better understanding of the technical and social arrangements through which electricity supply disruptions are managed in countries where disruptions are routine can provide insight that can help understand how adaptation to a more renewables-centric power system could work in industrialized countries.

A basic scheme of blackout coping mechanisms contains at least three components, each of which draws on combinations of social and technical elements. First, generators and other backup power systems are available. Due to the regular need to use such backup, they are (probably) more reliable than is the case in countries where power outages are rare. Second, there are alternative energy sources, such as kerosene and biomass, which can provide energy services when electricity is not available. This is not just a matter of substituting one fuel for another, but instead requires necessarily social systems of distribution as well as technological devices suited to these alternative sources. Non grid-tied local generation and micro-grids are also sometimes available, again an alternative to a centralized grid. Third there are short-term and longer term time—and resource—management strategies that distribute activity with respect to expected value. Kavishe (2015), for example, describes how industry manages blackouts through shifting working hours, using low-skilled low-cost labor, and investing in high productivity systems to take advantage of power when it is available, if there might not be power tonight, study now, and so on. Because power shortages can be expected, the consequences of power shortages can be expected. While on the

[2]There is a popular concept particularly in some formerly French-colonized countries in Africa of "*Système D*" where the D stands for *se débrouiller* (see Kaufmann 1985) and/or (according to some) *se démerder* (to get yourself out of shit), now referring to the informal economy in general, or otherwise to do or construct without the proper equipment, tools, or rules, mundanely a form of "do-it-yourself."

surface, this level of coping would hardly be viewed as practical or acceptable policy in the US, it is also not distant from the types of suggestions that are sometimes provided in demand response and conservation material directed to residential customers, as Carlsson-Kanyama and Lindén (2007) detail—e.g., to do one's dishes at night or refrain from using the clothes dryer.

Learning from Blackouts

The question is not whether society is flexible when forced to be, since clearly the answer is yes. Instead what is most interesting is the nature and consequences of flexibility and inflexibility, how flexibility might be increased, and how the consequences of flexibility and inflexibility can be made less costly—to individuals, organizations, governments—in terms of economic costs, financial costs, lost amenities, time and process synchronization, and other social costs such as stress, suffering, and psychic tolls.

Above I remarked on narratives about coping with blackouts as experienced in the United States. In discussing the role of narrative in conceptualizing how people can and do adapt to climate change events, Paschen and Ison (2014) argue that local narratives about adaptation help draw connections between otherwise seemingly unrelated socio-cultural and institutional aspects, as well as highlight individual and community local knowledge and coping capacity. This ability to capture complexity, context-specific adaptability, and emergent knowledge (Paschen and Ison 2014) as rendered in stories contrasts with a flatter, more top-down, focus on universals, discrete elements, and quantification (Moezzi 2015) characteristic of demand response-based view of demand flexibility.

In summary, blackouts reveal existing interconnections between different practices and potential domino effects of failures among them, many of which are difficult to predict without empirical evidence. These connections are in continual evolution. Second, blackouts also provide evidence and examples about how electricity shortage problems have been, or could be, coped with, and the potential consequences of such coping. Third, blackouts also provide an escape from electricity that may itself provide pleasure including different, and perhaps deeper, forms of sociality, and "regressive" snapshots of what everyday life could require.

An improved ability to cope with blackouts is not the central point of this argument, in part because seriously considering a future studded with blackouts is hardly a winning vision. Rather, it is that it is possible to configure society so that it is less dependent on having any amount of centrally provided power available at any time. For a simple example, shifting geographies of cooling, such as via "cooling centers," or more archaically, with theaters (Cooper 1998), shopping malls, grand cool churches and mosques, can reduce the amount of space that needs to be cooled. Cooling centers as a course of action are already included, to some extent, in municipal heat response plans (e.g., Bernard and McGeehin 2004) but less so as a matter of demand response itself.

Expanding Demand Response

Darby and McKenna (2012) argue that residential demand response, while potentially effective in reducing load, requires a relatively large societal effort. Given the small loads at play—a bit of WiFi power for a timer—for many particular end uses, many actors must participate in order to produce an adequate aggregate load reduction. Most demand response has been focused in the commercial and industrial sectors, and few aggregators have been interested in residential demand response (Darby and McKenna 2012). Current residential demand response mechanisms include energy efficiency and conservation programs; static time-of-use pricing; critical day pricing; critical peak pricing; peak time rebates; real-time pricing; demand-side bidding; and dynamic demand via equipment cycling (Darby and McKenna 2012). As noted above, so far the demand response literature has largely focused on getting people to "participate" in demand response and assessing the load reduction effects of this participation, rather than on exactly what people do to provide this load response, how to increase the capacity for this response (beyond price/utility ratios and direct load control mechanisms), and what the individual and social consequences of response might be. These issues normally lie beyond the scope of power utility interests. In thinking about how a high penetration of renewables can be integrated into power systems, a broader view is necessary.

Time Dependency and Utility Interests

The idea of time-dependent rates is as old as the electricity grid (Hausman and Neufeld 1984). At the turn of the twentieth century, peaks were in the early evening, driven by lighting, leaving large excess capacity for much of the rest of the day. Industry tried to encourage load in other hours by supporting the conversion of steam engines to motor, discouraging grid-independent power, implementing time-of-day rates, encouraging electric vehicle charging at night, and focusing on load factors. As noted above, this pressure to build load and shape demand is echoed in contemporary US arguments for further electrification. Despite the long history of time-of-use rates, they are rare in the US, with less than 1% of households using time-of-use rates in 2001 (King 2001). In UK only a quarter of households are on time of use (Higginson 2014). For residential end users, even current participants in current demand management schemes, demand response may often take quite a bit of effort (Burchell et al. 2016; Higginson 2014), in part because the array of alternatives to electricity is so poor.

The problem of trying to change load shape to economically fit supply capacity is usually rendered as demand-side management. In the US, demand-side management is usually considered to include (a) load reduction or conservation; (b) load shifting; (c) peak-clipping; (d) valley-filling; (e) load growth (Faruqui and

Table 13.3 Some types of demand classified with respect to their time flexibility

Category of demand	Examples
Time-dependent, linked to environmental circumstances such as weather, darkness	Cooling, heating, lighting
Time-dependent, linked closely to other schedules	Mealtime, working together, serving patrons, transportation
"Always on" load, such as many electronics, communication networks	Standby modes, WiFi (in some cases)
Baseload that can be off or cycled for a while, with little loss of amenity	Refrigeration, storage domestic hot water, charging, etc.
Load that technically might be very flexible as to time	Individualized/atomized operations (e.g., working alone); some industrial processes
Storage	Charging vehicles, electronics, pre-cooling, etc.

Chamberlin 1993); and (f) load shape flexibility. The first four, and especially the first two, of these are based on short-term definitions. From the perspective of thinking about societal flexibility of electricity use, this supply-side view of demand might be translated into a demand-side view of supply. Table 13.3 provides a provisional view, asking (a) what are current "baseline" levels, and thus the grounds for accounting for saving and shifting?; (b) how might changes work in the long-run versus in response to short term?; and (c) what are the levels of willingness and ability of end users to provide these shifts in support utilities and the grid?

Smart Grid

The smart grid is a not a centrally planned concept or implementation but rather an umbrella term that has taken on currency and a variety of definitions, courtesy of a wide variety of stakeholders. In aggregate, a great deal of promises have been made for this smart grid—including ability to manage supply and renewables intermittency and accordingly support greenhouse gas emissions goals and reduce fossil fuel dependence; providing two-way communications between producers and users; and managing end-use demand and costs, safety, security, and reliability (Tricoire 2015). Skjølsvold et al. (2015) note that this smart grid has become a sort of catch-all solution to the anticipated need for a more efficient, more manageable, mode of supply, and supply–demand linkages. It inscribes users themselves and encodes expectations about them, such as that they will engage with time-dependent home energy management, sell electricity, etc.

Most renditions of the smart grid are optimistic about the eventual participation of their customers, particularly residential customers, in demand management programs. There is rather slim evidence that many people really want to manage their electricity use in this way or that they would consider it "empowering" especially in the long term (e.g., Burchell et al. 2016; Darby and McKenna 2012;

Hargreaves et al. 2013; Higginson 2014; Peters et al. 2009). Experimental and small-scale programs do sometimes show savings, but not only are these programs often centered on volunteers, they are generally short-term rather than longer term. Leaving issues of equity and prioritization aside, here is also a question of what rates (and other rules) would be required to achieve the levels of demand response required—itself rarely discussed publicly.

One of the basic social discomforts expressed about the smart grid is ceding control to the utility and its agents. There has been little acknowledgment of this fear other than some expectation that it is a matter of short-term distrust and ignorance that will eventually be ceded. But historically, a combination of fear and fascination with robots and dumb agents runs very deep (Schelde 1993). Individuals, households, and communities can do change schedules and habits, sometimes very quickly and massively (e.g., during Ramadan, war, vacations, etc.) but the reasons for, and infrastructural support for, these changes matter. So it remains to be seen why exactly customers would find these demand response activities to be of value, other than the schematic economic argument where individuals balance electricity price with its utility. Table 13.4 summarizes some typical assumptions promoted in smart grid frameworks, along with corresponding

Table 13.4 Promoted expectations of a smart grid future, and reservations thereof

Promotional line	Reservations
Prices will be used to trade off personal costs and benefits of load shifting and conservation	Does not allocate across users; takes effort; does not address equity
Short-term demand response delivers capacity	Short-term capacity is important but longer term shifts in demand may also be necessary
Demand response can be scaled to any degree	May not scale well
Focus on discrete "conservation" actions rendered as relatively independent choices	Actions are integrated into practices which depend on material arrangements linking "people, organisms, artifacts, and things" (Schatzki 2015)
Transform consumers to engage with the smart grid and to value the practices and advantages that it enables	Little evidence that people want to co-manage their energy use in accordance with utility interests
Energy tariffs are otherwise logical indicators of the costs of supplying electricity	Current electricity prices are based on a variety of historic, social, regulatory, and profit motives; as in the case of, e.g., demand charges, they may have little relationship to cost of provision
Demand response is a positive way of giving control to users	Demand response can have a psychic toll and physical repercussions
The grid should be perfect	Expecting perfection creates dependencies may increase both vulnerability and electricity use
Electrification and smart grid to create a near-ubiquitous network offering power, management, and control	Control controls, and this control may often be unwelcome

reservations that arise when considering these assumptions from a broader, more practice-centered, perspective.

Even when customers are required to be on time-of-use rates, or can be enticed into them, demand response efforts traditionally focus on the minds of end users (Strengers 2013). Hence, visions of how the new flexibility in demand will be achieved seek solutions in terms of rational price response, knowledge, and cooperation from end users, including their acceptance of automated agents that choose to consume or not on the basis of some pre-arranged criteria. By focusing instead on practices as composing demand, a whole new set of strategies and potential points of influence comes into view (Strengers 2013). In the case of residential air conditioning, this could include, for example, reformation of housing construction to better allow passive cooling (Strengers 2013) as well as the cooling centers noted above. Having a variety of these alternatives available increases the ability of end users to reduce electricity demand, whether through curtailing energy services, shifting modes, or shifting timing.

Time Flexibility, Synchronicity, and Sociotechnical Capacity

In some regions of the world, electricity markets include capacity markets (Anders 2015; Creti and Fabra 2004), which are a type of futures market for electricity. The product on offer in these markets is the commitment to provide a given amount of electricity—through generation, demand response, or transmission improvements—at a given point in the future (e.g., three years). These markets can be configured in various ways; they are controversial but have been growing in popularity (Schlandt 2015) especially as linked to the growing penetration of renewable energy in electricity markets (Bothwell and Hobbs 2016). Capacity markets are intended to incentivize investments in future capacity (particularly generation) but they have also been criticized as potentially dissuading investment in renewables because of their intermittency, as opposed to the more predictable power available through thermal generation (Bothwell and Hobbs 2016).

In explicitly including demand response, capacity markets frame demand response as a future resource that has value and can be developed. Generally, demand response capacity is restricted to the realm of electricity markets. This chapter argues that the notion of demand response capacity can be expanded, both to include stakeholders that do not explicitly participate in electricity markets (e.g., government entities such as those involved in urban planning, emergency planning, codes and regulations, as well as technology developers, non-governmental organizations, etc.) and to include notions of capacity that are less rooted in narrow notions where price is the major incentive for, and determinant of, response.

The discussions of blackouts and demand response above provided examples of how one might increase the degrees of freedom available to provide demand reduction capacity, especially by promoting different *modes* of doing things that rely less on electricity. These modes depend not only on individual volition but also

Table 13.5 Some elements of sociotechnical capacity perspective on demand response

Dimension	Examples
Time flexibility	Shift timing of activities within days, weeks, or longer time period
Space flexibility	Shift the geography of activities so that distributed services serve more people (e.g., more communal spaces for lighting and environmental control) or that environmental conditions are more hospitable (e.g., avoid living in very hot areas)
Material flexibility	Buildings designed with operable windows or other features that enable reduced need for electricity-based cooling
Tailor services more closely to what is required or wanted	Variable speed drives; reduce overheating or overcooling
Manual backup modes	Integrate manual backup modes in technology design; conduct tests in place to ensure that these backup systems work
Reduce standby demand	Smart on/off; forms of technology memory that reduce the need for electricity outside of active-on modes
Storage	Storage on either side of the meter; expanded ways of thinking about storage
Retain/develop non-electric energy sources	This could include technologies such as solar thermal cooling, generators, etc.

on the availability of material configurations, whether dormant or active. For example, as Shove et al. (2012) argue in the case of bicycling, where relevant elements from a more bicycling-centric past persist, "including meanings, competences, and bicycling-related infrastructures," bicycling can more easily re-emerge as a viable mode of transport that is alternative to cars, for example. This "sociotechnical capacity" for reducing electricity dependence could draw on material configurations from the past, as in the bicycling example, or, for an electricity-centered example, on the ability to passively cool buildings. It could also draw on new material configurations that are potentially designed, in part, to provide alternatives to electricity dependence. Table 13.5 lists some conceptual dimensions of sociotechnical capacity, as well as some simple examples of how these capacities might be achieved.

With respect to the time dimension of sociotechnical capacity, a fair amount of evidence is already available. From a technical perspective, most people in industrialized countries have been able to "live like it is daytime in the dark and cold" for decades, even a century. Schivelbusch (1995) describes the night as "frontier," where the availability of industrialized lighting allowed colonization of the night. With the internet, it is now completely normal to communicate, shop, etc. in the middle of the night. Shaw (2015) argues that finally and slowly, the night has become clearly inhabited. What seems remarkable, however, is the fact that relatively little of this happens in public physical space. From the perspective of aggregate load, activity is quite synchronized. For most cities and for most households, rhythms are still remarkably linked to daylight. This has implications for space usage as well as electricity usage as well; streets and parking lots that are severely congested in the daytime may be nearly empty at midnight, and so on. In general, despite the possibilities of light and a 24-h society and despite the diversity of ways of living and the variability of energy use seen, for example, from one house to another (Lutzenhiser et al. 2016), load may be becoming more synchronized, rather than less. Mass actions, usually coincidental (such as a major soccer game) but sometimes intentional (such as Earth Hour) can have major impacts on load shapes (Olexsak and Meier 2014).

Next Steps

This chapter has aimed to pry open currently relatively narrow concepts of demand response and demand management, especially as they are being imagined in future high-penetration renewables scenarios. The expanded view urges a shift to focus on sociotechnical entities rather than on individual actors deliberating on choices and taking discrete actions. As Strengers (2013) notes, considering "minds," "actions," and "technologies" as separate entities in the bid for demand management is inadequate for the scope of the problem. A number of different research directions could help support this broader view and potentially help guide modes of adaptation to better suit the anticipated new shape of electricity:

- In current demand response schemes, what do users actually do to provide demand response? What enables, motivates, and restricts demand response? Are current expectations of electricity users as future providers of much higher levels of demand response reasonable?
- By viewing demand flexibility as a matter of sociotechnical capacity to change demand, including the material means and social experience to do so, can society develop better support for demand response as well as reduce the potential costs of electricity shortages? What can be learned from locales where high levels of flexibility are already required? How can current experience with blackouts and other electricity supply irregularities in both grid-dependent and less grid-dependent areas help both increase flexibility with respect to electricity supply and reduce vulnerability to shortages?
- While further electrification of energy supply systems is usually viewed as a path to clean, controllable, energy, what are the potential downsides of this dependence, and how can these risks be mitigated?
- How have demand shapes changed historically? What are recent trends in demand shape evolution? What factors affect these shifts and to what extent have load shapes been resistant to change? In what ways might load be reshaped over the long term to better suit the potential changes in the time availability of supply?

- What older practices (including their material configurations) might be especially valuable to revive or retain as alternatives to greater dependence on electricity? What policies, processes, or principles might help?

References

Aghaei, J., and M.I. Alizadeh. 2013. Demand response in smart electricity grids equipped with renewable energy sources: A review. *Renewable and Sustainable Energy Reviews* 18: 64–72.

Amin, S.M. 2011. US grid gets less reliable. *IEEE Spectrum* 48 (1): 80.

Anders, S. 2015. *Half-Empty Planes: Utilization Rates for California's Electric Grid Part II. The EPIC Energy Blog*. https://epicenergyblog.com/2015/07/14/half-empty-planes-utilization-rates-for-californias-electric-grid-part-ii/.

Anderson, G.B., and M.L. Bell. 2012. Lights out: Impact of the August 2003 power outage on mortality in New York, NY. *Epidemiology* 23 (2): 189–193.

Apt, J., L.B. Lave, and M.G. Morgan. 2006. Energy conundrums: Power play: A more reliable US electric system. *Issues in Science and Technology* 22 (4): 51–58.

Ballo, I.F. 2015. Imagining energy futures: Sociotechnical imaginaries of the future smart grid in Norway. *Energy Research & Social Science* 9: 9–20.

Bernard, S.M., and M.A. McGeehin. 2004. Municipal heat wave response plans. *American Journal of Public Health* 94 (9): 1520–1522.

Bothwell, C., and B.F. Hobbs. 2016. *Crediting Renewables in Electricity Capacity Markets: The Effects of Alternative Definitions Upon Market Efficiency*. Working paper, Department of Geography & Environmental Engineering, The Johns Hopkins University.

Burchell, K., R. Rettie, and T.C. Roberts. 2016. Time for change? The hard work of energy demand reduction. In *DEMAND Centre Conference*, 13–15 Apr 2016. Lancaster.

Burlando, A. 2010. *When the Lights Go Out: Permanent Health Effects of Transitory Shocks*. Job Market Chapter. http://people.bu.edu/barlando.

Carlsson-Kanyama, A., and A.L. Lindén. 2007. Energy efficiency in residences—Challenges for women and men in the North. *Energy Policy* 35 (4): 2163–2172.

Cooper, G. 1998. *Air-Conditioning America: Engineers and the Controlled Environment, 1900–1960*. John Hopkins University Press.

Creti, A., and N. Fabra. 2004. *Capacity Markets for Electricity*. Center for the Study of Electricity Markets, CSEM WP 125, University of California Energy Institute.

Darby, S.J. 2012. Metering: EU policy and implications for fuel poor households. *Energy Policy* 49: 98–106.

Darby, S.J., and E. McKenna. 2012. Social implications of residential demand response in cool temperate climates. *Energy Policy* 49: 759–769.

Denholm, P. 2015. *The Role of Storage and Demand Response. Greening the Grid Series*. USAID Office of Global Climate Change. NREL/FS-6A20-63041, Sept 2015.

Deumling, R. Thinking outside the refrigerator: Shutting down power plants with NAECA? In *Proceedings of the 2004 ACEEE Summer Study on Energy Efficiency in Buildings*. Washington, DC: American Council for an Energy Efficient Economy.

Deverell, E. 2009. Crises as learning triggers: Exploring a conceptual framework of crisis-induced learning. *Journal of Contingencies and Crisis Management* 17 (3): 179–188.

Economist. 2015. Unplugged: Rolling power cuts are fraying tempers. *The Economist*.

Faruqui, A., and J.H. Chamberlin. 1993. *Principles and Practice of Demand-Side Management*. TR-102556. Prepared for the Electric Power Research Institute (EPRI).

Hargreaves, T., M. Nye, and J. Burgess. 2013. Keeping energy visible? Exploring how householders interact with feedback from smart energy monitors in the longer term. *Energy Policy* 52 (1): 126–134.

Hausman, W.J., and J.L. Neufeld. 1984. Time-of-day pricing in the US electric power industry at the turn of the century. *The Rand Journal of Economics* 15 (1): 116–126.

Higginson, S.L. 2014. *The Rhythm of Life is a Powerful Beat: Demand Response Opportunities for Time-Shifting Domestic Electricity Practices*. PhD dissertation, Loughborough University.

[IEA]. International Energy Agency. 2005. *Saving Electricity in a Hurry*. Paris: OECD.

Jasanoff, S., and S.H. Kim. 2013. Sociotechnical imaginaries and national energy policies. *Science as Culture* 22 (2): 189–196.

Kaufmann, J.C. 1985. 'S'en sortir', 'pas de probèmes', 'faire avec': au coeur de la culture populaire. *Anthropologie et Sociétés* 9 (2): 195–201.

Kavishe, T.E. 2015. *Coping with Power Interruptions in Tanzania: An Industrial Perspective. A Case Study of One Small Scale Animal Food Processing Industry in Moshi Municipality*. Masters Thesis, University of Oslo.

Kimura, O., and K.I. Nishio. 2013. Saving electricity in a hurry: A Japanese experience after the great east Japan earthquake in 2011. In *Proceedings of the 2013 ACEEE Summer Study on Energy Efficiency in Buildings*. Washington, DC: American Council for an Energy Efficient Economy.

King, C.S. 2001. *The economics of real-time and time-of-use pricing for residential customers*. Berkeley, California: American Energy Institute.

Labanca, N., I. Maschio, and P. Bertoldi. 2015. Evolutions in energy conservation policies in the time of renewables. In *Proceedings of the 2015 ECEEE Summer Study on Energy Efficiency*. European Council for an Energy-Efficient Economy.

Lovins, A.B., and L.H. Lovins. 2001. *Brittle Power: Energy Strategy for National Security*. Amherst, MA: Rocky Mountain Institute. Black House Publishing Co.

Lund, P.D., J. Lindgren, J. Mikkola, and J. Salpakari. 2015. Review of energy system flexibility measures to enable high levels of variable renewable electricity. *Renewable and Sustainable Energy Reviews* 45 (May): 785–807.

Luther, D. 2014. The big blackout: Why I'm going low-tech to prep for an EMP. Blog post. In *The Organic Prepper*, 30 July 2014.

Lutzenhiser, L. 2014. Through the energy efficiency looking glass. *Energy Research & Social Science* 1 (1): 141–151.

Lutzenhiser, L., M. Moezzi, and A. Ingle. 2016 (forthcoming). *Advanced Residential Energy and Behavior Analysis*. Prepared for the California Energy Commission.

Marnay, C., and O.C. Bailey. 2004. The CERTS microgrid and the future of the macrogrid. In *Proceedings of the 2004 ACEEE Summer Study on Energy Efficiency in Buildings*. Washington, DC: American Council for an Energy Efficient Economy.

Matthewman, S.D., and H. Byrd. 2014. Blackouts: A sociology of electric power failure. *Social Space* (socialspacejournal.eu).

Michaelowa, A., H. Connor, and L.E. Williamson. 2010. Use of indicators to improve communication on energy systems vulnerability, resilience and adaptation to climate change. In *Management of Weather and Climate Risk in the Energy Industry*, 69–87. Springer Netherlands.

Moezzi, M. 2015. Numbers, stories, energy efficiency. In *Proceedings of the 2015 ECEEE Summer Study on Energy Efficiency in Buildings*. European Council for an Energy Efficient Economy.

[NAS]. National Academies of Sciences, Engineering, and Medicine. 2016. *Analytic Research Foundations for the Next-Generation Electric Grid*. Washington, DC: The National Academies Press.

[NERC]. North American Electric Reliability Council. 2004. *Technical Analysis of the August 14, 2003 Northeast Blackout: What Happened, Why, and What Did We Learn?* Report to the NERC Board of Trustees by the NERC Steering Group, Princeton, NJ.

Nkwetta, D.N., V.V. Thong, J. Driseen, and R. Belmans. 2007. Energy sustainability in sub-Saharan Africa. In *Proceedings of the 9th International Conference on Electrical Power Quality and Utilisation*. Barcelona.
Nye, D.E. 1990. *Electrifying America: Social Meanings of a New Technology 1880–1940*. Cambridge, MA: The MIT Press.
Nye, D.E. 2013. *When the Lights Went Out: A History of Blackouts in America*. Cambridge, MA: The MIT Press.
Olexsak, S.J., and A.K. Meier. 2014. The electricity impacts of Earth Hour: An international comparative analysis of energy-saving behavior. *Energy Research & Social Science* 2: 159–182.
Onishi, N. 2015. Weak power grids in Africa stunt economies and fire up tempers. *International New York Times*.
Paschen, J.A., and R. Ison. 2014. Narrative research in climate change adaptation—Exploring a complementary paradigm for research and governance. *Research Policy* 43 (6): 1083–1092.
Pasquier, S.B. 2011. *Saving Electricity in a Hurry. 2011 Update*. Paris: International Energy Agency.
Peters, J., M. Moezzi, S. Lutzenhiser, J. Woods, L. Dethman, and R. Kunkle. 2009. *PowerChoice Residential Customer Response to TOU Rates*. Report to the California PIER Demand Response Research Center.
Powells, G., H. Bulkeley, S. Bell, and E. Judson. 2014. Peak electricity demand and the flexibility of everyday life. *Geoforum* 55: 43–52.
Roscoe, A.J., and G. Ault. 2010. Supporting high penetrations of renewable generation via implementation of real-time electricity pricing and demand response. *IET Renewable Power Generation* 4 (4): 369–382.
Rupp, S. 2016. Circuits and currents: Dynamics of disruption in New York City blackouts. *Economic Anthropology* 3 (1): 106–118.
Schatzki, T. 2015. Spaces of practices and of large social phenomena. *EspacesTemp.net. Travaux.*
Schelde, P. 1993. *Androids, Humanoids, and Other Folklore Monsters: Science and Soul in Science Fiction Films*. NYU Press.
Schivelbusch, W. 1995. *Disenchanted Night: The Industrialization of Light in the Nineteenth Century*. University of California Press.
Schlandt, J. 2015. *Capacity Markets Around the World*. Berlin: Clean Energy Wire Factsheet.
Shaw, R. 2015. Night as fragmenting frontier: Understanding the night that remains in an era of 24/7. *Geography Compass* 9 (12): 637–647.
Shove, E. 2015. Infrastructures and practice: Networks beyond the city. In *Beyond the Networked City: Infrastructure Reconfigurations and Urban Change in the North and South*, ed. J. Rutherford, and O. Coutard. Routledge.
Shove, E., M. Pantzar, and M. Waston. 2012. *The Dynamics of Social Practice: Everyday Life and How It Changes*. Sage Publications.
Skjølsvold, T.M., M. Ryghaug, and T. Berker. 2015. A traveler's guide to smart grids and the social sciences. *Energy Research & Social Science* 9: 1–8.
Smith, D.C. 2003. Power cuts: Risks and alternatives to the current transmission system. *Refocus* 4 (6): 22: 24–25.
Stephens, J.C., E.J. Wilson, T.R. Peterson, and J. Meadowcroft. 2013. Getting smart? Climate change and the electric grid. *Challenges* 4 (2): 201–216.
Strengers, Y. 2011. Peak electricity demand and social practice theories: Reframing the role of change agents in the energy sector. *Energy Policy* 44: 226–234.
Strengers, Y. 2013. *Smart Energy Technologies in Everyday Life: Smart Utopia?* Palgrave Macmillan.
Trentmann, F. 2009. Disruption is normal: Blackouts, breakdowns, and the elasticity of everyday life. In *Time Consumption and Everyday Life: Practice, Materiality, and Culture*, ed. E. Shove, F. Trentmann, and R. Wilk. Oxford: Berg.
Tricoire, A. 2015. Uncertainty, vision, and the vitality of the emerging smart grid. *Energy Research & Social Science* 9: 21–34.

Trovalla, E., and U. Trovalla. 2015. Infrastructure as a divination tool: Whispers from the grids in a Nigerian city. *City* 19 (2–3): 332–343.

[White House]. Executive Office of the President. 2013. *Economic Benefits from Increasing Electric Grid Resilience to Weather Outages*. Washington, DC: The White House.

Williamson, L.E., H. Connor, and M. Moezzi. 2009. *Climate-Proofing Energy Systems*. HELIO International.

World Energy Outlook. 2015. *WEO 2015 Electricity Access Database*. http://www.worldenergyoutlook.org/resources/energydevelopment/energyaccessdatabase/.

Chapter 14
Energy Systems and Energy-Related Practices

David S. Byrne and Françoise Bartiaux

Abstract In this chapter we attempt to synthesize two relevant bodies of social theory which can be used to understand how human beings—consumers, distributors, producers, and regulators—act in relation to energy systems. The two key words are actions and systems. Practice theories deal with how social life is constituted by practices, or is a product thereof, and with how "people perform the actions that compose practices" (Schatzki 2015: 27). Complexity theory is a general framework of reference which deals with systems which are emergent in character: that is to say they cannot be understood by an analytical programme which seeks to explain them in terms of the properties of their components taken alone. Our approach here is to begin with two sections which in somewhat brutal summary outline the essentials of social theories of practices and complexity theory. We then continue with a discussion of practice and action to show how they are interrelated into a web of interconnected practices. In a similar vein we develop a complexity theory founded discussion of the constraining and enabling role of systems. We then proceed to attempt a synthesis of practice theory and complexity theory with specific reference to how such a synthesis can help us to understand and shape the whole emergent complex system which incorporates institutions and humans and is reconstructed or reshaped by the interaction of all of these entities in daily life. On the basis of this synthesis we will try to make some policy recommendations which will really be about how policy makers should understand what they are trying to influence because such an understanding is foundational to effective intervention.

D.S. Byrne (✉)
School of Applied Social Sciences, Durham University, Durham, UK
e-mail: dave.byrne@durham.ac.uk

F. Bartiaux
Institute of Analysis of Contemporary and Historical Societies,
Université Catholique de Louvain and National Fund for Scientific Research,
Louvain, Belgium

N. Labanca (ed.), *Complex Systems and Social Practices in Energy Transitions*,
Green Energy and Technology, DOI 10.1007/978-3-319-33753-1_14

Introduction

In this chapter we attempt to synthesize two relevant bodies of social theory which can be used to understand how human beings—consumers, distributors, producers, and regulators—act in relation to energy systems. The two key words are actions and systems. Practice theories deal with how social life is constituted by practices, or is a product thereof, and with how "people perform the actions that compose practices" (Schatzki 2015: 27). Complexity theory is a general framework of reference which deals with systems which are emergent in character: that is to say they cannot be understood by an analytical programme which seeks to explain them in terms of the properties of their components taken alone. It is useful perhaps to consider practice theories as departing from Sociology's traditional dichotomy of structure and agency, and then to see complexity theory as mode of thinking which helps us to engage with social structures. In terms of the title of a classic article we are trying to integrate "The Two Sociologies" (Dawe 1970).

One of Marx's mostly widely quoted dictums asserts that: "Men make history, but not in circumstances of their own choosing." Of course the totality of the social world is constituted by social actions but we cannot understand the social world without reference to the emergent components of social reality which are the product of historical action and which do in fact have agentic properties of their own. The expression used most generally for these taken together is social structure. As Westergaard remarked " … structure is only a metaphor, but useful to denote persistence and causal force" (2003: 2). Indeed the word "structure" is perhaps an unfortunate linguistic choice since it implies a degree of rigidity which is inappropriate in considering the character of systems which are inherently dynamic, even if their system state remains functionally constant for considerable periods of time.

Our approach here is to begin with two sections which in somewhat brutal summary outline the essentials of social theories of practices and complexity theory. We then continue with a discussion of practice and action to show how they are interrelated into a web of interconnected practices. In a similar vein we develop a complexity theory founded discussion of the constraining and enabling role of systems. Of particular significance here is the role of institutions, considered as collective entities having the properties of complex systems in themselves and also being effective social actors. We can see all the collective actors in an energy system, producers, distributors, regulators, legislative bodies, judicial bodies, social movements, indeed even perhaps terrorist organizations—as institutional complex systems which act in creating and shaping the ongoing character of energy systems in relation to the social practices of all these actors, including individuals and households as consumers (producers) of (renewable) energy.

We then proceed to attempt a synthesis of practice theory and complexity theory with specific reference to how such a synthesis can help us to understand and shape the whole emergent complex system which incorporates institutions and humans and is reconstructed or reshaped by the interaction of all of these entities in daily life. On the basis of this synthesis we will try to make some policy recommendations which will really be about how policy makers should understand what they are trying to influence because such an understanding is foundational to effective intervention.

Short Account of Social Theories of Practices

Social practice theories mainly draw on Bourdieu's works on practices and habitus (1972, 1980, 1994), on Giddens' structuration theory (1984)—namely the key role of routines in structuring societies—and on Wittgenstein's works. In his seminal book, Schatzki (1996) acknowledges these influences and others while he develops his own practice theory, which was further enriched by Schatzki himself and other scholars (Schatzki et al. 2001; Schatzki 2002), and by Reckwitz (2002). Though really important, as exemplified for example by the word "turn" in the title "The Practice Turn in Contemporary Theory" (Schatzki et al. 2001), these publications came in the spotlight in 2005 when Warde summarized them (mainly Schatzki 1996; Reckwitz 2002) and underlined their potential for consumption studies (Warde 2005). This article has inspired many empirical studies and has been cited more than 1300 times according to Google Scholar (May 2016). Nearly one decade later Warde (2014) synthetises the origins and developments of this "practice turn" and shows the remaining problems to be addressed.

In 2005 too, Shove and Pantzar (2005), and later the same ones with Watson presented a "deliberately slim-line version of practice theory" (Shove et al. 2012: 82) where they theorize practice change. For them, practices change by "the distribution and the circulation of materials, competences and meanings," the reconfiguration of "relations between practices," the reshaping of "the careers of practices and those who carry them" or the "forging and breaking some of the links, relationships, networks and partnerships involved" (Shove et al. 2012: 163).

To acknowledge these different influences and theories, the line of thought devoted to the study and/or theorisation of social practices is usually referred to in its plural form as "the social practice theories." This short account of social practice theories (SPT) is primarily based on the works of Schatzki and Reckwitz. It defines a practice and shows that a practice approach transcends the micro/macro opposition.

Practice: Performance and Entity

Schatzki (1996: 89) defines a practice as both a coordinated entity and a performance. A practice as a coordinated entity is a "temporally unfolding and spatially dispersed nexus of doings and sayings. Examples are cooking practices, voting practices, industrial practices" with major key components of the nexus. He first envisions that "[t]hree major avenues of linkage are involved: (1) through understandings, for example, of what to say and do; (2) through explicit rules, principles, precepts, and instructions; and (3) through what we will call 'teleoaffective' structures embracing ends, projects, tasks, purposes, beliefs, emotions and moods" (1996: 89). In other words of the same author, practices are "open spatial-temporal nexuses of doings and sayings that are linked by arrays of understandings, rules and end-task-action combinations (also emotions and even moods) that are acceptable for or enjoined of participants" (Schatzki 2015:15). Reckwitz (2002) later adds the material arrangements as a fourth key component. There is no hierarchy between these components coordinating the doings and sayings in order to constitute a practice, and according to the practice studied, their relative importance may be different. In several researches on energy-related practices (e.g. Gram-Hanssen 2010; Bartiaux et al. 2014) four linking components are considered: material arrangements, know-how and routines, institutionalized rules, and teleoaffective structures. It can be noted that in further work, Schatzki differentiates practices from material arrangements by which he means "collections of people, artefacts, organisms and things that are linked by such matters as contiguity, causality, and physical connections" (Schatzki 2015: 15). Practices and material arrangements are interconnected: "[w]hile practices effect, use, react to, bestow meaning on and are inseparable from the entities that compose linked arrangements, arrangements induce, channel, prefigure and are essential to practices." (Ibidem: 15).

Practice as a performance is carrying out a practice as just defined, i.e. a coordinated nexus of doings and sayings. According to Schatzki (1996: 90), practice as a performance "denotes the do-ing, the actual activity or energization, at the heart of action. (…) It designates the continuous happening at the core of human life *qua* stream of activity and reminds us that existence *is* a happening taking the form of a ceaseless performing and carrying out."[1]

Both notions of practice refer to each other as underlined by Schatzki (1996: 90): "Each of the linked doings and sayings constituting a practice is only in being performed. Practice in the sense of do-ing, as a result, actualizes and sustains practices in the sense of doings. For this reason, a general analysis of practices *qua* spatiotemporal entities must embrace an account of practice *qua* do-ing; in more standard language, it must offer an account of action."

Both Schatzki and Reckwitz underline the collective character of practices. As clearly stated by Reckwitz (2002: 250), "[t]o say that practices are 'social practices' then is indeed a tautology: A practice is social, as it is a 'type' of behaving and understanding that appears at different locales and at different points of time and is carried out by different body/minds."

Schatzki (1996) acknowledges Wittgenstein's influence in departing from the usual dichotomies in social sciences such as holism/individualism, or macro/micro. He thus distances himself from individualist theories that "problematize any construal of social existence as simply interrelations among individuals" and "all give

[1]In all citations, italics are from the author(s) quoted.

theoretical pride of place to the actions, strategies, mental states, and rationality of individuals, the cooperation, negotiations, and agreements reached among individuals, the rules, norms, and threats governing people's behavior, and the unintentional consequences of behavior that often extend beyond actors' purview." (Schatzki 1996: 4, 6).

Which place is given to the individual in practice theories? Reckwitz (2002: 249–50) answers by explaining that "[t]he single individual—as a bodily and mental agent—then acts as the 'carrier' (Träger) of a practice—and, in fact, of many different practices which need not be coordinated with one another." This is well a "secondary role" respectively to practices as Shove (2010: 1279) has made it clearer.

Indeed, Shove (2010: 1279) underlines that practice theories are irreconcilable with psychological models based on attitudes (A), behaviours (B) and individual choice (C)—that she somewhat mockingly calls the ABC model—given "the incommensurability of these contrasting paradigms, and hence about the impossibility of merger and incorporation": theories of practice "emphasise endogenous and emergent dynamics" and view individuals as "carriers of practices" whereas many psychological theories—namely behavioural and psycho-social theories—view them "as autonomous agents of choice and change" and "focus on causal and external drivers" (see also Shove et al. 2012: 143–164). Shove's (2010) position is endorsed by many social scientists working on energy-related topics (her article entitled "Beyond the ABC" is cited more than 750 times according to Google Scholar in May 2016.)

Regarding so-called social wholes, Schatzki emphasizes that "social" "means pertaining to social coexistence" (1996: 170), coexistence being defined as "the hanging-together of human lives" (Ibidem, 171). And he concludes his book by saying that "[s]ocial life is an intricate nexus of practice, thus centrally a complicated weave of constellations of normativized understanding and intelligibility articulated through action." (Schatzki 1996: 210). He later writes that "the social is a field of embodied, materially interwoven practices centrally organized around shared practical understandings." (Schatzki 2001: 3).

Summarizing, Schatzki (2015) refers to one of his earlier publication (2002) where he "argued that social life, or human existence, transpires as part of bundles of practices and material arrangements." (Schatzki 2015: 15)

By making obsolete the debate on micro/macro, SPT opens the consideration of another representation of the links between individuals and societies than concentric circles. (Note that this representation is well in line with our anthropocentric system of thought whereas other systems exist (Descola 2012)). Recently, Schatzki (2015: 26) makes it very clear by using the word "bankrupt": "this multi-level perspective [the one of Rip and Kemp, 1998] is bankrupt: the phenomena it assigns to the micro, meso and macro levels are in fact laid out on the single plane of the practice-[material] arrangements plenum." This single plane refers to the "flatness of social life" (Ibidem: 17) and to his "flat social ontology."

Complexity Theory—A Brief Guide

The term "complexity theory" is widely employed but as Castellani and Hafferty (2009) note, the word theory is not the best one to describe what is implied by the complexity turn. That turn is a rejection both of the reductionist programme which argues that all larger scale entities can be understood in terms of the properties of the components which make them up and of simplistic holism which asserts that entities can only be understood as themselves alone—somewhat on the lines of the phenomenological emphasis on back to the things themselves. We might better think of it as involving a new frame of reference which hinges around the notion of "emergence" which can be summed up by the old expression to the effect that the whole is greater than the sum of its parts, so long as we recognize that the parts and interactions among them play a role in the constitution of the whole. Indeed we cannot think of entities in isolation. All are influenced by and influence other entities which also have a complex character. The idea of emergence as an explicit position can be dated back to Lewes (1875) and was very much a consequence of sophisticated reflection on the implications of Darwin's understanding of species in interaction in what we now call ecological systems.

And that brings in the word "system." The complexity turn in contemporary science runs on from the development of systems theory, both in the Cybernetics movement and in the form of General Systems Theory as outlined by Bertalanffy (1968). Cybernetics had its origins in engineering mathematics and the application of those methods to the solution of problems in whole engineering systems, although as Prigogine and Stengers (1984) have observed, the development of heat engines in the nineteenth century and the consequent need for physics to engage with thermodynamical systems had already initiated a turn from Newtonian reductionism towards understandings based on consideration of the properties of systems as a wholes. This led to developments in mathematics through attention to nonlinearity, which in this context means something which cannot be described by an equation of a form which through differentiation can be reduced to a linear form. In practical terms this means that the effects of inputs into the system cannot be predicted by linear law—they may be disproportionate to any differentiable equation based description of the system behaviour—the essence of Chaos Theory. The usual, if somewhat erroneously expressed, way of describing this is in terms of the butterfly effect. It is not that the flapping of a butterfly's wings can cause a hurricane, but that rather in the models describing weather systems as they develop over a time a difference in the specification of the initial parameters of that tiny order can generate a qualitatively different outcome in a relatively short time—the difference between a flat calm and a hurricane. There have been a number of developments in mathematics which have attempted to generate ways of describing nonlinear systems. These include nonlinear difference and differential equations and splines, although splines really just describe linear phases with breaks in the linearity.

The development of computing power is another contributor to the complexity turn, particularly in relation to the possibility of agent based modelling.

A well-known precursor of this is Conway's game of life, a cellular automaton in which the state of cells in a grid depends on the state of neighbouring cells according to a set of specified rules. If the game is driven forward through numerous iterations in which the rules are applied, then quite complex forms emerge from the very simple initial specification. Agent based models are similar although the agents interact in more complex ways. They have become an important part of the quantitative programme in the social sciences since we can build such models and see what emerges from interactions among agents following a given rule set. Again as with the game of life we see emergence. Macy and Willer (2002) argued that the development of agent based models means that the social sciences can turn from factors i.e. analytical accounts based on the identification of causal entities through factor analysis of quantitative descriptions of social entities, to actors, i.e. to agents acting to create the social world.

Those actors recreate or reshape or change the social pre-existing world through their actions is the essential proposition of all social science. And of course this resonates absolutely with the approach of practice theories. What is much more contentious in pure micro-emergent positions, as elegantly but in Byrne's view erroneously argued by Sawyer (2005) is that there is nothing there other than the product of contemporary action. In other words there is no historically created social structure, an emergent social system, which operates to constrain and shape the potential of agents as actors. Indeed, as Room (2011) notes in pure micro emergence, there is no place for social institutions, crucial collective entities with agentic powers of their own, e.g. corporate entities with an independent legal existence, let alone for a general over-arching social system. Here we might turn to Reed and Harvey's (1992) elegant and persuasive synthesizing of complexity theory as a scientific ontology with critical realism as a philosophical ontology. This has been developed with reference to the work of Archer (2003) in way which allows for agency in context but sees the social world as composed of complex systems at all levels interacting and nesting with no hierarchical ordering of causal powers. Human social actors, whether individual or collective have agentic power but in specific socio-historical contexts—people make history but not in circumstances of their own choosing. We might add that Actor Network Theory (ANT) actually gives non-human entities agentic power and although this theme was developed in relation to the production of scientific knowledge it has been more widely applied, not least in relation to the power of ecological systems. Practice theory, drawing in particular on Bourdieu's concept of habitus, emphasizes the routine non-agentic character of social practices but when people are confronted with profound changes in the context within which they perform practices then things which had not been subject to conscious reflection come to the forefront, for example in the formation of class practices in a context of deindustrialization.

So individual human actors are complex systems which form, for example, emergent complex systems of households—the crucial unit of social consumption some of which of course contain just one individual. These in turn live in neighbourhoods with different spatial characteristics within localities and regions and these exist within nation states which themselves in the contemporary world are part of trans-national government systems. Think of a person in a modal nuclear family household with a partner and two children living in a working class multi ethnic neighbourhood in a post-industrial municipality in a post-industrial region in France in the EU in the globalized world system. Cross-cutting this set of nested geographical systems, all of which have causal powers with implications for all the other systems, the essence of practice theory is that the practices of individual human participants thereto have causal powers for all social systems but those actions happen within constraints deriving from the causal powers of other systems, systems including corporate entities and any emergent institutional form.

In essence the complexity frame of reference resonates well with the central traditions of social theory—the social has causal powers but actors act within that social not as programmed automatons but with some autonomy, although that autonomy is shaped by all that is social around them and that which they have internalized as part of the social world. Key here are the discussions on the social constitution of mind (see also Schatzki 1996: 55 *et sq.*) as well as Bourdieu's conception of habitus. Wilhite (2012) on his part has stressed the embodiment of social practices. What we have now through the complexity frame of reference is a way of thinking about practices in context and of deploying a whole range of powerful techniques to describe and understand those practices.

Practice and Action

As already introduced above in the "Neither micro nor macro" section, social practice theories rests on a "flat ontology" of social life (Schatzki 2015: 17). And as clearly demonstrated by Shove (2010) they are thus impossible to merge with other theories grounded in the view of individuals being autonomous agents choosing their behaviours—like in economic and some psychological theories. Here there are versions of Complexity Theory which agree with Practice Theory, notably those which derive from the work of Deleuze as developed by DeLanda (2006). However, other versions which relate to Bhaskerian realism work in relation to a layered ontology of causation—see Harman (2010). These certainly see social action as being in large part socially determined but also generally allow for a degree of autonomy.

How then do social practice scholars speak about action, which they do much more often than speaking about actors and agents? But Reckwitz (2002: 256) speaks of agents as carriers of practices: "In practice theory, agents are body/minds who 'carry' and 'carry out' social practices. Thus, the social world is first and foremost populated by diverse social practices which are carried by agents. Agents, so to speak, 'consist in' the performance of practices (which includes—to stress the point once more—not only bodily, but also mental routines). As carriers of a practice, they are neither autonomous nor the judgmental dopes who conform to norms: They understand the world and themselves, and use know-how and motivational knowledge, according to the particular practice." So agents are carrying and carrying out social practices and these social practices are the focus of the analysis.

Indeed, for Schatzki (2001: 3), practice approaches' flat social ontology "contrasts with accounts that privilege individuals, (inter)actions, language, signifying systems, the life world, institutions/roles, structures, or systems in defining the social. These phenomena, say practice theorists, can only be analysed via the field of practices. Actions, for instance, are embedded in practices, just as individuals are constituted within them. (…) institutions and structures are effects of them [practices]." The field of practices is "the total nexus of interconnected human practices (Ibidem: 2).

Earlier, in his seminal book, Schatzki had explained his conceptions of action and behaviour: "actions consist of (or, in the case of omissions, in the absence of) particular behaviors in particular circumstances. This indicate, notice, that behaviors must be distinguished from actions. A behaviour is a bodily doing or saying, a type of action. Actions are either bodily doings or sayings or something a person carries out by way of performing a bodily doing or saying in a specific circumstances. (…) Indeed, the performance of an action consists not only in a bodily doing or saying (basic action), but also in sensations and images that accompany that behaviour." (Schatzki 1996: 38–39).

A further quote may be useful to better understand how behaviours and actions are embedded in practices: "[w]ithin practices (1) intelligible or paradigmatic patterns of behavior, combinations of conditions, and situational relevancies are laid down and lived through; (2) people's behaviour becomes informed by these patterns, combinations, and relevancies; (3) people come to understand everchanging patterns, combinations, and lines of relevancy, as well as the conditions of life that bodily activity expresses on their basis." (Schatzki 1996: 37).

In deconstructing what Schatzki (2002: 207) calls "the fanfare attending the 'discovery' that people are assemblages or networks" as made in Actor Network Theory (ANT), Schatzki (2002: 208), acknowledges that both types of theories (ANT and practice theories) agree that "(1) humans actors are composed of active physical subsystems that maintain causal relations among themselves and with the environment outside their skin, and (2) what people are capable of doing depends in part on the people organisms, things, and artifacts around them." In this book, Schatzki (2002: 189–256) discusses at length how his practice theory differs from other social theories (including ANT). Of special interest in our discussion on practice and action is where Schatzki (2002: 240–241) locates intentionality as it relates to human agency; but first he distinguishes two processes that bring about practice change: "Practice organizations are not static. The understandings, rules and teleoaffective structures that organize integrative practices frequently change. So, too, do the doings and sayings that constitute these practices. These two processes can be called "reorganization" and "recomposition." I reiterate that the constant flow of human and nonhuman doings, in addition to altering practices, maintain them. (…) The reorganization of rules and teleoaffective structures is an occasional and largely intentional process. By contrast, both recompositions of practices and shifts in their practical understandings are continual and largely unintentional events."

So up to now, practice theories differ from complexity theory in their different views on individuals (a secondary role in SPT/agents if not actors in complexity theory) and these views rest on different social ontology (flat/multi-layered). But they do match very well in at least four ways: (1) in acknowledging that the whole—being it called the total nexus of practices versus a system—is greater than the sum of its parts (i.e. the practice's components versus the different actors), and that the parts and interactions among them play a role in the constitution of the whole; (2) in seeing the multiple and variable influences in systems (of practices); (3) in showing that a historical approach of the social system (of practices) is quite desirable: and (4) in considering the constraining and enabling role of systems, as explained below.

The Constraining and Enabling Role of Systems

We act, autonomously and routinely, with social structures which both enable and constrain our actions. Let us illustrate this fundamental principle by drawing on the life experience of David Byrne as an energy user. As a child and adolescent growing up in a coal mining region—the northeast of England—where coal was the dominant domestic fuel although town gas and electricity were also available, and where coal miners even received part of their wages in the form of deliveries of coal, a very important part of his social practices were about fuelling fires which also heated a boiler which provided hot water and about cleaning up the dirt which was created both within the household by burning coal and externally by the prevalence of coal smoke in the air. The region was the earliest locale of the carboniferous capitalism which dominated the industrial world for most of the last two centuries and coal, coal dust and coal smoke were everywhere. The availability of town gas and electricity, important then for cooking and lighting but not otherwise, were themselves products of coal—coal used to make town gas in coke-works and coal used to generate electricity in power stations. Indeed, much of the basic science and engineering science fundamental to the getting and using of coal had been developed exactly in this region.

Important daily social practices included lighting fires, banking them up to keep them going over night, cleaning grates, and disposing of ash for removal. Laundering clothes was a constant activity done in the late 1940s by heating water on a gas fired boiler and then pounding clothes and bedding in a poss tub with a poss–stick. To Poss was the north of England vernacular expression for beating clothes in a tub to clean them. Housewives often had shoulders, to use a local expression, like coal heavers.

All this changed as domestic appliances transformed or eliminated much of the heavy domestic labour of women. Electric powered washing machines progressively reduced the task of laundry. The replacement of open coal fires with smokeless systems, and particularly with the availability of North Sea Gas with gas fired central heating, eliminated all the labour associated with coal fires.

In David Byrne's last house in urban Tyneside where he carried out all the domestic labour he could control his gas heating on line and did laundry with an automatic machine which reduced the labour to drying and ironing—not that he ever ironed much. Now David Byrne lives in a house in a deep rural area in the Scottish borders. There is no gas supply to the village—it was too far from the nearest urban settlement and at the time it might have been available many households had installed oil fired systems and wanted to retain them—a decision which has led to many regrets as oil has become so much more expensive than gas. Byrne's space and water heating and cooking for 70% of the year is provided by a large multi-fuel stove which mostly burns Furnacite—a semi smokeless fuel. This is much easier to manage than an open fire and much more efficient but it has reintroduced the practices of fuelling and ash removal into his daily routine. It is also less dirt generating than an open fire but much more so than gas so laundry is now a much bigger element in his daily practices.

In all this we can see systems acting to constrain the nature of practices. The original driver for the shift away from coal burned on open fires was a public health recognition of the damaging effects of smog. This led to a system which combined prohibitory regulation with subsidy. People had to get rid of open coal fires but were given grants to provide replacement systems, often fairly basic ones in the early years. There was in fact opposition from miners and their families although this was bought off with the substitution of solid smokeless fuel for coal in that element of remuneration in kind. The opening up of the North Sea gas fields made relatively clean gas not only widely available but also relatively cheap which in an era of rising real incomes made efficient gas based domestic central heating the norm in the urban UK. Solid fuel became largely a rural energy source for households although it has only just ceased to be the largest single fuel in electricity generation. Privatization of energy supply with the abolition of nationalized gas, electricity and coal institutions has supposedly led to a market based systems of delivery of energy to households. People in the Scottish borders actually do look keenly at prices for coal based fuels and wood, and at the quality of wood in terms of dryness, but do not seem any more active in changing electricity suppliers than the UK norm.

So we have institutions and systems which enable and constrain practices. These include energy providers, energy regulators and all components of the energy system. Local government plays a role. Whilst in Scotland there is a general strong planning regime support for the development of sustainable energy sources there are also rules relating to conservation of the status of older buildings. So Byrne is not able to install solar photovoltaic panels on the south facing roof of his rural home, something he had in his urban house, because they would be visible from the street in a conservation area. His practice in relation to PVPs was only to check and return his usage on a quarterly basis but they were an important component of his domestic energy system. At the same time there is a whole local network of stove suppliers, wood suppliers and coal merchants which constitutes a system enabling the use of relatively efficient solid fuel as the basis of much of domestic energy use.

What this vignette shows is the interaction of systems at a whole range of levels. Individuals are complex systems in and of themselves with agentic powers involving both routinized practices and innovation, e.g. in relation to personal commitments to sustainability and therefore contextually constrained options about domestic fuels and heating systems. Multi member households are emergent complex systems which can function differently depending on practices performed by each member. Households are embedded in social, market, and governance systems—often populated by institutions which in and of themselves are complex emergent systems—which both constrain and enable the character of social practices. Byrne's house, built in the 1880s, could function effectively as a residence in a post-apocalyptic context in which gas and electricity ceased to be available since he can burn wood, although wood supply would be a real issue. Modern dwellings in larger urban settlements often lack chimneys and could not. Building regulations permit such dwellings. Building regulations change but last for a long time before changing. Think complex, think interactions.

Energy Systems and Energy Practices: Policy Recommendations

This section is intended to giving policy recommendations in the fields of energy retrofit and renewable energies by integrating and applying some of the main ideas and concepts briefly sketched above and relating to complexity theory and to practice theories.

A first important point to stress is that Shove's (2010) critique of the ABC model is still quite relevant, even after several years. As she puts it: "Given that the ABC is the dominant paradigm in contemporary environmental policy, the scope of relevant social science is typically restricted to that which is theoretically consistent with it. At this point it is important to acknowledge that the ABC is not just a theory of social change: it is also a template for intervention which locates citizens as consumers and decision makers and which positions governments and other institutions as enablers whose role is to induce people to make pro-environmental decisions for themselves and deter them from opting for other, less desired, courses of action." (Shove 2010: 1280). This self-deterrence by governments and other political institutions already points to the relevance of analysing as well political practices in the field of energy policy. However this research agenda has not been much implemented so far (see below).

Shove's (2010) critique resonates very well with Zélem's (2012) one about the notion of social acceptability, that she sees as placing the burden of high energy consumption, and the change thereof, solely on individuals in such a way that policy makers and their political parties have no other responsibility than to device "acceptable" policy. She proposes instead the notion of sociotechnical feasibility that obliges to question both the techniques' meanings and the political choices related to energy.

This is not simply a conceptual discussion but a case for a "turn" in energy policy at most levels: European, national, regional if competent, and local. Indeed, the notion of "social acceptance" has flourished in the literature, especially the one devoted to the renewable energy and with a psychological orientation[2] while the notion of "social acceptability" is widely used in other disciplines, including sociology.

Social practice theories have been operationalised in energy research, especially on energy consumption. Even energy retrofit by home owners is studied with this theoretical framework as critically synthesized by Karvonen (2013). One of the questions raised is whether energy retrofit forms one practice in the way that practices are defined by social practice theories. Bartiaux et al. (2014) answer by the negative drawing from their qualitative survey released in 2009–2010. Even if their answer is justified by arguments related to political procedures and institutionalized rules (above all, the Energy Performance Certificate) as well as to craftsmen and professionals' practices, their analysis is based on in depth interviews realized only with homeowners, as also done by Judson et al. (2014). Diversifying the stakeholders and integrating their narratives in the analysis would give a more complete (and more complex!) view of the system of interconnected practices carried on and by political institutions, business, and home owners. In this regard an interesting analysis on infrastructure is reported in Shove et al. (2015).

Political instruments could thus push forward such scientific practice. Another scientific practice that deserves more support from energy policy makers is to fund evaluation studies of energy-policy instruments with the methodologies developed in the framework of complexity theory (for a state of the art, see Byrne and Callaghan 2014). To impose on Member States (or subnational entities if relevant) to realize such research in a comparative way and providing them with a mandatory and very detailed research protocol is not an unusual practice for the European Commission. An example of this is the Commission delegated regulation on the energy performance of buildings by establishing a comparative methodology framework for calculating cost-optimal levels of minimum energy performance requirements for buildings and building elements (European Commission 2012).

Regarding energy policy instruments Bartiaux et al. (2014) conclude their comparative study (between Denmark, Latvia, Wallonia, a Region of Belgium, and the Coimbra area in Portugal) by the following sentences: "this analysis based on social practice theories shows the way for novel recommendations on energy policy that are directed to the social context of energy-related renovations rather than to individual house owners. As outlined above, these recommendations are of at least three types: subsuming and integrating practices that are otherwise disparate; strengthening the components linking doings and sayings; and reinforcing the synergies between two or more components" (Bartiaux et al. 2014: 536). Many examples are given there on these three types of recommendations that stress the

[2]See for example http://climatepolicyinfohub.eu/social-acceptance-renewable-energy#footnote1_00q6zyd

connexions within and between practices' components (such as for the institutionalized rules, between the EPC and other policy instruments such as fiscal ones), as well as the interrelations already existing and/or to be improved between craftsmen's practices and homeowners' practices.

Furthermore, how to device novel energy policy instruments by integrating the many contributions of complexity theory with the ones drawn from the operationalization of social practice theories?

Both complexity theory and practice theories envision the social as complex adaptive systems or as the total nexus of interconnected human practices: by sharing this systemic approach and the importance given to interactions and interconnections between parts of the system, both groups of theories stress complexity and the insufficiency of top-down policies namely in the field of energy (as information given through energy labels on appliances, energy performance certificates on buildings, sensitisation campaigns, and so on): people are not empty reservoirs waiting to be filled with information, they act in relation to their understanding, know-how and motivational knowledge as well as in relation to how systems bound them.

And to progress on this integration, there is no need to first give an answer to the difference outlined above between complexity theory and practice theories namely in their opposite views on individuals (agents if not actors in complexity theory/a secondary role in SPT) and in their different social ontology (multi-layered/flat). Indeed, complexity is more a frame of reference which can allow for a range of theories of the social than a specific theory per se. As we have indicated there are versions of a complexity way of thinking about the social which are founded in a flat ontology, as with practice theory, and others which very much challenge the notion that we should work with a flat ontology but rather see the social, and in particular social causation, as layered with complex and multi-directional chains of causation running between, among and across these layers. For us in practical terms this is not a problem. Rather we think that practice theory and the complexity frame of reference are both useful tools for thinking about real policy issues. They can be deployed in relation to other useful perspectives, and we would mention particularly the approaches being developed in institutional economics. Complexity thinking does have an advantage in that it has been associated with forms of quantitative and calibrated modelling which have some predictive capacity (see Allen 1997). These are tools which are both familiar in form to engineers and planners and can be used by them. What practice theory does is constantly remind us of the actual forms of social life which surround any planning or policy process. So for us what we have are tools for thinking about policies in the process of forming those policies which suggest a way of making those policies work in practice. That has been what we have attempted to demonstrate in this chapter.

References

Allen, P.M. 1997. *Cities and Regions as Self-Organizing Systems*. Amsterdam: Gordon and Breech.

Archer, M. 2003. *Structure, Agency and the Internal Conversation*. Cambridge: Cambridge University Press.

Bartiaux, F., K. Gram-Hanssen, P. Fonseca, L. Ozoliņa, and T.H. Christensen. 2014. A practice-theory approach to homeowners' energy retrofits in four European areas. *Building Research & Information* 42 (4): 525–538.

Bertalanffy, L. 1968. *General System Theory: Foundations, Development, Applications* (revised edition 1976). New York: George Braziller.

Bourdieu, P. 1972. *Esquisse d'une théorie de la pratique précédée de trois études d'ethnologie Kabyle*, Genève: Droz. (Updated and reorganized in Bourdieu, P. (1977) *Outline of a Theory of Practice*. Cambridge: Cambridge University Press.

Bourdieu, P. 1980. *Le sens pratique*. Paris: Éditions de Minuit. (Bourdieu, P. 1990. *The Logic of Practice*. Cambridge: Polity Press).

Bourdieu, P. 1994. *Raisons pratiques. Sur la théorie de l'action*. Paris: Seuil.

Byrne, D., and G. Callaghan. 2014. *Complexity Theory and the Social Sciences. The State of the Art*. London and New York: Routledge.

Castellani, B., and F. Hafferty. 2009. *Sociology and Complexity Science*. Berlin: Springer.

Dawe, A. 1970. The two sociologies. *British Journal of Sociology* 21 (2): 207–218.

DeLanda, M. 2006. *A New Philosophy of Science*. London: Continuum.

Descola, P. 2012. Beyond nature and culture. Forms of attachment. *HAU: Journal of Ethnographic Theory* 2 (1): 447–471.

European Commission. 2012. Commission delegated regulation (EU) No 244/2012 of 16 January 2012 supplementing Directive 2010/31/EU of the European Parliament and of the Council on the energy performance of buildings by establishing a comparative methodology framework for calculating cost-optimal levels of minimum energy performance requirements for buildings and building elements. Available on http://www.eib.org/epec/ee/documents/comparative-methodology-epbd.pdf.

Giddens, A. 1984. *The Constitution of Society: Outline of the Theory of Structuration*. Cambridge: Polity Press.

Gram-Hanssen, K. 2010. Residential heat comfort practices: Understanding users. *Building Research and Information* 38 (2): 175–186.

Harman, G. 2010. *Towards Speculative Realism*. London: Zero Books.

Judson, E.P., U. Iyer-Raniga, and R. Horne. 2014. Greening heritage housing: Understanding homeowners' renovation practices in Australia. *Journal of Housing and the Built Environment* 29 (1): 61–78.

Karvonen, A. 2013. Towards systemic domestic retrofit: A social practices approach. *Building Research & Information* 41 (5): 563–574.

Lewes, G.H. 1875. *Problems of Life and Mind (First Series)*. London: Trübner.

Macy, M.W., and R. Willer. 2002. From factors to actors: Computational sociology and agent-based modeling. *Annual Review of Sociology* 28: 143–166.

Prigogine, I., and I. Stengers. 1984. *Order Out of Chaos: Man's New Dialogue with Nature*. Boulder, CO: New Science Library.

Reckwitz, A. 2002. Toward a theory of social practices: A development in culturalist theorizing. *European Journal of Social Theory* 5 (2): 243–263.

Reed, M., and D.L. Harvey. 1992. The new science and the old: Complexity and realism in the social sciences. *Journal for the Theory of Social Behaviour* 22: 356–379.

Room, G. 2011. *Complexity, Institutions and Public Policy*. Cheltenham: Edward Elgar.

Sawyer, R.K. 2005. *Social Emergence*. Cambridge: Cambridge University Press.

Schatzki, T. 1996. *Social Practices: A Wittgensteinian Approach to Human Activity and the Social*. Cambridge: Cambridge University Press.

Schatzki, T. 2001. Introduction: Practice theory. In *The Practice Turn in Contemporary Theory*, ed. Schatzki, T.R., K. Knorr Cetina, and E. von Savigny. New York: Routledge.
Schatzki, T. 2002. *The Site of the Social. A Philosophical Account of the Constitution of Social Life and Change*. University Park: The Pennsylvania State University Press.
Schatzki, T. 2015. Practices, governance and sustainability. In *Social Practices, Intervention ane Sustainability—Beyond Behaviour Change*, ed. Strengers, Y., and C. Maller. London and New York: Routledge.
Schatzki, T., K. Knorr Cetina, and E. von Savigny. 2001. *The Practice Turn in Contemporary Theory*. New York: Routledge.
Shove, E. 2010. Beyond the ABC: Climate change policy and theories of social change. *Environment and Planning A* 42 (6): 1273–1285.
Shove, E., and M. Pantzar. 2005. Consumers, producers and practices. Understanding the invention and reinvention of Nordic walking. *Journal of Consumer Culture* 5 (1): 43–64.
Shove, E., M. Pantzar, and M. Watson. 2012. *The Dynamics of Social Practice. Everyday Life and How it Changes*. London: Sage.
Shove, E., M. Watson, and N. Spurling. 2015. Conceptualizing connections: Energy demand, infrastructures and social practices. *European Journal of Social Theory* 18: 274–287.
Warde, A. 2005. Consumption and theories of practice. *Journal of Consumer Culture* 5 (2): 131–151.
Warde, A. 2014. After taste: Culture, consumption and theories of practice. *Journal of Consumer Culture* 14 (3): 279–303.
Westergaard, J. 2003. Interview with John Westergaard. *Network* 85: 1–2.
Wilhite, H. 2012. Towards a better accounting of the role of body, things and habits in consumption. In *COLLeGIUM: Studies Across Disciplines in the Humanities and Social Sciences: The Habits of Consumption*, ed. Warde, A., and D. Southerton, vol. 12, 87–99. Helsinki: Helsinki Collegium for Advanced Studies. Available at: http://www.helsinki.fi/collegium/journal/volumes/volume_12/index.htm.
Zélem, M.-C. 2012. "Les énergies renouvelables en transition: de leur acceptabilité sociale à leur faisabilité socio-technique" Revue de l'Énergie, n ° 610. Available on http://www.global-chance.org/Les-energies-renouvelables-en-transition-de-leur-acceptabilite-sociale-a-leur-faisabilite-sociotechnique.

Part IV
Summary and Conclusions

Chapter 15
Key Messages from the Authors of the Book

Nicola Labanca, Paolo Bertoldi, Isabella Maschio and Daniele Paci

Abstract The main research and policy indications produced by the authors of the book are summarised in this final chapter. Formulated research and policy indications reflect the positivist and the constructivist approaches adopted by these scholars to study the ongoing transition to renewables. Contributing authors conducting their studies from a positivist perspective mainly stress the need for researchers and policy makers to give priority to a complex system approach when addressing this transition. Rather than just focusing on decarbonisation of countries, some of them highlight, in particular, the relevant effects that can be generated by *de-olification* of societies and by a close monitoring of the environmental impact of giant corporations. Others discuss the urgency of going beyond purely neoliberal approaches to energy sustainability and of complementing energy efficiency with energy conservation policies. On the other hand, scholars representing the constructivist perspective point to the need for researchers and policy makers to go beyond approaches informed by complexity by avoiding, among others, constantly taking literally the metaphors developed around the energy and information concepts. At the same time, they urge to move the focus of policy and research agendas from an abstract notion of energy to social practices and to develop technical tools enabling the dynamic match between energy end-uses and available renewable energy sources entailed by radical transitions to renewables. In addition, they highlight the importance of achieving a deeper understanding of the concept of smartness and of expanding current approaches to demand response and demand management by focusing on the role played by people's practices and on how existing disparities on energy generation, transmission and use can be reduced. The chapter finally draws some main conclusions concerning how the above-mentioned complementary approaches have allowed framing the problem of energy sustainability in this publication.

N. Labanca (✉) · P. Bertoldi · I. Maschio · D. Paci
European Commission, Directorate-General Joint Research Centre,
Unit C.02 Energy Efficiency and Renewables,
Via E. Fermi 2749, 21027 Ispra, VA, Italy
e-mail: Nicola.labanca@ec.europa.eu

N. Labanca (ed.), *Complex Systems and Social Practices in Energy Transitions*,
Green Energy and Technology, DOI 10.1007/978-3-319-33753-1_15

The complementary perspectives adopted by the authors of the chapters included in this book has allowed them to discuss a series of key aspects concerning the on-going energy transition which are often neglected by researchers and policy makers having to deal with it. This chapter is dedicated to synthesizing these key aspects and highlighting the main conclusions achieved by the various contributors as well as formulating some indications for research and policy making based on what they have presented. Summarized indications have been grouped under two separate chapter sections depending on whether they reflect what is, in our opinion, either a positivist or a constructivist perspective.

Indications for Research and Policy Making from a Positivist Perspective

Facing the Challenges of the On-Going Energy Transition by Giving Priority to a Network Perspective

If the present level of complexity of societies and economies has to be maintained, the on-going energy transition will necessarily entail an increased complexification of energy (notably electricity), information, and transport networks. Ruzzenenti and Fath point out in Chap. 3 that this increased complexification will generally be caused by the necessity of fulfilling existing energy needs through highly distributed, intermittent and low intensity renewable energy sources. Moreover, they highlight the possibility of a dramatic increase in the electricity demand caused by fuel shifts towards electricity caused by this transition and by an ever increasing automation of human activities. According to Ruzzenenti and Fath, besides this very likely and difficult to manage increase in electricity demand, the main impediments to a complete transition to RES are represented by the inadequacy of the existing electricity and transport networks, by the existing obstacles to make power (i.e., energy produced per unit of time) rather than energy the main commodity traded on the markets and by the intrinsic unpredictability of RES availability. They explain that the current energy transition is mainly a policy problem that needs to be faced in a systemic way by giving priority to the creation of the technical and institutional conditions for the establishment of a market of power. In this respect, priority should be given to the development of technical and institutional solutions enabling the purchase and sale of energy and power anywhere in the grid on the condition that there exists a free and suitable electric meter (and inlet/outlet point) to enter the grid. Overall, the current energy transition would hence be accelerated if policy measures stimulating the diffusion of end-use technologies using RES (e.g., electric vehicles) would be accompanied and subordinated to the implementation of measures more focused on improving existing electricity network and stimulating a transition from an energy to a power market.

Focusing on De-Olification of Societies

In Chap. 4 Ruzzenenti and Fath illustrate instead how oil shapes the energy metabolism of current economies. They discuss how economic growth, the international division of labor, the development of global value chains, stock markets, currency rate exchanges, markets of commodities, and the balance of payments are all globally correlated to oil price. Oil has increased the scale of economy and the energy metabolism of countries and cities by decoupling primary energy production from land (like any other fossil fuel), by increasing the density of energy production and use and by dramatically enhancing the transport system and thereby allowing the integration of the production structure globally. The authors conclude therefore that crude oil is largely responsible for the present energy metabolism of societies and current levels of greenhouse gas emissions. Rather than a general de-carbonization, *de-olification* of economies should hence become the main target for a massive transition to RES. The transport sector would need to be addressed in particular to achieve this end, because oil is still non-fungible in this sector. On this line of reasoning, they also argue that a transition to RES can reduce the scale of the energy metabolism of countries and cities, facilitate governance of energy systems and monetary and financial processes, and ease the implementation of a circular economy. In this respect, they point out that low oil prices associated with present oil-economy foster the establishment of international outsourcing which represents a major impediment to the governance of these processes. The dismantling of the oil-economy could however also have negative effects on a global scale, as oil price is a *control variable* that could be used for global governance purposes. By downsizing the scale of energy metabolism, a transition to RES may however partially restore the link with land, as, even in a globalized economy, the ultimate source of energy would be bounded to land surface. All in all, the points discussed by these authors prove that the ways in which oil shapes the present economy and society are still very little understood and that a deeper understanding of this mutual interaction can provide important insights concerning the feasibility and the impacts of a massive transition to RES.

Monitoring the Environmental Impacts of Giant Corporations

In Chap. 5, Ruzzenenti concludes that competition for survival in nature and in the domain of the economy is driven by power output maximization and that the creation of hierarchies is functional for this maximization and for the associated scaling up of metabolic processes within natural and socio-technical complex systems. Following this line of reasoning this scholar shows that the gigantism and complexity achieved by international corporations is, under a thermodynamics and cybernetics point of view, consistent with the general evolution pattern followed in nature and consisting in maximizing power output production through the creation of hierarchies. Paradoxically, however, this gigantism of corporations and of the productive structure becomes also the main factor responsible for the instability of natural cycles, the main threat for human beings' survival and a major impediment to a low-carbon transition. Ruzzenenti highlights therefore the urgency of widening the objective of present energy studies by including firms among the categories usually considered when studying the dynamics of energy consumption. As already proved by existing life cycle analysis studies, firms have necessarily to be considered as an elementary unit of analysis as they can represent one major determinant of large-scale energy consumption dynamics that cannot be captured by present energy accounting systems developed around the categories of consumers/sectors/cities/countries. The inclusion of this category within existing energy accounting systems could indeed represent a first and important step for energy governance that could be performed at the global level and that would be highly needed to achieve a low-carbon transition.

Going Beyond Purely Neoliberal Approaches to Sustainability

Thomas Bauwens explains in Chap. 6 how a polycentric governance approach can greatly enhance energy systems capability to adapt to the ever-changing environmental conditions that can affect their performances within a low-carbon transition. In particular, he provides case studies showing how community-based energy initiatives, as part of large polycentric systems, can effectively increase this capability by reducing possible incentives to free-ride on the constructive behavior of others, by overcoming the obstacles represented by a lack of trust in conventional energy actors and by existing vested interests in preserving the current system leading to institutional rigidity. The author of this chapter draws attention to how the implementation of a polycentric approach for the governance of future low-carbon complex energy systems requires that policy makers create favorable conditions for the self-organization of local communities. In particular, he discusses how these stakeholders are supposed to ensure a coordination in the whole systems of initiatives by guaranteeing the establishment of an overarching set of common rules and sanctions for noncompliance. Moreover, he underlines how these stakeholders should also play the role of facilitators managing negations and situations of conflicts among lower level governance units while representing neutral sources of information for energy systems functioning. The necessity to go beyond purely neoliberal and competitive approaches to sustainability is stressed also by Labanca in the conclusions of Chap. 2. He for example discusses how, rather than by exports, the resilience of an economy in a country is enhanced in the long term by the presence of as many as possible self-feeding return loops whereby energy and material flows generated by this economy are constantly redirected back into the country to maintain internal productive capacities and processes. The same resiliency principle would also indicate that healthy economies have to constitute intricate networks wherein human expertise, material infrastructures, and cultural systems grow together and play a mutually supportive role.

Complementing Energy Efficiency with Energy Conservation Policies

Ruzzenenti and Bertoldi illustrate in Chap. 7 how natural and human made systems exhibit at a global level a positive correlation among energy efficiency improvements, energy consumption, and complexity growth and discuss the implications of this correlation for policies that can be implemented to foster a low-carbon energy transition while avoiding the unwanted increases in the absolute values of final energy consumption. In particular, they discuss the need to revisit the role of energy efficiency policies by integrating these policies with measures aiming at an absolute reduction in the energy demand. By focusing on the European context, these experts discuss how a first and preliminary step to be made in this direction consists in the production of harmonized and consistent statistics on energy consumption by countries. Then they underline the need to intensify information measures on energy performances of technologies based also on life cycle assessments. Moreover, they stress the need for a reinforcement of R&D activities in the fields of energy efficiency and renewable energies without neglecting research on the impacts of present and future energy efficiency programmes and associated rebound effects. In addition, they suggest designing policy measures where minimum energy efficiency requirements are introduced in conjunction with requirements concerning maximum allowed total consumption of technologies (e.g., besides achieving given minimum energy efficiency requirements, cars may be required to not emit more than a given amount of CO_2 equivalents per year, refrigerators may be required to not consume more than a given amount of KWhs per year, houses could not consume more than a given amount of primary energy per year, etc.). They also hypothesize that a maximum CO_2 budget per person/household/building could be introduced in the long run by leaving choice on how to meet it and using money to be paid for possible budget infringement for energy efficiency investments. Energy price signals are finally also seen as a policy measure that could be used to counteract energy consumption increase, e.g., by establishing additional taxes on energy consumption and using the additional money so collected by governments for investments on energy efficiency.

It may be worth mentioning that the necessity of complementing energy efficiency with energy conservation policies is stressed by Ruzzenenti also in Chap. 3 of this book. In this chapter, he explains how power output maximization drives the evolution of societies and economies at the global level and then shows how energy efficiency improvements are functional to this power output maximization. Labanca in Chap. 2 and Shove in Chap. 9 also point to the need for expanding present policy approaches mostly exclusively informed by the energy efficiency paradigm. The alternative approaches they indicate do not however concern policy measures whereby a cap on total energy consumption can be established. As partly illustrated in the reminder of this chapter, they are instead mostly informed by a constructivist perspective and are focused on how social practices responsible for energy consumption can be changed.

Indications for Research and Policy Making from a Constructivist Perspective

Complementing Possible Policy Approaches Informed by Complex Systems

Nicola Labanca illustrates in the first two chapters of this book how the complexification of energy systems that can be associated with the on-going energy transition can be considered to be the result of a particular social construction. This social construction relies on and constantly validates the assumption that functions accomplished by people within societies can be reproduced and sustained through underlying networks wherein abstract units of energy, matter, information, and monetary values circulate. In doing so, he discusses how policies exclusively informed by the abstract principles regulating the evolution of these networks, and focused, e.g., on promoting an increase in networks' output production while improving the efficiency of this production, may cause serious problems due to how these policies cannot take into account the specificities of the contexts and of the people they are applied to. In particular, he points out how the implementation of technical solutions exclusively informed by these principles (e.g., solutions abstractly aiming at diversifying mobility options in a city by promoting the diffusion of more energy efficient and less polluting vehicles) can represent a way to bypass any political consideration concerning the actual utility of these solutions. He therefore underlines the necessity of combining this policy approach with a more genuinely political one based on collective decisions for a reorganization of specific systems outputs (e.g., decisions concerning how the need for vehicles in a city can be reduced by reorganizing existing pathways or by promoting mobility practices not relying on vehicles) that can have beneficial results for the community while reducing unwanted energy and environmental impacts. This necessity is made more urgent by the fact that the on-going energy and technological transition makes existing energy and communication networks more brittle due to the increased coupling being established among their nodes, the progressive reduction to a single energy type (i.e., electricity) circulating among these nodes, the more intermittent character of used energy sources, the increased burden on these networks determined by how they tend to progressively integrate within them all human activities, and the impossibility of creating control systems that can completely secure from networks breakdown caused by their increasing complexification. While the previously mentioned policy approach would reinforce an integration within existing complex systems, the latter approach would rely on developing solutions which reduce dependency on these systems while allowing a more active involvement of citizens in the design and implementation of policies enhancing energy sustainability and reducing unwanted environmental impacts of technologies. It may be interesting to note how the relevance of this latter policy approach is confirmed also by the conclusions achieved by Moezzi in Chap. 13 in relation to how reduce energy grid dependencies.

Escaping the Literal Interpretation of Energy and Information Metaphors

Labanca discusses in Chap. 8 how the complex systems constituted by energy and information technologies tend to disengage from any form of social control while generating artificial dynamics of consumption growth which can have extremely negative and large-scale impacts on persons and natural resources. He illustrates how one main social factor contributing to the creation of these perverse dynamics is to be found in a huge process of homogenisation and reification generated by the constantly literal interpretation of metaphors developed around the energy and information concepts. The literal interpretations of these metaphors would basically lead to a reorganization of societies based on the assumption that persons and existing socio-technical systems can be constantly identified with motors and information processors. He illustrates how these literal interpretations reinforce each other within present competitive market settings while being validated by technicians, scientists and the huge technological apparatus whereby complex systems are being socially constructed. In addition, he discusses how these interpretations increase our dependence on the supply of abstract and standardized resource units which are constantly mistaken for entities actually populating everyday life. He then concludes that possible ways out of the social entrapment generated by this phenomenon can only be found by acknowledging the axiomatic incompatibility of social organizations to the descriptions and reductions operated by scientists (including economists) and the fundamental value of people's practical knowledge in the definition of solutions to increase their well-being and the sustainability of their ways of life. The presence of this social phenomenon gives more relevance to the alternative policy approach already illustrated by Labanca in the conclusions of the second chapter of this book. He explains that the literal interpretation of energy and information metaphors can only be escaped in the field of energy policies by subordinating the implementation of solutions reducing technologies energy inputs or increasing technologies outputs to collective and political decisions taken by people in relation to how to change the social practices they reproduce in the context of their everyday life. This change of perspective can indeed disclose a huge variety of additional options whereby people can actively improve the quality of their lives by reorganizing their practices while reducing the impacts of their activities on the environment and available natural resources. The author of Chap. 8 remarks that this change of perspective is particularly needed with the on-going energy transition given the way in which it can trigger dramatic increases in energy demand and the risks of systems crashes associated with the progressive and particular complexification that this transition entails. He points out that it is not so unrealistic to assume that the actual accomplishment of this transition depends on the possibilities that societies will be given to temporarily escape the literal interpretation of energy and information metaphors through the development of ways of life allowing them to temporarily and comfortably survive while disconnected from current energy, information, and monetary systems. This condition of temporary disconnection would be enabled by the huge variety of context dependent solutions that people can elaborate to increase their well-being while reducing their dependence on these systems by relying on their practical knowledge.

Moving the Focus of Policy and Research Agendas from an Abstract Notion of Energy to Social Practices

Elizabeth Shove discusses in Chap. 9 how the dominant paradigm presently reproduced in research agendas, journal articles, reports, and policies around the world consists in stripping energy out of the context where it is produced and used and in conceptualizing it as an oil equivalent resource. She points out that this produces an understanding of policy action based on abstract ideas of efficiency, consumption, and behavior which is progressively disconnected from what people do. This dominant paradigm is however being called into question, as, e.g., the on-going transition to renewable energy increasingly requires that energy is studied in relation to sociotemporal rhythms of generation and use, as energy efficiency agendas are being subject to increasingly powerful critiques based on the presence of long-term counterproductive rebound effects, as the results of efforts to change consumers' behaviors are proving to be limited and as utilities' business models are increasingly oriented to the provision of energy services that are clearly more tightly focused on how and when energy is used. Elizabeth Shove discusses the urgency of developing a new paradigm anchored in social practices that depends on the possibility of taking an abstract and reified notion of "energy" out of energy policy. This new paradigm would allow acknowledging that many different areas of policy making contribute to configure and shape social practices impacting on available natural resources and the environment. Rather than making research and policy agendas which focus on substitutions (e.g., with renewables), efficiency or minimal behavioral changes that do not actually modify present ways of life, this new paradigm challenges researchers and policy makers to imagine and try to establish configurations of technologies, practices, and sociotemporal orders which are really different from the present and which are more compatible with renewable energies and reduced energy demand.

Facing Radical Energy Transitions by Matching Energy End-Uses to Available Renewable Energy Sources

Allen et al. discuss in Chap. 10 why renewables cannot maintain the metabolic rate that is being presently generated within societies through fossil fuels. The reasons for this are mainly found in the fact that renewable energy sources need to be processed in order to be transformed into a fuel and that this often involves extensive transportation. While fossil fuels are found practically ready to use, renewable energy sources like, e.g., biofuels, require extensive gathering activities, water, fertilizers, and time to grow before they can be used as fuels. This fundamental qualitative difference between fossil fuels and renewables means that it is not possible for the latter energy sources to maintain the present energy metabolism of economies on the large scale. A massive transition to renewables would hence entail a radical reorganization of societies, as it would affect food production, products manufacturing, and living standards and would require a higher portion of human activity in the energy sector. They therefore see a radical transition to renewables as a gigantic translation problem consisting of fitting the characteristics of specific energy carriers to specific energy users at specific times and in specific situations. Moreover, they suggest that this problem could be handled through a digital platform using information describing the characteristics of energy carriers and the contexts of energy users (a) to fit specific sources to users and (b) to aggregate all these specific matches to higher levels of matching for insights about policy. The problem of matching the different processes producing energy carriers from the various renewable energy sources to the different needs/abilities of energy end-users constituting a society would become in this way a huge congruence problem between society as a whole and its parts. These scholars explain that the solution to this problem would require a multilevel and fine-grained analysis and specific grammars for translation that could be performed through a scalable web platform whereby decision making could be supported.

Taking into Account Sociotemporal Rhythms of Social Practices

In Chap. 11 Palm and Ellegård highlight the importance of understanding energy consumption in relation to people's activity patterns in order to develop suitable policies for more efficiency energy consumption and climate change mitigation. By focusing on households, they explain how time diaries from a population can be used to reveal differences in the energy consumption associated with aggregate activity patterns generated by people of different generations or by men and women. These types of analysis represent the necessary prerequisite for any policy discussion and intervention related to modifications in social practices that can enable a transition to renewables while allowing energy conservation. In addition, time diaries can be useful for energy advisors and other experts giving individual advice on how to reduce energy consumption and associated costs. The authors of this chapter also point out that, when it comes to understanding whether and how energy consumption patterns of households can be changed, it becomes important to understand how activities of various members are interlinked: they claim that suitably designed time diaries help achieve this result.

Achieving a Deeper Understanding of the Concept of Smartness

In Chap. 12, Padovan and Arrobbio point out that, contrary to views generally adopted by researchers and policy makers, smart grids cannot be simply defined and studied in terms of series of technical characteristics supposed to mark their smartness. Smart grids should rather be seen as a new type of apparatus aiming to reduce the disparities represented by the asymmetric lines of power, knowledge, information, decision-making, intensity, and artifacts that constitute any energy grid. Researchers and policy makers involved in the building up of smart grids are therefore invited to adopt a radically new perspective. As any transition toward smart grids must start from existing infrastructures, these stakeholders should first of all start by understanding these infrastructures in relation to embedded constraints and opportunities for the development of the previously mentioned type of smartness. Then they should consider that, rather than through the linear addition of technical operations, a transition to smartness can only be achieved through a co-evolution of varying lines and strata of practices, technics, and discourses according to patterns that remain ultimately uncontrollable. In relation to energy conservation or peak shaving targets to be achieved by smart grids, researchers, and policy makers should then take into account that smart feedback devices encourage end-users to reduce their level of consumption only to a limited extent and a wider policy and market support is generally needed to achieve significant energy consumption reductions. Moreover, they should consider that achieving these targets depends on existing possibilities to change social practices generating energy consumption and that these possibilities can be markedly increased by improving communication and reciprocal interpretation between human practices and technical systems. The authors of this chapter also underline the point that the inevitable struggle and harmonization of interests among different human and not human agents involved in a transition to smartness should always be considered.

Expanding Current Approaches to Demand Response and Demand Management

In Chap. 13 Mithra Moezzi highlights how scenarios regarding the future high-penetration of renewables require us to expand the present concepts of demand response and demand management. In particular, she urges a shift of focus—concentrating on comprehensive socio-technical entities rather than on individual actors deliberating on choices and taking discrete actions. Individual minds, actions and technologies cannot indeed be considered as separate entities in the bid for demand management. Mithra Moezzi indicates a series of research directions that could support this broader view. These directions generally point to the need for considering a wider spectrum of demand response options while understanding whether current demand response expectations are reasonable. To further this aim, it could prove extremely useful to assess to what extent existing material means and social experience can improve demand response while reducing costs associated with electricity shortages. Experience from countries and regions where high levels of demand flexibility are already required could be useful in this respect. Other insights could then come from studying how demand profiles have changed historically, which factors have affected these shifts and which load profiles have been resistant to change. It would then be important to assess which older practices (including their material configurations) might be valuable to revive and/or become valid alternatives to greater dependence on electricity and which policy measures might be implemented to exploit these alternative options. Current experience with blackouts and other electricity supply irregularities could also indicate additional possibilities to both increase flexibility with respect to electricity supply and reduce vulnerability to shortages. Moezzi also underlines that it would be important to study potential downsides of increased dependence on electricity associated with the on-going energy transition and how associated risks can be mitigated.

Integrating Ideas and Concepts of Complexity and Social Practice Theories into Policy Making

In Chap. 14, Byrne and Bartiaux highlight that both complexity theory and social practice theories prompt a systemic approach to policy making which is mostly incompatible with the top-down approach typically adopted in the field of energy today. Rather than as the result of unidirectional and linear causation mechanisms that can be activated by policy makers, change is indeed seen by these theories as an emergent phenomenon resulting from the mutual interactions and interconnections existing among all parts of the system. In order to understand how and whether policy action can trigger some type of change, it would hence be necessary to consider how this action can affect the interactions among all the parts of the system. To exemplify this, they suggest that the impacts of policies implemented to stimulate residential building energy retrofits can only be understood when all involved stakeholders, related narratives, and the whole system of practices carried on by political institutions, business companies, and building owners are taken into account. Rather than just targeting individual building owners, these policies should hence address the whole social context involved in energy retrofits and, in doing so, they should aim to properly integrate all involved practices, strengthen the components linking people doings and sayings and reinforce synergies between two or more components. Byrne and Bartiaux argue for policy instruments that can stimulate the adoption of the principles of complex systems and social practice theories within policy making related to energy. They suggest, for example, that policy makers might start by supporting the realization of evaluation studies of energy policy instruments with the methodologies developed in the framework of complexity theory. At the same time, they share the critiques raised by a series of scholars in relation to the notion of social acceptability of policies and the problem that this notion places the burden of reducing energy consumption solely on individuals while leaving policy makers and their political parties just with the responsibility of devising acceptable policies. In their opinion, the notion of socio-technical feasibility would instead be more appropriate to assess energy policies in so far as it obliges analysts to question both technological meanings and political choices related to energy.

Conclusions

After having summarized, some of the main research and policy directions described in the various chapters, it is time to conclude this book by briefly drawing some main conclusions and mentioning some of challenges faced during the writing of this publication.

Rather than just focusing on the technological challenges associated with the on-going energy transition and discussing in detail the technical innovations that can enable it, the authors of this publication highlight the interactions between involved social and technological aspects under a complex system and social practice perspective. This approach makes it possible to identify and address fundamental questions and frames the issue of the energy sustainability of this transition in a way that has not so far received the attention it deserves within research and policy agendas being adopted worldwide.

The contributions included in this publication have showed that this transition has to be inscribed within complex systems dynamics of growth in energy consumption. Phenomenological evidence so far collected indicate that socio-technical systems whose intricacy and density of energy flows increase have better chances of surviving in the long term compared to lower density systems, provided these systems manage to balance this growth with adequate energy efficiency improvements. When applied to the energy systems that will result from the on-going massive shift to renewables, this insight suggests that energy efficiency has a still fundamental although complementary role to play in ensuring the survival of these systems. In the long term, energy efficiency improvements can help increase energy systems resilience against ever changing environmental conditions. This increased resilience results from an augmentation in the diversity of systems outputs and/or in the pace at which these outputs will be produced thanks also to these energy efficiency improvements.

Concerning the way in which future energy systems resilience can be increased, it is then worth pointing out that complex system theories provide a quite robust theoretical basis for arguing that community-based and collaborative approaches to renewable energy production, management and use can represent a very effective way to increase systems resilience compared to approaches devised within current competitive market settings.

The above-mentioned evidence also, and somewhat paradoxically, shows how relevant it becomes that new energy conservation policies are implemented to counteract expected growth in renewable energy consumption. Despite the rhetoric developed around renewable energy sources—especially that which suggests that these sources are infinitely abundant and that the on-going shift to renewables is just a technical issue that can be solved without significantly affecting how our lives are organized, the above-mentioned dynamics of growth will indeed always render available energy sources scarce, not to mention their persistent, unwanted, and negative environmental impacts. As previously mentioned, energy efficiency improvements in the utilization of available energy inputs and policies that can be devised to foster these improvements do not necessarily help solve this issue at the system level, because they often create the conditions for an increase in systems outputs production and therefore contribute to the creation of a situation of increased energy scarcity. These policies must therefore be accompanied by a new generation of policies capable of curbing total systems consumption. As some of the authors of this book argue, new types of policies aiming at energy conservation can in principle be represented by energy taxation measures where associated increases in governments' revenues are used for energy efficiency improvements or by suitable command and control policy measures establishing maximum levels of energy consumption for specific sectors of the economy, technological applications or energy end-uses. However, alternative policy approaches exist—and can be further developed—based on the insights provided by social practice theories. These theories and the work accomplished by researches applying them to study the dynamics of consumption lead us to conclude that very effective energy conservation policies can be designed and implemented by focusing on a reorganization of energy outputs. Rather than by reducing energy inputs of human activities, a reduction in energy consumption can also be achieved by taking the possible reorganization of social practices and human activities reproduced in a given context as a starting point. In other words, energy conservation targets can be effectively achieved also by improving the ways in which people organize, e.g., their transportation, their work, their leisure or the ways in which they prepare their food, they purchase what they need, etc., without taking the reduction of energy inputs of these various activities as the only target. Interestingly, this approach to policy making can in principle be more democratic in so far as it enables the more active involvement of people in the design and implementation of associated solutions. Moreover, contrary to what typically happens in the field of energy policy, it allows for considering solutions for energy conservation which are not merely related to the production of new less energy consuming technologies.

One of the main objectives of this book has been to illustrate the extent to which people and their socio-technical capacities represent a huge reservoir of possibilities and opportunities for identifying very effective context dependent solutions to energy conservation which can also enhance the resilience of future renewable energy systems, this aspect certainly not being a minor point.

This book has then also allowed us to discuss why renewable and nonrenewable energy sources have to be considered as completely different types of energy and to explore some of the main implications of this difference for the on-going energy transition. In particular, it has been possible to illustrate why the complexification of energy systems entailed by renewable energy sources might entail a different relationship with time, also highlighting the point that social practices theories have a fundamental role to play in the study of this aspect and in the design of policy instruments fostering the current energy transition by addressing the time dimension of practices. The possibility that people can cope with the increasingly intermittent character of the energy supply that can be expected in the future can for example only be assessed by focusing on how these practices are distributed, how they are interconnected and how they can be possibly shifted in time. Finally, this publication has allowed us to identify and consider some of the dangers associated with the reification processes associated with entities like energy. The implications of the reification processes whereby researchers and policy makers aggregate very different natural and technical energy transformation processes and reduce them to a question of production and consumption of same energy units are absolutely important. As explained in various chapters, the very perverse ways in which these reification processes obscure relevant issues at stake with the on-going energy transition while reinforcing dependency on the supply of abstract resource units can hardly be overestimated.

With this book we have tried to indicate research avenues and elements of understanding in relation to the above-mentioned aspects. The task has not been easy for several reasons. One of these reasons is that this is an interdisciplinary challenge. To engage with these issues it has been necessary to deal with topics concerning at least the research fields of energy policy analysis, economics, biology, ecology, physics, and sociology. Another reason relates to our conviction that any serious attempt to understand how the targeted complex systems dynamics evolve and can be possibly affected implies their disaggregation into the myriad of social practices generating them. This has obliged us to propose a method of analysis combining a positivist with a constructivist perspective, despite the fact that we are aware that accounts and explanations produced under these two perspectives can in some respects be completely different (notably in relation to the historical character of notions like energy, information, time and in relation to how the nature of these entities can be differently perceived and interpreted in different historical periods). Clearly, we are also aware that the experts who endorse one of these perspectives might in some case not agree with the conclusions achieved by experts endorsing the other.

A further reason and source of possible misunderstandings relates to the point that complex systems theories can in principle be invoked to explain the social construction of the dynamics under study and to devise suitable policy instruments that can affect them, while our main tenet is that complex systems themselves are the result of a social construction whose impacts have to be investigated by avoiding the reductions inevitably associated with any theory laden approach.

We hope to have managed to satisfactorily cope with these issues and to illustrate why it is so important that policy making and research activities dealing with the current energy transition take complex systems dynamics into account and become more and more informed by complex systems theories. At the same time, we hope to have managed to explain why social practices represent an essential field of study, and how this field of study can indicate completely new research avenues and policy approaches to address this transition. The topics discussed and the conclusions drawn in this book provide several clear suggestions about how this might be concretely realized.

MIX
Papier aus verantwortungsvollen Quellen
Paper from responsible sources
FSC® C105338

If you have any concerns about our products,
you can contact us on
ProductSafety@springernature.com

In case Publisher is established outside the EU,
the EU authorized representative is:
Springer Nature Customer Service Center GmbH
Europaplatz 3, 69115 Heidelberg, Germany

Printed by Libri Plureos GmbH
in Hamburg, Germany